PRELIMINARY EDITION

MAKING SENSE OF INTERMEDIATE ALGEBRA

Models, Functions, and Graphs

Instructor's Version

Judith Kysh
University of California at Davis

Karen Wootton
Indiana State University

Jon Compton
University of New Mexico

Derek Lance
Chattanooga State Technical Community College

CPM Educational Program
College Preparatory Mathematics: Change from Within
directed by Judith Kysh, Tom Sallee, Elaine Kasimatis, Brian Hoey

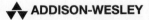

ADDISON-WESLEY

An imprint of Addison Wesley Longman, Inc.

Reading, Massachusetts • Menlo Park, California • New York • Harlow, England
Don Mills, Ontario • Sydney • Mexico City • Madrid • Amsterdam

Sponsoring Editor: Jason Jordan
Development Editor: Elka Block
Project Editor: Kari Heen
Production Supervisor: Kathleen Manley
Design Supervisor: Susan Carsten
Text Designer: Ellen Pettengell
Cover Designer: Susan Carsten
Composition: Patricia Brown
Technical Artist: Techsetters, Incorporated
Illustrator: Dan Clifford
Marketing Managers: Andy Fisher and Liz O'Neil
Manufacturing Supervisor: Ralph Mattivello

Cover photo: © Superstock
The bridge icon on the cover and title page is a registered trademark of CPM.

ISBN 0-201-76802-X — Instructor's Version

ISBN 0-201-49995-9 — Student Version

1 2 3 4 5 6 7 8 9 10—CRW—0100999897

CONTENTS

OUTLINE OF MATHEMATICS CONTENT

All chapters include problems that require practice, use, or extension of what has been developed in previous chapters; therefore, only the material introduced in each chapter is listed below.

PREFACE

NOTE TO THE STUDENT

THIS BOOK IS DIFFERENT. It is probably quite different from any other mathematics text you have used in the past. First, it is all problems. Some are explorations, some are investigations, some are based on realistic situations, and some are purely mathematical exercises. The big ideas of the course are developed in and through the problems you will be doing during class while working in small groups with other students and with the assistance of your instructor. Only a part of the class time will be lecture. Much of the time will be used for working on the problems through which you and your group, assisted and guided by your instructor, will develop and verify mathematical generalizations and rules.

IT'S OK NOT TO UNDERSTAND SOME THINGS AT FIRST. The approach of this text is to develop the important ideas over time, rather than covering them exhaustively in one chapter, testing, and moving on. The new material introduced in early chapters is developed and used throughout the course. At the end of a particular chapter you may not think you completely understand the material presented in that chapter, but several chapters later, after you have had the opportunity to use those ideas in a continuing sequence of problems, you will acquire a deeper understanding than you could possibly achieve in the short term. Problems designed for assignments to be done outside of class provide continuing review and practice of the work done in earlier chapters.

A GRAPHING CALCULATOR WILL BE AN IMPORTANT TOOL. Many of the problems that you will encounter during class will be based on the use of a graphing calculator. The calculator makes it possible to see a lot more of the patterns and relationships among graphs and between graphs and their equations. It is a wonderful tool that enables you to become a "triple threat" problem solver, able to analyze and solve problems numerically, algebraically, and graphically. You will be able to understand a lot more mathematics because you can do that.

UNDERSTANDING

We assume you are taking this course to prepare for a future college course or technical field that will require you to be able to use the mathematics you have learned. The skills you learn in this course will be useful to you in many college courses such as chemistry, economics, psychology, and zoology as well as other math courses. What this means is:

Your goal for this course should be *understanding*.

Only you will know if you understand something. We all know how easy it is to memorize something and even do well on a test without having any real idea about what is going on. If you settle for just performing well, you will be cheating yourself because later, when you need to use some part of the mathematics you have studied, you will not have a clue as to where to begin.

We have done our best to design a course that makes it easy for you to try to learn *and* to understand what you have learned. But no matter what — learning is *hard work*. As the commercial says, "*No (brain) pain, no gain.*" So if you want to do well in this course, there are four things which you will have to do:

- Attend class regularly.
- Make *understanding* the mathematics your highest goal.
- Discuss the questions with your group.
- Do the homework.

Let's look at each of these points more closely.

ATTEND CLASS REGULARLY. This course will be different. You should expect to be actively involved in your learning *every* day. You will have the opportunity to discuss new ideas, practice these ideas, and get guidance and assistance from your group members and instructor. In this course, your instructor will not be doing most of the talking. You will! You will be developing your own definitions and examples rather than searching for rules and examples in the book. The group discussion and problem solving will allow you to build understanding of mathematical ideas as you learn and use new mathematical language. *Full participation in class is essential.*

MAKE UNDERSTANDING THE MATHEMATICS YOUR HIGHEST GOAL. If you want to understand, you will need to be willing to spend time thinking and trying out alternative approaches. Often you will not be able to come up with the right answer on the first try, so you will need to be willing to stick with it. At this level there are generally several ways to think about a topic, and you should try to see more than one of them.

DISCUSS THE QUESTIONS WITH YOUR GROUP. Most job situations today demand that you work with others, discussing ideas, listening, testing, and putting the good parts of one person's idea with the good parts of someone else's to get a solution. Many of the problems in this book will ask you to discuss your ideas with your group and to listen to other people's ideas. This is an important skill to learn, so do not skip over that part of the assignment.

Most importantly, we want the mathematics to make sense to you. Mathematics should not seem like a random and arbitrary set of rules. The problems in this book have been structured so that you and your group, with support from your instructor, can construct, and therefore understand, much more mathematics than you could from being given a rule and assigned a bunch of exercises that all look the same. Understanding mathematics is seeing its relationships and interconnections. There are a few, (but not very many) rules that should be memorized.

DO THE HOMEWORK. Finally, you need to do your homework. No one expects basketball players to become good (or even decent), if they just watch others play. They have to get in there and practice. But they also need to know what to practice. In mathematics, you will often get stuck and not know what to do when you have no friend there and have no one to call to ask for help. In that situation write down what you do know about the problem, write down what you tried, and figure out what your question is. Then write down your question about the problem. This should be enough to convince your instructor that you have really tried to do the problem, and you will be prepared to learn what you need to in order to do the problem at your next class meeting.

Doing this kind of work is what we mean by "doing homework." In fact, the homework questions to which you should give the most attention are the ones you are not sure how to do. The questions you can easily answer do not help you learn anything new; they are just practice. Getting stuck on a problem is your *big opportunity* to learn something. Analyze the problem to find out just what the hard part is. Then, when you do find out how to do the problem, ask yourself, "What was it about that problem that made it hard?" Answering that question will get you ready to handle the next difficult problem so that maybe you won't get stuck, and you will have learned something.

A very useful way to help yourself learn is to use problem-solving strategies such as: finding a subproblem, guess and check, looking for a pattern, organizing a table, working backward, using manipulatives, drawing a graph or diagram, writing an equation, or finding an easier related problem. These strategies are useful not only for solving problems, but for *learning* as well. Much of what you learn in this course will not be brand new, but will build on mathematics you already know, even if it has a new name. So you can use one of the problem-solving strategies reviewed in Chapter 1 to try to go from what you know to what you need to know.

Three of your major goals for this course should be:

- to become a better problem solver
- to become better at asking questions
- to become better at explaining your reasoning.

NOTE TO THE INSTRUCTOR

This text is different. First, it is all problems. Some are developmental, some are investigations, and some are exercises. Some are applications, some are pure mathematics, while others are pure fantasy. Every problem requires students to do something: to organize, to draw, to represent, to graph, to solve, to calculate, to conjecture, to reason, to generalize, to explain, or to justify. To learn mathematics requires doing mathematics. To enjoy mathematics requires persevering through some frustration. Almost everything here is presented as a problem.

Because the basic text is different, there will be other differences too. Some of them are discussed below. In addition we have included extensive notes for instructors based on the experience of those who have used the book so far.

GRAPHING CALCULATORS Students will need to use graphing calculators throughout this class. The materials are designed so that one per pair of students, available for use during class time, will suffice. The homework problems generally require only a scientific calculator, not the graphing calculator. However, the more calculator access the better.

PORTFOLIO ASSIGNMENT Throughout the book, we offer suggested portfolio assignments. Additionally, there are notes to students stating that certain projects and/or lab reports should be written, revised, and placed in the portfolio. This will provide a way to track growth over the course. If you choose to use portfolios as an assessment item, it will be important to discuss with the students what a portfolio is and what it will be used for. If the approach of this course is new to you, you may want to hold off on trying portfolios until the second time around.

GROWTH-OVER-TIME PROBLEMS We have included two growth-over-time problems. One appears toward the end of Chapter 2, again at the end of Chapter 5, and finally at the end of Chapter 7. The other appears at the end of Chapters 3, 6, and 8. These are problems for students to do and put away in their portfolios. After the third try, students should compare all three and reflect on the growth in their knowledge over the course. Some instructors have suggested having students try the first problem sooner than the end of Chapter 2 so they will not be applying what they have learned in the chapter when first trying to solve it; then, when they work on the problem again, their growth will seem all the more impressive.

REVISION Another important practice that has been underemphasized in the past is revision. One way to encourage students to revise their work is to assign grades as follows: A, B, C, or revise (or A, B, revise). Once you see that a paper is below a C in quality, you can just write "revise" on it. Thus, if the student wants credit, he or she must reconsider and revise the complete assignment. A key part of changing the way we teach is to change the way we assess. Students will not believe that we want them to understand the mathematics if we continue to ask only routine exercise-level questions on exams. You can ask more challenging, in-depth questions when you know students will have the opportunity to revisit the assignment.

HOMEWORK For homework, when students are working individually and may not always be able to do every problem, we suggest that they be encouraged to do as much as they possibly can, write what they do know, and then write what it is they need to know to finish the problem. Within each day's set of problems you will find a variety of topics. There are problems about probability and geometry as well as algebra. Some of these may be new to your students. You could skip some but don't be too eager to skip them all. Many instructors report that their students can do the probability problems without any instruction.

Avoid falling into the trap of standing in front of the class "going over the homework." Instructors who have used the texts recommend starting each class with that day's new work and saving any homework review until the end of the class time, the last ten minutes. One idea is to write out complete solutions to a few key problems and give one copy to each group. The group should be able to answer most of each other's questions on the routine problems while you circulate among the groups to deal with the hard problems.

GROUPS Since this is a new course and the methods of instruction, particularly group work, might be new to you, we suggest that you not try to do everything at once. The most important thing to begin with is getting your groups to work efficiently. In a course like this, we expect students to be spending a majority of the class time working in groups. You should *not* be lecturing on the previous night's homework, nor preparing lengthy discussions of what the students will be doing each day. Many sections suggest beginning class by having the students in their groups start directly to work on the problems while you circulate. Don't worry about the fact that you will be answering the same question for several different groups, but be sure to keep moving from group to group. The students must be active learners.

SPIRALING AND TESTING The problems are spiraled. That is, once a concept is introduced, it never goes away. For example, what is developed in Chapter 1 is practiced, used, and developed further in Chapters 2 through 8. Similarly, the focus of each chapter is an investigation or lab that the problems in the preceding chapter have been building toward. This spiraling has important implications for individual testing. Students will not necessarily have mastered an idea by the end of the chapter in which it is introduced. They may not, for example, be good at writing exponential equations (Chapter 4) until the end of Chapter 6 or later. One way to deal with this delay is to make chapter tests cumulative, and test about 60 percent on material from preceding chapters and only 40 percent on the current chapter's content. Another way is to give more problems than you grade and select, for example, each student's 10 best out of 13. That way they can get credit for what they currently know and continue to work on what they don't. Holistic scoring can make grading papers more efficient and encourage revision.

PREPARATION FOR TEACHING The best preparation for each class is to write out the solutions to all the problems you expect your students to do. Not only does this allow you to anticipate where problems and questions might arise; it also gives you solutions that can be given to the students as follow-up. In fact, a good way to deal with homework is to give a solution sheet to each group and ask the students to take a *few* minutes to discuss the problems in their groups. This should not be a lengthy activity, but it does get the students focused on "talking" mathematics as they deal with homework questions. They can start this discussion before class begins if you hand out the homework solution sheet as they are coming in and getting into their groups.

WHAT DO THE BARS AT THE BEGINNING OF EACH CHAPTER REPRESENT?

The six bars on the opening page of each chapter represent the major areas in which students should be making progress throughout the course. At the beginning of each new chapter students should take a look at each of the bars and ask, "How am I doing? What are some of the things I now know for sure in this area? What do I need to do some more work on?"

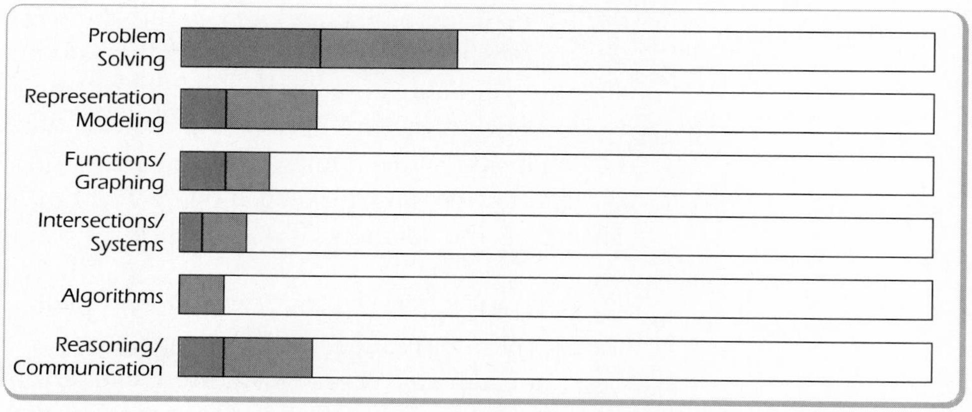

PROBLEM SOLVING A major concern of this course is problem solving. Students should be developing their ability to integrate the use of basic problem-solving strategies such as guess and check and making organized lists with the use of algebraic procedures and graphing.

REPRESENTATION/MODELING Learning to represent situations with diagrams, graphs, equations, and other models has become a much more important aspect of mathematics now that technologies for carrying out algorithmic procedures have been developed. Learning to represent situations in several ways and to move back and forth between representations such as graphs, tables, and equations is emphasized throughout this course.

FUNCTIONS/GRAPHING This course is mostly about functions: linear, exponential, quadratic, other polynomial, logarithmic, rational, and radical plus others that arise out of specific situations. Students will learn to represent functions with tables, graphs, and equations and for each one-to-one function to find an inverse function.

INTERSECTIONS/SYSTEMS Intersections (and nonintersections) of graphs of functions lead to systems of equations (and inequalities), which sometimes can be solved by algebraic means and sometimes cannot. Students will build on what they know about solving linear and quadratic equations to solve 3-by-3 systems of linear equations, 2-by-2 pairings of lines, parabolas, and other polynomials that can be solved algebraically, and some pairings of these with exponential and logarithmic equations for which they will have to use estimation and graphing or some combination of methods.

ALGORITHMS As students gain familiarity through practice with some types of problems that appear routinely as subproblems in other problems, they will need to develop facility with some routine procedures, including the tried-and-true tools of algebraic reorganization, rules of exponents, and others they will identify as they work through the course. Procedures that they should learn so well that they are almost automatic are identified throughout the course, often with a note to include them along with examples in the students Tool Kit (see below).

REASONING/COMMUNICATION Developing the abilities to give clear explanations, to make conjectures, and to develop logical mathematical arguments is an important part of this course. Throughout, students will practice articulating and justifying ideas both orally and in writing, in both mathematical and standard language.

RESOURCE PAGES

The next to a problem indicates that there is a Resource Page for that problem. Resource Pages give additional information in relation to the problem or provide a convenient format for working on the problem.

You may want to make multiple copies of some Resource Pages; others can be used in the book; a few might be copied as overhead transparencies. For example, the Lab Report Write-up Outline (first referenced in EF-1) provides information on writing up lab reports that will be useful for the Sharpening Pencils Lab and several others. Tool Kit blanks (first referenced in PS-29) can be duplicated for use throughout the course. Two copies of the special axes for graphing sequences (BB-16) will make this problem much easier for students and allow them to see the main point of the problems. The Resource Page for EF-21 can be used directly by students or copied as a transparency and used with the whole class. There is also an unreferenced page in the Resource Pages for Chapter 2 that contains graphing grids for use under regular paper in case students need to draw just one graph and don't want to use a whole piece of graph paper. All the Resource Pages are located at the end of both the instructor and student versions.

Resource Pages on finding regression equations are included only in instructor's version so the instructor can decide whether to encourage the use of the graphing calculator for this purpose. Some instructors have found that too-early introduction of use of the calculator to determine best-fit equations detracts from students understanding the ways in which they can develop equations of lines, parabolas, and exponential functions from their graphs.

THE TOOL KIT

The  marks the problems that contain information students should make note of, the tools of algebra that will be useful throughout this and other courses. Encourage students to use the Tool Kit Resource Pages or develop their own format. Some instructors copy the Tool Kit Resources Pages for students on yellow paper and then refer to them as the student's "Yellow Pages". The Tool Kit is a reference students should develop and revise as the course progresses. Many instructors allow the use of the Took Kits during exams. Some restrict the number of pages so students will be forced to revise and condense their notes. Blanks for the Tool Kit are in the Chapter 1 Resource Pages, and blanks for the Calculator Tool Kit entries are in the Chapter 2 Resource Pages.

ANSWERS

Many of the answers for the Extension and Practice problems are in the back of the student text, as well as in the back of this text for your reference. Sometimes we did not give answers but gave suggestions for getting started. In some cases, we gave the problem set-up and a few steps, for others we gave a reference to a similar problem, and for some, where there were several similar exercises to be done, we did one example.

We did not include answers to the problems that are to be done in groups in class in the student text because those should be resolved through group discussion and your guidance. In addition, answers to all the problems (answers, not solutions), except most of those which are graphs, are given with the problems in this instructor's version.

ACKNOWLEDGMENTS

First, we want to thank the high school teachers who used these materials in their classrooms through the developmental stages, for their contributions to the program on which this text is based.

Second, we want to recognize and thank the college instructors who agreed to class-test the first rough adaptation of the high school version. Their discussion and comments have been a tremendous help in preparing of this preliminary version.

Lindsey Bramlett-Smith
Santa Barbara City College; Santa Barbara, California

Jon Compton
University of New Mexico; Los Lunas, New Mexico

Sarah Donovan
Solano Community College; Suisun, California

Maggie Flint
Northeast State Technical Community College; Blountville, Tennessee

Dorothy Hawkes
Solano Community College; Suisun, California

Brenda Jinkins
State Technical Institute; Memphis, Tennessee

Kenneth Johnson
Sierra College; Rocklin, California

Roberta Lacefield
Waycross College; Waycross, Georgia

Derek Lance
Chattanooga State Technical Community College; Chattanooga, Tennessee

Mike Mallen
Santa Barbara City College; Santa Barbara, California

Karla Martin
Walters State Community College; Morristown, Tennessee

Elizabeth Mefford
Walters State Community College; Morristown, Tennessee

Dick Phelan
Sierra College; Rocklin, California

Jim Ryan
Madera Center Community College; Madera, California

Karen Wootton
Indiana State University; Terre Haute, Indiana

Finally, we want to thank, in advance, the college instructors who will be using this preliminary version, for their willingness to work with the text while it is still undergoing revision and for their anticipated suggestions for making it better.

J.K.	K.W.
T.S.	J.C.
E.K.	D.L.
B.H.	

CHAPTER ONE

PROBLEM SOLVING

Getting Started

IN CHAPTER 1 YOU WILL HAVE THE OPPORTUNITY TO:

- review problem-solving strategies and begin to see them as learning strategies;
- recall and use skills that you have learned in previous classes in new contexts of modeling and solving larger problems;
- establish good routines for group work;

BY THE END OF THIS CHAPTER YOU WILL HAVE REVIEWED SOME BASIC ALGEBRA SKILLS AND SHOULD BE ABLE TO:

- solve equations;
- find mathematical models to fit real-life data;
- interpret the meaning of the slope and y-intercept of a linear graph;
- draw graphs of lines and parabolas.

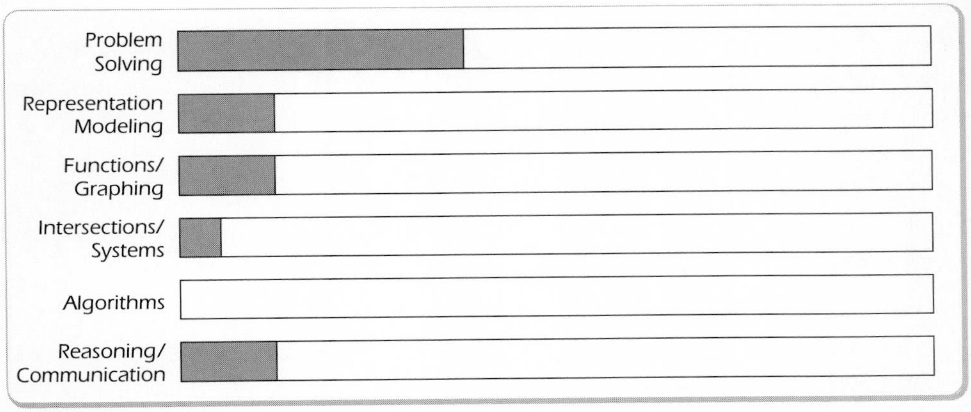

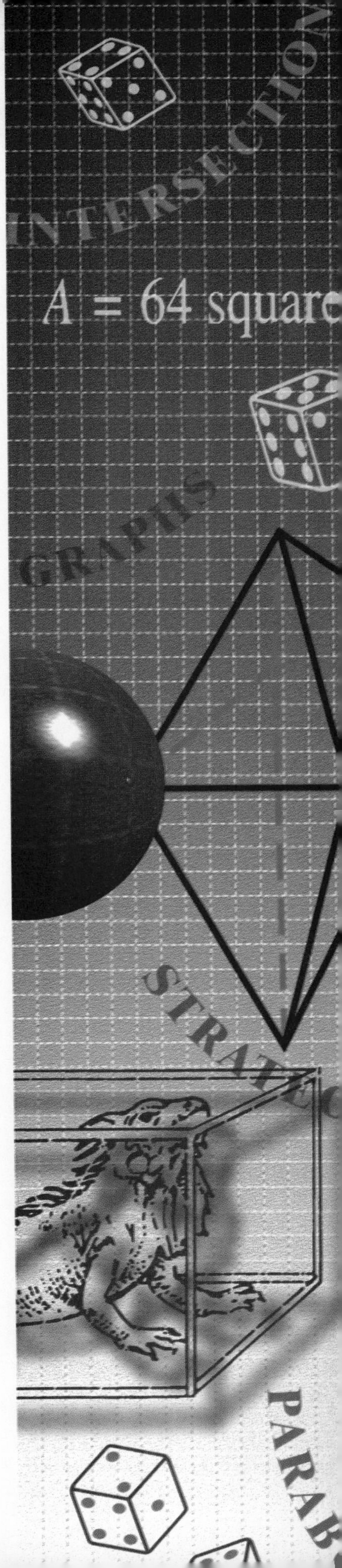

MATERIALS	CHAPTER CONTENTS		PROBLEM SETS
Eight-foot loops of yarn, one for each group	1.1	Getting Started: Guess and Check	PS-1 – PS-12
Function-walk axes	1.2	Data Walk and Graphs	PS-13 – PS-27
Tool Kit Resource Page	1.3	Equations for Graphs, Tool Kit	PS-28 – PS-38
PS-39 Resource Page String, cardboard, and two pushpins for each group	1.4	Subproblems with Digger the Dog, Function Machines	PS-39 – PS-51

1.1 GETTING STARTED: GUESS AND CHECK

Since these materials require that a major portion of class time be devoted to small-group work and since this will probably be a new approach for most students, you will need to discuss the value and reasons for using this approach. However, it is best to begin the class with the Yarns activity (PS-1) and very little (or no) discussion about the course. As the students ***experience*** *that the course is different, discuss the reasons for the differences. The Yarns activity is included as the first problem (it can be on the overhead projector if students don't have books yet) to get students to jump into cooperative group work immediately.*

After the students have done the Yarns activity, then you can have a brief whole-class discussion about the course and the role of group work. Emphasize the fact that, just as the Yarns problem is easier when done with others, so will be the mathematics we learn. You can also highlight key points from the introduction. At this point you may wish to discuss your course syllabus, grading, and so forth. We don't expect you to spend too much time on these "business" items. The first day should be a learning day. Although this is an introductory chapter, important ideas are introduced here. In particular, we are developing problem-solving strategies that will be used throughout the course. We do ***not*** *expect mastery of these ideas by the end of this chapter; hence there is no need to give an individual test for this chapter.*

A guess-and-check approach to word problems and writing equations is the focus for this section. Model PS-2 for the whole class, drawing as much information as you can from the students. PS-2 does need to be led since it is a dynamic process.

The purpose of PS-1 is to give the students an opportunity to get acquainted as they work together in groups. Each group needs an 8-foot length of yarn with the ends tied together, making a loop. This is a group-building activity. Don't skip it.

PS-1. Yarns In your group, make each of the following shapes with the yarn provided by your instructor. Show the shapes to your instructor as you complete them.

a. square

b. 5-pointed star

c. tetrahedron

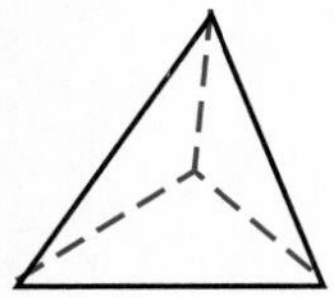

d. square-based pyramid

e. octahedron

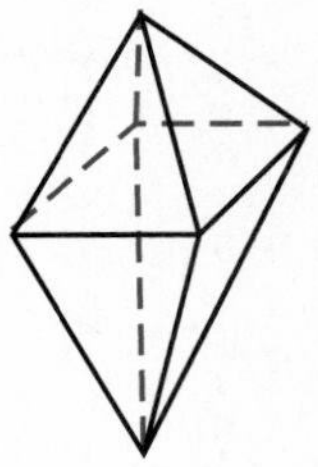

f. cube

We recommend modeling a systematic guess-and-check approach as laid out below in PS-2. Start with no table, just the guess, and build the table across, step-by-step as you do each necessary calculation to check the guess. It is important not to get to the solution too soon, so be sure to start with a couple of numbers that do ***not*** *work. Three guesses are usually enough for students to see the pattern of the relationship, so they should make at least three guesses. After modeling problem PS-2 for the whole class, have students do PS-3 the same way in their groups.*

As you move from group to group, ***insist*** *that they use guess and check and that they each record and check at least three guesses. It may seem strange to insist that they guess when they may want to write an equation, but the point here is to learn a strategy that will be useful for many different problems, not just for getting the answer to this problem. Be sure students organize their guesses and the checking into a table as they go. They should* ***not*** *try to set up the whole table before starting—and neither should you when you model this strategy!*

When PS-2 and PS-3 are complete, call the groups back together and suggest "guessing x" as the next guess in the PS-2 table. Then work your way across the table to show how you can arrive at the equation, which students (not you) should then also solve just for verification. Circulate as students attempt to solve this equation, and assess their elementary algebra skills by seeing how much or how little help they need.

PS-2. Example of Using a Guess-and-Check Table to Solve a Problem The length of a rectangle is 5 centimeters more than twice the width. The perimeter is 60 centimeters. Use a guess-and-check table to find how long and how wide the rectangle is.

STEP 1 Start a table. Why is the width a good thing to guess?

Guess Width	

STEP 2 Make a guess:

Guess Width	
10	

Then test your guess by writing down the steps you take to check it. These steps are your column descriptions.

STEP 3 Calculate the length:

Guess Width	Length
10	25

STEP 4 Find the perimeter. Remember, rectangles have two "width" sides and two "length" sides. Drawing a picture of the rectangle is a good organizer here.

Guess Width	Length	Perimeter
10	25	70

STEP 5 Check the perimeter against 60, and identify it as correct, too high, or too low.

Guess Width	Length	Perimeter	Check P=60
10	25	70	too high

STEP 6 Go back to Step 2 and make a new guess. Repeat this process until you find the solution.

STEP 7 Write a sentence stating the solution.

PS-3. Solve the following problem by using a guess-and-check table similar to the one in PS-2. Your purpose in making and checking the guesses is to understand the mathematical relationship, so make at least three guesses even if you guess right on the first or second try. Be sure to write a sentence stating the solution.

One number is 7 more than a second number. The product of the numbers is 2958. Find the two numbers. ***51 and 58; −51 and −58 also work, but we don't expect these answers at this point.***

PS-4. Now go back to the end of your table for problem PS-2 and guess x for the width. Fill in each column of the table using x, and write an equation that represents this problem. Do the same for PS-3. Then see if you can solve the equations you wrote to get the same results. ***$2x + 2(2x + 5) = 60$; $x(x + 7) = 2958$***

PS-5. Write an equation to help you solve the following problem. Doing a guess-and-check table first will help.

A cable 84 meters long is cut into two pieces so that one piece is 18 meters longer than the other. Find the length of each piece of cable. ***$x + (x + 18) = 84$; pieces are 33 and 51 meters.***

EXTENSION AND PRACTICE

PS-6. Write an equation for the following problem:

A monkey stores coconuts and bananas in a small cave. He has 17 more coconuts than bananas and a total of 53 coconuts and bananas. How many coconuts does he have? A guess-and-check table can help you solve this problem. ***18 bananas, 35 coconuts; $2x + 17 = 53$***

PS-7. At a recent gathering of a local chapter of the Veggie Lover Society, Carlos ate 12 more turnips than Judy and 16 more than David. If the three of them ate a total of 83 turnips, how many did Carlos eat? Find the solution, and also write an equation. ***Carlos ate 37 turnips; $x + x - 12 + x - 16 = 83$***

PS-8. Find the error in this problem. Explain what the error is, and show how to do the problem correctly.

$$3(x-2)-2(x+7)=2x+17$$
$$3x-6-2x+14=2x+17$$
$$x+8=2x+17$$
$$-9=x$$

Error: negative sign is not distributed; $x = -37$.

PS-9. Solve each of the following equations showing all work. This should be review, so we have included the answers for you to check. For a reminder of how to solve these equations, read and work through the examples in Appendix A.

a. $3x+6=-45$ — ***$x = -17$***

b. $2x-\frac{2}{5}=x+8$ — ***$x = \frac{42}{5} = 8.4$***

c. $\frac{3}{4}(x+1)=9$ — ***$x = 11$***

PS-10. There will be elementary probability problems throughout the text because some students may take an elementary probability and statistics course after this. You should be able to figure out these probability problems by using common sense or by consulting with your group.

Uyregor has a collection of normal, fair dice. He takes one out to roll it.

a. What are all the possible outcomes that can come up? *1, 2, 3, 4, 5, 6*

b. What is the probability that a 4 comes up? *1/6*

c. What is the probability that the number that comes up is less than 5? *2/3*

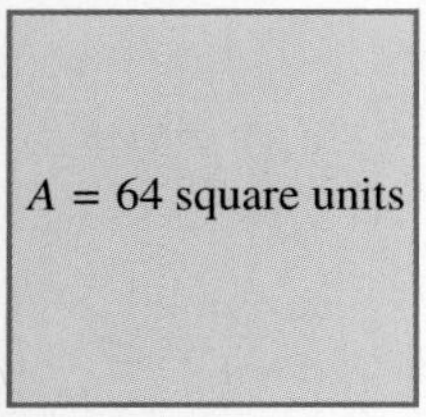

PS-11. The accompanying figure shows a square. If the area of the square is 64 square units, what is the length of one side of the square? Explain how you got your answer.

PS-12. Read or reread the introduction to the course. Focus on the goals of the course, and then write a paragraph that outlines your goals. Include information about how this course will assist you in fulfilling your goals. Be prepared to turn it in the next time your class meets.

1.2 DATA WALK AND GRAPHS

There is a lot to do today, so no time should be spent going over homework or conducting business items at the start of class. If the activity moves smoothly enough, you will have time to handle administrative items later, while the groups are discussing and working. It may seem as if we are dealing with a lot of deep topics today (asymptotes, domain, range, etc.), but we don't expect the students to master these concepts at this point. We are trying to give them some realistic situations in which to start thinking about these ideas and to refer back to later.

Today we take the students outside for a Data Walk. The xy-coordinate system should be set up or marked off before class. Some instructors have used two pieces of rope marked with colored tape to indicate the units. You could also use chalk to mark and label the axes on the pavement. You will use only the first quadrant, but we hope students will ask about the other quadrants. Whichever method you use to produce the axes, the units should be about 30 inches apart to give students enough room to stand comfortably side by side along the x-axis. For all the graphs, 0 to 10 on the x-axis and 0 to 16 on the y-axis will be adequate .

Before going outside, give each group an integer (0 to 8) and five different-colored index cards. Some students may get more than one. Tell everyone to bring his or her calculator, pencil, and paper outside. Instead of a function rule, you will be giving students situations to graph. The five situations are described in PS-13. When you go outside, have those students with a red card (there should be one from each group) stand side by side on the x-axis, facing straight ahead into the first quadrant with the mark for their number between their feet. The students who are observing should stand behind them facing the same direction because it corresponds to the standard orientation we use when graphing. Start by telling them, "Be sure you are standing on the horizontal axis with the mark that corresponds to your number between your feet. This is the x-axis. Your number represents gallons of milk." Then read Situation 1 and say, "Calculate your cost using your calculator (conferring with your group members, if necessary). When I say 'go,' take that many steps forward. Ready? ***Go!****"*

Mistakes will be made. Some students may have difficulty figuring out that they have to estimate and stop at 1.8 for $1.80 or 3.6 for $3.60, but resist telling students where to stand. Let them correct their mistakes. Each should walk a path parallel to the y-axis, but there will be some who try to turn toward it. Be sure the observers record as much detail as they can about the resulting graph. Have the students with the remaining colors do Situations 2 through 5. The fifth one, the area of the rectangular pen, is difficult, but we want the students to see a parabola.

After the walk, your students should have enough understanding to answer the questions in PS-14 through PS-18.

PS-13. Today your instructor will be leading you in an activity called a **Data Walk**. Your instructor will give you an integer and an index card. Write down the integer and don't forget it. You will need this card, your calculator, a pencil, and paper for this activity. Your results for this problem will be used again and again throughout this chapter and Chapter 2. Keep your results handy.

Once the problem is set up, your instructor will give you five situations to graph. When your card color is called, you should line up on the x-axis as illustrated.

Situation 1: Milk costs $1.80 per gallon. Your integer is the number of gallons of milk you are going to buy. What is the cost?

Situation 2: Driving at a constant rate on the freeway, your car consumes 2 gallons of gas each hour. You start with 13 gallons. Your number is the number of hours you have been driving. How many gallons are left in your tank?

Situation 3: For Davis's birthday, his grandma gives him $5.00, which he quickly puts into his piggy bank. Each week, Davis adds one quarter to the bank. Your number is the number of weeks that have passed since his birthday. What is the amount of money in his piggy bank?

Situation 4: James frequently travels from his house to his girlfriend's house 10 miles away. Sometimes he drives or takes the bus, other times he rides his bike, occasionally he runs, or even walks. Your number is the number of hours it takes him to make this trip. What is his average speed?

Situation 5: Miguel needs to make a rectangular pen for his pet iguana out of 16 feet of fencing. Your number is the length of *one* side of the rectangle. What is the area of the pen?

As a group, re-create the graphs you made. Plot on graph paper the points the people represented. Make sure each group member has a copy. Later, you will need these graphs and the answers to the following questions.

PS-14. For the first graph you created in the Data Walk, each person's integer represented a number of gallons of milk. What did the y-axis (the output) represent? Clearly label the axes as "number of gallons" and "total cost." Label the axes appropriately for the rest of the graphs you created. *Situation 2, no. of hours driving vs. no. of gallons of gas left; 3, no. of weeks since Davis's birthday vs. amount of money in the bank; 4, no. of hours vs. speed; 5, length of one side vs. area of pen.*

*The next problem asks the students to label and interpret the y-intercept for each graph. The graph in Situation 4, which is time versus speed, does not have a y-intercept. If there is time, you may want to touch on the idea of an asymptote, but **only** when the questions start to come up. Do not give a formal description at this point. As the students ask "What should we do if there is no y-intercept?" you can respond with, "Can you tell me **why** there is no y-intercept?"*

PS-15. For each graph in PS-13, label the point where the graph crosses the y-axis by giving its coordinates. This point is called the **y-intercept**. In the graph for Situation 1, the y-intercept has the coordinates (0, 0). Verify this! This point represents the relationship: "zero gallons of milk cost $0.00." For each of the other graphs, write the coordinates of the y-intercept and say what this point represents. *Situation 2, (0, 13); before you start driving on the freeway (and using gas), there are 13 gallons of gas in your tank. 3, (0, 5); on his birthday (0 weeks after his birthday) Davis has $5.00 in his bank. 4, there is no y-intercept because James can't make the 10-mile trip in 0 hours no matter how fast he travels. 5, (0, 0); a pen with one side length of 0 has zero area.*

PS-16. Choose one *other* point (not the y-intercept) in each of the situations in PS-13, and interpret the point. By **interpret** we mean write out in words the real-world meaning of what the numbers represent.

PS-17. For each of the graphs in PS-13, you plotted only a few points, namely when $x = 0, 1, 2, \ldots, 8$. Would it make sense to use other numbers for x? For instance, in Situation 1, could x be negative? How about $x = 0.5$? Explain. *We hope students will realize that for each situation, negative values are not allowed, but decimals (or fractions) are. If you can lead them to see that it makes sense to connect the points, great. If not, don't worry about this now. We will come back to it.*

PS-18. The graphs for Situations 1, 2, and 3 in PS-13 are all lines. How does the steepness of each line and its direction (up or down) relate to the situation? Explain completely. *We are hinting at the slope concept here, but they may not get it. We will deal with it in the next lesson. We hope students will notice that one graph represents a decreasing quantity and that the line of data points "slants downward."*

EXTENSION AND PRACTICE

PS-19. In the debate over whether or not to raise the speed limit, some people bring up the amount of time a driver saves by driving faster. On your paper, draw a set of axes similar to the one shown here.

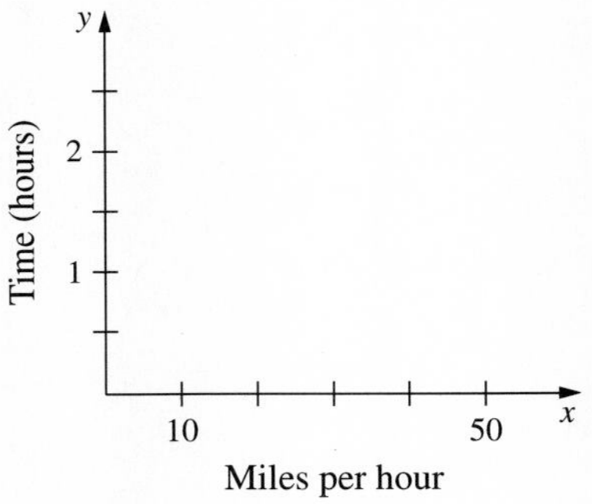

a. How much time will a 50-mile trip on the freeway take if the driver is averaging 25 miles per hour? 40 miles per hour? 50 miles per hour? Represent these points on your graph. *2 hr, 1.25 hr, 1 hr*

b. How long does the trip take if the driver averages 75 miles per hour? Represent this on your graph. *40 min*

c. How much time is saved by driving at an average speed of 75 miles per hour rather than 55 miles per hour on this 50-mile trip? *At 55 mph, it takes approximately 55 min, so about 15 min would be saved.*

PS-20. Find the error in this problem. Explain what the error is, and show how to do the problem correctly.

$$4x - 2 = 5(x + 7)$$
$$4x - 2 = 5x + 35$$
$$x - 2 = 35$$
$$x = 37$$

Subtracting 5x from both sides should give −x; x = −37.

PS-21. Solve each of the following equations, showing all work. If you've forgotten how to do any of these, check Appendix A for examples.

a. $8 - x = 3(x + 2)$ *x = 1/2*

b. $5(4 - x) = 38$ *x = −18/5 = −3.6*

c. $\frac{x}{5} = \frac{5}{x}$ *x = 5 or −5*

PS-22. Solve the following problem and write an equation. You may want to start with a guess-and-check table.

Howie has 23 bills in his wallet. Some are \$10 bills and some are \$50 bills. If the wallet has \$550, how many \$50 bills does he have? *8*

PS-23. What is the x-coordinate for the y-intercept of any graph? Explain why it is this value. *0*

PS-24. If it takes 11 yards of canvas to make three tents, how many yards of material will it take to make 30 similar tents? *110 yd*

PS-25. Solve each of the following for x:

a. 42 percent of x is 112. *≈ 266.67*

b. 42 is x percent of 112. *37.5%*

c. 27 is x percent of 100. *27%*

d. 27 percent of 500 is x. *135*

PS-26. What is this graph telling you? Explain completely. *It shows a rapid increase in the number of coffee shops starting from mid-1995.*

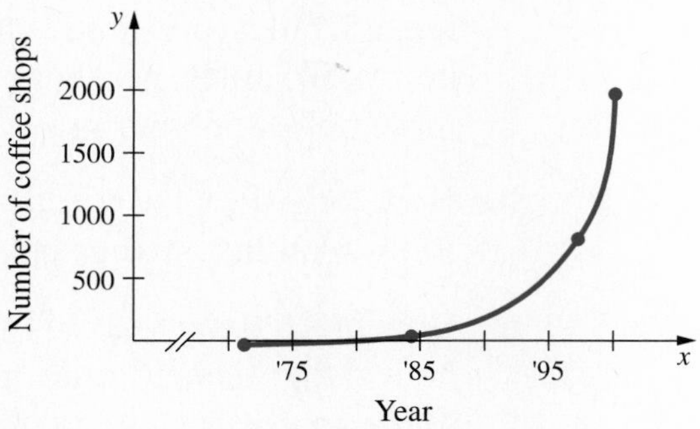

PS-27. Rank the following lines from steepest to least steep. Justify your order.

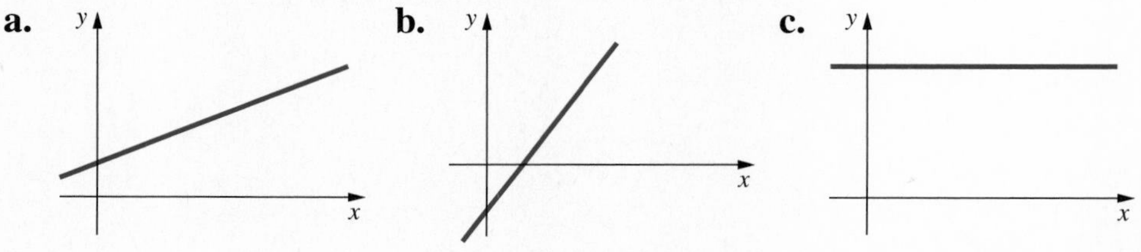

b, a, c

1.3 EQUATIONS FOR GRAPHS, TOOL KIT

Today you will develop equations for your graphs from the Data Walk. For now, we deal only with the three graphs that turned out to be lines. We will also introduce the idea of a Tool Kit and suggest your first entry.

PS-28. In the Data Walk in PS-13, the first three graphs were all **linear** (straight lines). Find your three linear graphs from the Data Walk. You should have the y-intercepts and axes clearly labeled. Parts a–d spell out what you need to do for the "gallons of milk" graph.

a. Find your data points for the first graph. Make a table on your paper like the one shown here, and record your points in the table as shown.

The first column is sometimes called the input; the second is the output.

Number of Gallons of Milk	Cost in Dollars
0	0
1	
2	
3	
4	
⋮	

b. For the "milk" graph, write out in words how to find the cost of 2 gallons of milk, the cost of 3 gallons of milk, the cost of 5 gallons of milk, and the cost of x gallons of milk. *The costs are $1.80 \cdot 2$, $1.80 \cdot 3$, $1.80 \cdot 5$, and $1.80 \cdot x$.*

c. We can represent this data relationship with an equation. If x represents the number of gallons of milk and y represents the cost, write an equation to represent this data. The variable x is referred to as the **input** and y is the **output**. Using your equation, verify that each of the inputs (0 through 8) gives the correct output.

d. Lines can usually be represented by equations in the form $\boldsymbol{y = mx + b}$. Compare your equation from part c with $y = mx + b$. What number does the $\boldsymbol{m}$ correspond to in the "milk" equation? Describe in words what this number represents? What number corresponds to $\boldsymbol{b}$? What does it represent? *1.8 represents dollars per gallon; $b = 0$, the cost of 0 gallons of milk.*

Some groups may have difficulty with part e because we don't walk them through the development of writing the equation as we did in part b. If your students need the extra guidance, remind them to relate each input value 0 through 8 to its output, and then after completing the table for 0–8, to use an x-value as the input value. Prompt them, "After 8, use x. Then what goes in the y column?"

e. For each of the other two linear graphs, "freeway driving" and "Davis's savings," use an approach similar to what you did for parts a through d above, and write equations (using x and y) that represent the data. For each equation, state what x and y represent. *(2) x = number of hours of driving, y = gallons of gas left; $y = 13 - 2x$. (3) x = number of weeks since Davis's birthday, y = amount of money in the bank; $y = 0.25x + 5.00$.*

NOTE When you see the symbol beside a problem, that means there is a Resource Page for use with that problem. The Resource Pages are included at the end of the text. Some instructors will duplicate them for their classes, but if that is not the case, you may want to make a copy or, in some cases, several copies for yourself. The Resource Pages are designed to make your work easier.

PS-29. This is the first of many **Tool Kit** problems. Throughout the text there will be problems in which we recommend you stop and make a note of some important and useful information. These notes are for your own use. Write them so they make sense to you, and revise them as you learn new related information. Including specific examples is often a good idea.

You may design your own Tool Kit sheet, or you may want to make copies of the one included at the end of the text. It is a good idea to keep several extra blank copies handy. Since you will probably be able to use it on tests as well as on assignments, you will want to keep it up to date. Tool Kit problems are marked with the symbol .

For your first entry, write down what you know so far about linear equations. A suggestion would be to include an equation, its graph, and as much as you now know about each of the following:

- $y = mx + b$
- slope
- x-intercept
- y-intercept

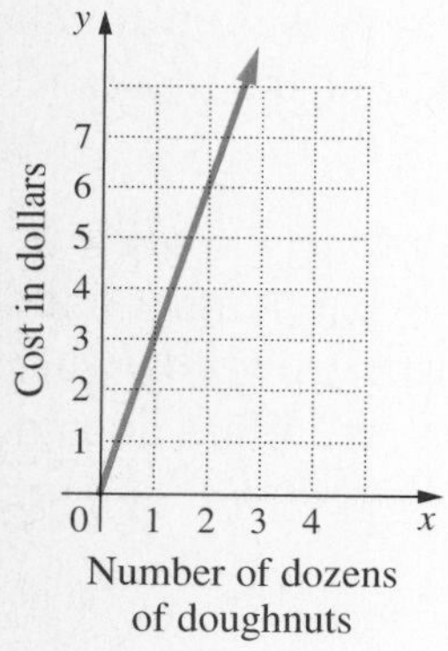

PS-30. Consider the graph shown at the left.

a. What does the graph represent? ***the cost in dollars for a certain number of dozen doughnuts***

b. What does the y-intercept represent? ***Zero dozen doughnuts cost $0.00.***

c. Write an equation that represents this graph. State what x and y represent. ***x = number of dozens of doughnuts, y = cost of doughnuts; $y = 3x$.***

d. If doughnuts were $4.50 per dozen, would the line be steeper? Explain why or why not.

e. Write an equation for day-old doughnuts that sell for $1.75 per dozen. How would the steepness of this graph compare with the graph in part a?

EXTENSION AND PRACTICE

PS-31. For each equation in parts a and b, what is y when $x = 2$? What is y when $x = 0$? Where would the graph of each equation cross the y-axis?

a. $y = 3x + 15$ ***21, 15, (0, 15)***

b. $y = 3 - 3x$ ***–3, 3, (0, 3)***

c. Now, for each equation in parts a and b, what is x when $y = 0$? ***–5, 1***

d. Where does each graph cross the x-axis? ***(–5, 0), (1, 0)***

PS-32. Find the error in the problem. Explain what the error is and redo the problem showing a correct solution.

$$\frac{5}{x} = x - 4$$
$$x \cdot \frac{5}{x} = x - 4$$
$$5 = x - 4$$
$$9 = x$$

Error in line 2; multiply the expressions on both sides of the equation by x, $x = -1, 5$.

PS-33. The perimeter of a triangle is 76 centimeters. The second side is twice as long as the first side. The third side is 4 centimeters shorter than the second side. Draw a picture, and then write an equation. Solve the equation to find the length of each side. *16 cm, 32 cm, 28 cm*

PS-34. Solve each of the following equations. Show all your steps or explain your solution.

a. $3x + 5 = 8x - 72$ *15.4*

b. $7x + 2(3x + 1) = 5(6x - 3)$ *1*

c. $6x - 7(x + 4) = 5 + 2(x + 3)$ *13*

PS-35. Graph the equation $y = 7 - x$ by setting up a table of values and choosing the input (x values) to be 0, 1, 2, and 3. For each input value x, calculate the output y and plot the points. Write a description of a real-world situation that this equation might represent.

PS-36. *Help*! Dogbert just ate Dilbert's new project! Dilbert had spent several hours in the lab building and measuring the lengths of the edges of three different cubes. He left the cubes in the lab and just brought home the edge lengths and the volumes. Dogbert's bite removed the side lengths, and all that remained were the volumes. Help Dilbert out by finding the lengths of the edges of the cubes. Explain how you found these lengths.

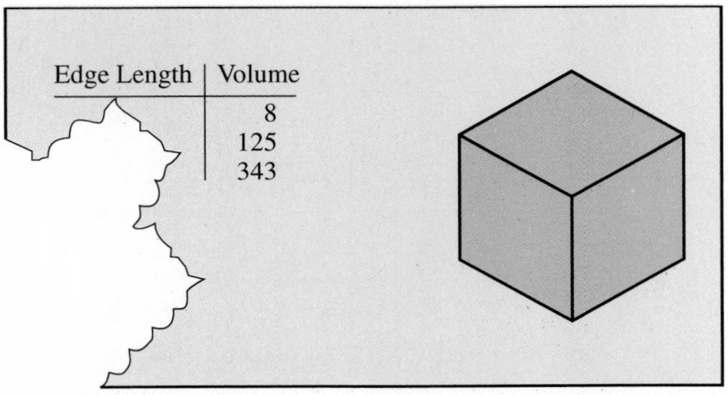

Students who have not taken a geometry course recently may not know how to start the following problem. We include it to provide practice in using ratios and writing equations, and the students' answer section shows how to do part a. However, if you think students will need an introduction to similar triangles, then it would be better to skip the problem rather than to digress.

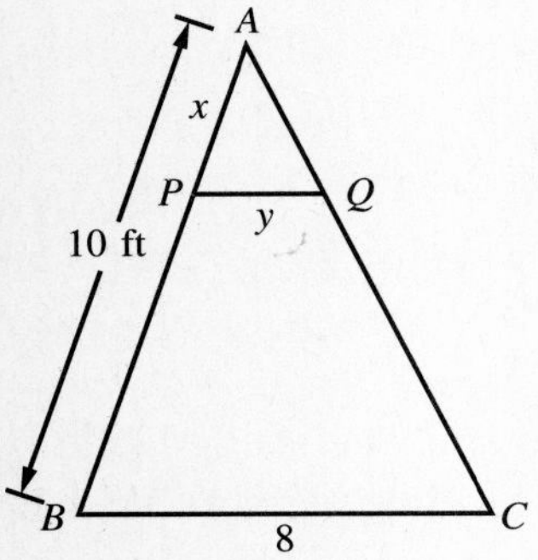

PS-37. In ΔABC shown here, $\overline{PQ}$ is drawn x units down on side $\overline{AB}$, and is parallel to $\overline{BC}$. $\overline{AB}$ has a length of 10, $\overline{BC}$ has a length of 8. To answer the following questions, it might be helpful to redraw the small triangle, ΔAPQ, and the large triangle, ΔABC, separately. Show all work.

a. If $x = 1$, what is the length of y? *4/5 = 0.8*

b. Suppose $x = 2$. Now what is the length of y? *8/5 = 1.6*

c. Suppose $x = 3$. Now what is the length of y? *12/5 = 2.4*

d. Write an equation relating x and y using the form "$y =$." *y = 8x/10, y = (4/5)x, or y = 0.8x*

e. Use your equation to find y when $x = 7$. Explain in words what this means. *y = 5.6*

f. If you graphed your equation, what do you think the graph would look like? Explain how you know. ***a line passing through origin with a positive slope***

g. The relationship described in this problem is often called **direct variation**. Why does this description make sense?

PS-38. A card is drawn from a well-shuffled deck of 52 playing cards. What is the probability that a king is drawn? *4/52 = 1/13*

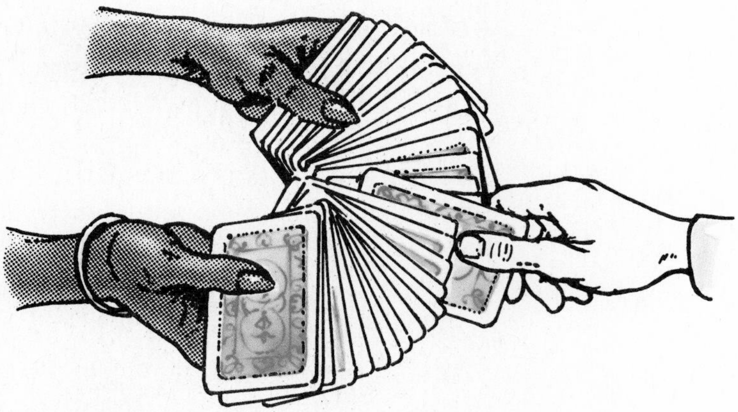

1.4 SUBPROBLEMS WITH DIGGER THE DOG, FUNCTION MACHINES

Identifying subproblems is an important and useful problem-solving strategy we will revisit. Most students at this level can identify and solve subproblems even if they don't know the name of the strategy. The point of giving the strategy a name, however, is so that students can call the technique explicitly to mind when they need to use it.

It is very important to name and discuss problem-solving strategies as students use them throughout the course.

Strategies we expect students to use include guess and check, which we have already seen, organized lists, patterns (which we have also seen to some extent), making diagrams, working backward, eliminating possibilities, using a manipulative, solving a simpler or related problem, or making a model. Students should come to see problem-solving strategies as strategies for building understanding, and for learning. Guess and check, for example, is a valuable numeric strategy for getting inside and analyzing, but the next step is organizing and ***abstracting****—a major goal of this course.*

In this next problem, we ask the students to identify the subproblems involved. Seeing how the rope will bend around the shed is difficult for most students. Necessary visual tools for each group are string for the rope, a flat piece of cardboard and pushpins to mark the corners of the shed. Don't underestimate the students' capabilities. Instructors have reported that even those without any geometry experience can do this problem. You may have to remind them that the area of a circle is πr^2 and offer some hints as to how to find the area of the quarter of the circle, but they ***can*** *do the problem.*

A copy of the diagram for this problem is included as a Resource Page, so students can draw in the regions Digger can reach.

Most problems that you will meet in this course, in other mathematics courses, and in life, are made up of smaller problems whose solutions you need to put together in order to solve the original problem. These smaller problems are called **subproblems**. For example, you cannot solve the next problem directly. First you must solve the subproblems, which are not stated explicitly but which are necessary to solve in order to solve the larger problem.

Remember,  means there is a Resource Page for the next problem.

PS-39. Logan's dog, Digger, was constantly digging his way under the fence, escaping from the backyard, and making a nuisance of himself in the neighborhood. Digger was also in big trouble with Lula Mae, Logan's wife, who enjoyed her leisure time away from her law firm by taking care of her flower garden. They decided they would have to keep Digger tied up while they were away during the day. After a long discussion about how much of the backyard Digger should be allowed to destroy, they agreed that at least two-thirds of the yard should be safe for planting flowers and other plants. They tied Digger to the corner of the shed, at point A, with a 25-foot rope. Will this satisfy Lulu Mae? Answering the questions below may help you answer this question. ***Digger can reach 1492.26 sq ft, about 30% of the yard.***

a. Draw the region in which Digger is free to roam. In order to do this, it will help to place the Resource Page (see the diagram on the next page) on top of a piece of cardboard and place a pushpin at the two corners of the shed marked A and B. Anchor a string at point A to represent Digger's leash. Discuss with your group members why you should do this.

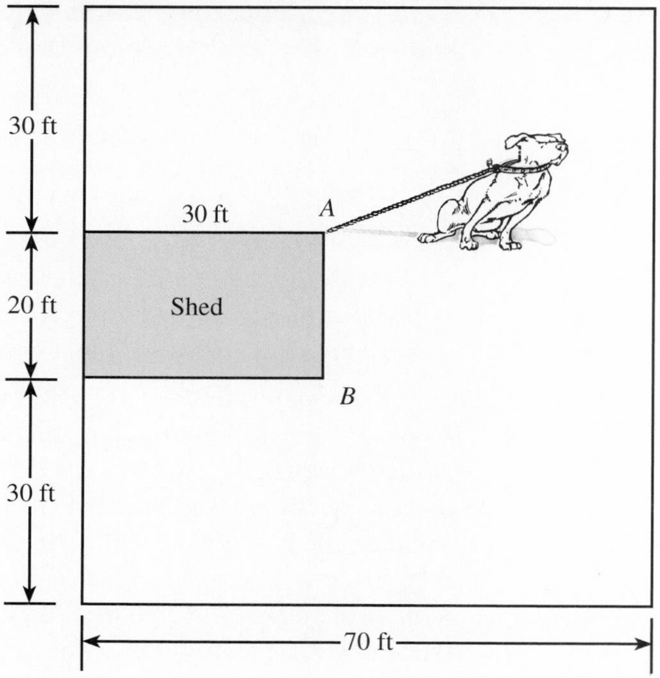

We have told the students to set up a model, using string and a pushpin for each corner. Have students use a piece of string to see how the rope wraps (bends) around the building. The pushpins should help students to see the corner at B as a new center.

b. Before actually working out any answers, write down all of the subproblems you need in order to solve the problem.

c. Now solve each subproblem and answer the question.

We want all students to think about what the subproblems are. Circulate from group to group to see whether they are getting started with these problems, and help those who don't have a clue. Responding to their questions with another question—such as "You want to know how to find the area of that ***quarter*** *of a circle?" or "Can you find the area of the whole circle? How much of it do you have?"—can clear up questions. Be sure everyone answers the question in the original problem. If some students need a challenge, you might ask, "What is the longest rope Digger can have and still leave two-thirds of the yard for flowers?* ***It's about 26.3 ft; students would probably guess and check.***

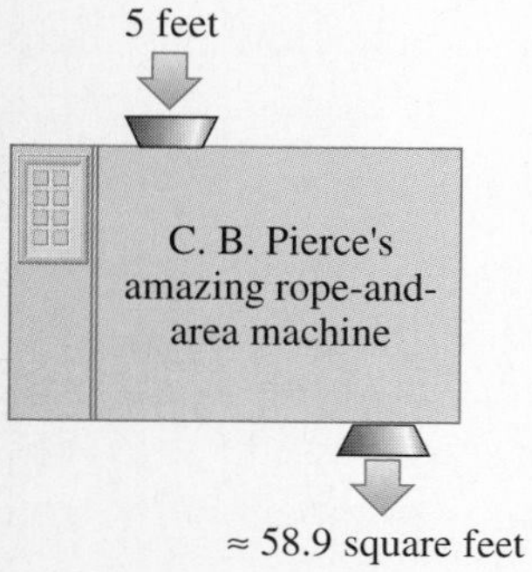

PS-40. How will your solution to the previous problem change if we lengthen Digger's rope? Explain. ***The longer the rope, the more area Digger gets into.***

PS-41. Carmichael B. Pierce, mathematician extraordinaire, has invented a fascinating machine that he hopes to sell to Logan and Lula Mae to cut down on the amount of work they are doing in calculating the destruction area of their dog, Digger. All you have to do is drop in a specific rope length, and the machine works out the area.

If this machine really works and Lula Mae puts in a specific rope length, is it possible for two different answers to fall out? That is, if she puts in 25 feet today, and puts in 25 feet again two weeks from today, will it give her the same answer? Explain. *No, yes*

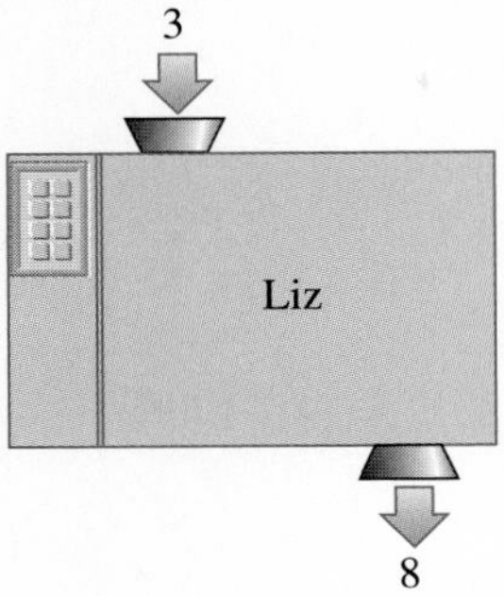

PS-42. Elizabeth's assignment for her math class was to make up a rule for a **function machine**. (She called the machine "Liz.") If we put 3 into her machine, the output is 8. If we put in 10, it gives us 29, and if we put in 20 it gives us 59.

a. What would her machine do with 5? with −1? with x? A table may help. *14, −4, 3x − 1*

b. Write a rule for her machine. *Multiply by 3 and then subtract 1.*

EXTENSION AND PRACTICE

PS-43. Stanley made a different machine. Here are four pictures of the same machine. Find its rule. *square*

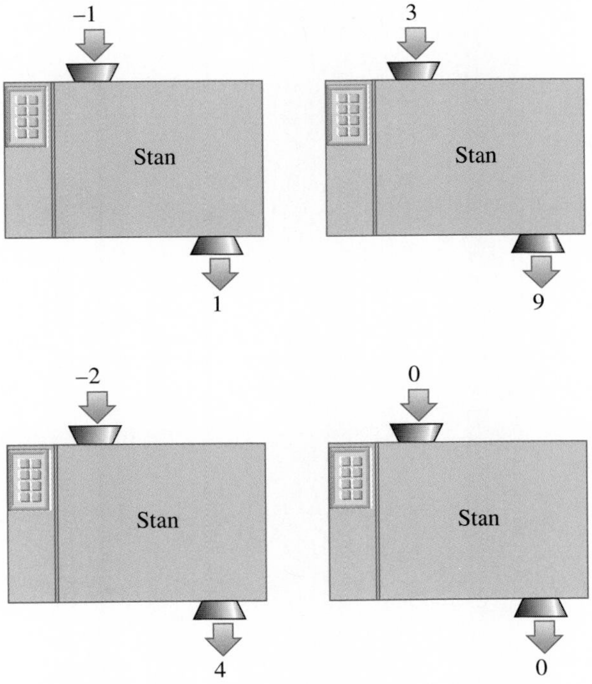

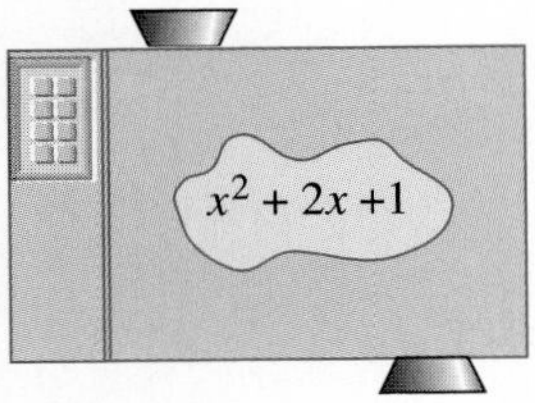

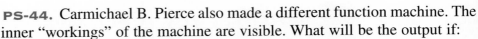

PS-44. Carmichael B. Pierce also made a different function machine. The inner "workings" of the machine are visible. What will be the output if:

a. 3 is dropped in? Show the steps to calculate the output. *16*

b. −4 is dropped in? Again show the steps. *9*

c. −22.872 is dropped in? *478.384384*

Students may not be able to solve the next problem. If they can get the answer(s) by guess and check, that is fine as long as they really do check.

PS-45. If 1 is the output from Carmichael's function machine, how can you find out what number was dropped in? Find the number or numbers that could have been dropped in. ***0 and –2***

PS-46. Find the error in the following problem. Describe the error, and show how to solve the problem correctly:

$$x^2 - 10x + 21 = 5$$
$$(x-7)(x-3) = 5$$
$$x - 7 = 5 \quad \text{or} \quad x - 3 = 5$$
$$x = 12 \quad \text{or} \quad x = 8$$

Equation is not set equal to zero; x = 8, 2

PS-47.

The Quadratic Formula	An important tool that you might have already used is the **quadratic formula**. This formula is particularly helpful in finding the x-intercepts of parabolas. The formula states: $$\text{If } ax^2 + bx + c = 0, \quad \text{then} \quad x = \frac{-b \pm \sqrt{b^2 - 4ac}}{2a}.$$

For example, suppose we wanted to find the x-intercepts of $y = 2x^2 - 3x - 3$. First we would let $y = 0$ (write an explanation in your Tool Kit for why we let $y = 0$) to get this equation:

$$0 = 2x^2 - 3x - 3$$

Then, since this does not factor easily, we must use the quadratic formula to solve for x. First identify a, b, and c. Since $a = 2$, $b = -3$, and $c = -3$, the quadratic formula gives us these solutions in ***radical form***:

$$x = \frac{-(-3) \pm \sqrt{(-3)^2 - 4(2)(-3)}}{2(2)}$$
$$x = \frac{3 \pm \sqrt{9 + 24}}{4}$$
$$x = \frac{3 \pm \sqrt{33}}{4}$$

To find the solutions in **decimal form**, use your calculator to take the square root of 33:

$$x \approx \frac{3 \pm 5.745}{4}$$

So, $x \approx \dfrac{3 + 5.745}{4}$ and $x \approx \dfrac{3 - 5.745}{4}$

$x \approx 2.186$ $\quad$ $x \approx -0.686$

Be sure you include all the information you need about the quadratic formula in your Tool Kit.

PS-48. Now solve these equations. When it makes sense to do so, give the solutions in both radical and decimal form. Show all your work.

a. $x^2 - 5x + 3 = 0$ *$(5 \pm \sqrt{13})/2$; 4.30, 0.70*

b. $x^2 + 3x - 3 = 0$ *$(-3 \pm \sqrt{21})/2$; –3.79, 0.79*

c. $3x^2 - 7x = 12$ *$\left(7 \pm \sqrt{193}\right)/6$; 3.48, – 1.15*

d. $2x^2 + x = 6$ *–2, 3/2*

PS-49. Write an equation, two equations, or use a guess-and-check table to help you solve the following problem.

A rectangle's length is four times its width. The sum of two consecutive sides is 22. How long is each side? *$l = 4w$, $l + w = 22$, $l = 17.6$, $w = 4.4$*

PS-50. Consider the equation $4x - 6y = 12$.

a. What do you suppose the graph of this equation looks like? Justify your answer.

b. Solve the equation for y (using the $y =$ form), and graph the equation. *$y = (2/3)x - 2$*

c. Explain how to find the x- and y-intercepts algebraically. Be thorough enough so someone who is still having trouble with the algebraic method can understand your explanation. ***Substitute x = 0 and solve for y. Substitute y = 0 and solve for x.***

d. Which form of the equation is best for finding intercepts quickly? Why?

e. Use the intercepts you found in part c to graph the line. Did you get the same line you got in part b? Should you? Explain. *Yes*

PS-51. Have you ever wondered why so many equations are written with the variables x and y? Even if you haven't, suppose you were going to reach into a bag that contained the alphabet and pull out one letter at random to use as a variable in equations. What is the probability that you would pull out an x? Let's assume you got the x. What is the probability that you would pull out the y if you chose again? *1/26, 1/25*

CHAPTER TWO

SHARPENING PENCILS

Exploring Functions

IN THIS CHAPTER YOU WILL HAVE THE OPPORTUNITY TO:

- start learning what it means to investigate a function;
- become familiar with function notation and begin to see many familiar equations and graphs as functions;
- develop your ability to determine the domain and range of a function;
- continue developing your ability to model data graphically, numerically, and algebraically;
- learn how to use the graphing calculator;
- continue developing a greater understanding of the slope and y-intercept of a linear equation;
- continue developing your ability to solve a system of equations.

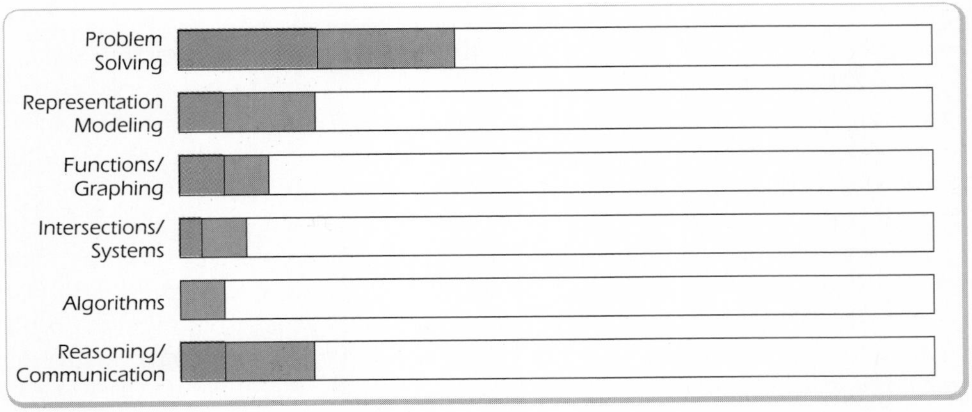

MATERIALS	CHAPTER CONTENTS		PROBLEM SETS
Scales, pencils, rulers, sharpeners, Resource Page–Directions for Lab write-up	**2.1**	Sharpening Pencils Lab	EF-1 – EF-18
Graphing calculators, Transparency–EF-21 Resource Page	**2.2**	Investigating a Function with a Graphing Calculator Domain and Range	EF-19 – EF-33
Graphing calculators, Resource Pages–Extra Practice and Calculator Tool Kit	**2.3**	Using a Graphing Calculator	EF-34 – EF-48
Graphing calculators	**2.4**	Intersections of Graphs	EF-49 – EF-64
Graphing calculators	**2.5**	Solutions, Graphs, and the Graphing Calculator	EF-65 – EF-77
Graphing calculators, Resource Page–Handy Graphing Grids	**2.6**	Summary	EF-78 – EF-90

2.1 SHARPENING PENCILS LAB

Today is the first lab that allows the students to gather some data through measuring, to plot it in reference to appropriate axes, and to use their graph to write an equation of good fit. You will need some very specific equipment — in particular, fairly accurate scales (yes, more than one if possible), pencils, pencil sharpeners, and rulers. If you don't want to sharpen pencils during class, then sets of presharpened pencils for each group will work.

If you really have a problem getting science instructors to give their scales up for a day, or you can't borrow an office letter scale, or you can't find an accurate food scale, then you can change this first investigation to use a roll of pennies. Since pennies are heavier, a small food scale will work. You will have to adjust some of the questions. For instance, you will be graphing the number of pennies remaining versus weight of remaining roll. Leaving the pennies in the paper roll still allows you to ask the y-intercept question. The slope question needs to be changed. Be sure to try out the weighing yourself first to see how it will work out. Using sharpened pencils produces a graph with a nice, natural y-intercept when the relationship explored is weight based on the length of the painted part.

This lab assignment is the first assignment we tell students to write up as a lab report, and it might be a good candidate for their portfolios. This is a good time to hand out a description of what you expect in lab reports. There is a description of the components of a lab report in the Resource Pages that can be used for this or any of the other labs done this year. It would be best to have this lab write-up due a couple of days (or more) after the lab itself. This will allow students the opportunity to refine their answers as they learn more in the chapter.

Note: If there is a problem getting the scales for more than one day in class, you may want to have students gather data for the Targets Investigation in Chapter 5 at the same time. If you are really strapped for time, have two students in each group do the measuring and weighing for this lab while the other two do the measuring and weighing for the Targets Investigation from Chapter 5.

At this point we don't expect students to use their graphing calculators to find the equation of the line of best fit, nor use formulas that some of the students might have seen (and forgotten) at some point. We would like them to place some thin straight edge such as their pen or pencil on the graph in the position that best fits the points. Once the straight edge is placed, have them trace the line on the graph paper. The students should then find the slope and y-intercept of the line they drew.

In this investigation you will be comparing weights of pencils of different lengths. While this activity, on the face of it, may not seem very inspiring or potentially enlightening, we have chosen it because it will lead to the development of mathematical models and concepts that are central to what you will be learning in this course. The term **mathematical model** might be new to you. One possible model is an algebraic equation used to represent a real-world situation. Knowing an equation that represents a situation allows you to make predictions based on the model. Although this activity does not quite represent a real-world situation, the fact that a large group of people can do it fairly quickly in a classroom will allow you to focus on the mathematical model this experiment will generate.

Be sure that each person in your group does a neat and accurate graph, answers all questions, and shows all work. Some instructors require a portfolio that includes examples of your work and problem-solving skills. Your lab write-ups of these investigations may be the first pieces to put into your portfolio.

EF-1. Sharpening Pencils Lab (EF-1 through EF-7) In this investigation, you are going to sharpen pencils. Throughout this investigation, remember not to use the eraser for erasing. The goal of this lab is to find a relationship between the length of the pencil and the mass of the pencil when the length varies. Here's how it works:

a. Start with a fresh pencil. Sharpen it just enough to be usable. Measure, in centimeters, the length of the *painted* part (excluding the metal and eraser part), and then weigh the pencil. Record both quantities in a table with two headings: "length of painted part (cm)" and "mass (g)" ("cm" and "g" are abbreviations for centimeters and grams).

b. Before you begin gathering further data, make a prediction about the relationship between the length of the painted part and the mass of the pencil. Will this relationship be linear (a straight-line graph) or nonlinear (not a straight-line graph)? Why? Be sure to record your prediction. It will become your lab hypothesis.

c. Now break off the point and sharpen the pencil. Measure the *painted* part and weigh it. Record your data in your table.

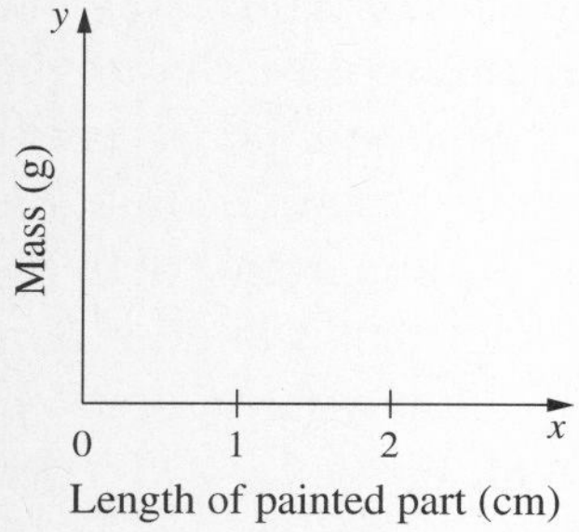

d. Plot the data on a graph where the x-axis represents the length of the painted part in centimeters and the y-axis represents the mass in grams. Use a full sheet of graph paper.

e. What should happen to the mass of the pencil as you sharpen it? How does this relate to the graph? Be clear in your written explanation. Collect at least 10 data points — the last few should be with the pencil down to a little nub. Don't forget to measure, weigh, and record after each sharpening!

f. Graph your data on your graph paper. Write a sentence comparing the graph to your predictions from part b.

In the following problems students should find the equation of the ***line of best fit****—by estimating or counting the slope and y-intercept. Using the graphing calculator to do linear regression is a wonderful use of the calculator, but at this point in the students' understanding it is best not to complicate this simple problem with a sequence of complicated calculator instructions. Instructors have reported that when they have introduced linear regression at this point, the students focus on making the calculator work rather than on understanding the slope-intercept relationship between the equation and its graph.*

EF-2. Lay a ruler on its edge on your graph so that it represents **a line of best fit**—the one that best fits all the points. The line will probably *not* go through every point on your graph. Try to come as close as possible to the points. Use the ruler to draw your line of best fit on the paper.

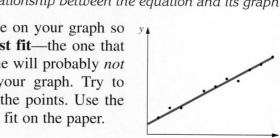

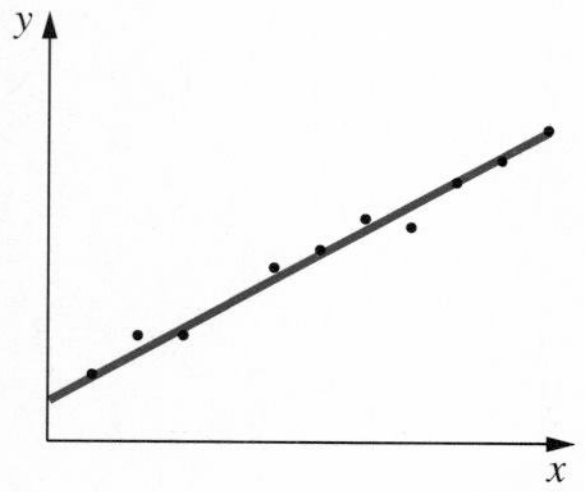

EF-3. Suppose you sharpened the pencil all the way down, so that no paint remained.

a. How much would this little pencil weigh?

b. Do you need to sharpen the pencil all the way down and weigh it to answer this question, or could you predict the weight of the stub from your graph? Explain. *This weight is the y-intercept.*

EF-4. How much change in mass is there for a one-centimeter change in length?

a. For example, use your graph to estimate the mass when the painted part is 3 cm long. Then estimate the mass when it is 4 cm long.

b. For a change in length of 1 cm, what is the change in mass?

c. Should the mass per centimeter of the painted part always be the same no matter which centimeter-long piece you examine?

d. How does the amount of mass per centimeter affect the graph? If the amount of mass per centimeter were larger, how would that change the line?

EF-5. Now, answer each of the following in complete sentences.

a. Explain what the y-intercept and slope of the line correspond to in relation to the pencil. *Note*: Be sure to state what the vertical change and the horizontal change of the slope represent. *Same as previous two problems: vertical change is change in mass; horizontal change is change in length.*

b. Explain why the data points for this lab lie in a straight line, and write an equation for your line.

c. Think of and describe a different situation, one that doesn't involve weighing, that would produce data points that lie in a straight line.

d. Explain what your new situation has in common with sharpening a pencil.

EF-6. The x-axis on your graph represents possible lengths for the painted part of the pencil. What are acceptable values for x in this problem? Explain. These "acceptable values" are known as the **domain**. Add this term to your Tool Kit. *From zero to the length of a new pencil.*

EF-7. What do the y-values represent? What are the acceptable y-values? These are known as the **range**. Add this term to your Tool Kit. *Weight from zero to the weight of a new pencil.*

Be sure to save your data and the answers to the previous questions for your lab report. The directions for this lab write-up (as well as future write-ups) are included in the Resource Pages. The purpose of this lab is to show how you can create an algebraic representation for the patterns you find within the data you collect.

EXTENSION AND PRACTICE

In Problem EF-8, students will forget the ± in part c. You might ask them to find all values of x for which $x^2 = 25$.

EF-8. Rearrange the following equations by solving each of them for x. Write each equation as $x =$ ________ (y will be in your answer).

a. $y = \frac{3}{5}x + 1$ $\quad$ *$x = \frac{5(y-1)}{3}$*

b. $3x + 2y = 6$ $\quad$ *$x = \frac{-2y+6}{3}$*

c. $y = x^2$ $\quad$ *$x = \pm\sqrt{y}$*

EF-9. Imagine that we add water to beakers A, B, and C, shown here. Sketch a graph to show the relationship between the volume of water added and the height of the water in each beaker. Put all three graphs on one set of axes. You may want to use colored pencils to distinguish the graphs or label each graph with its beaker letter. *x-axis: volume of water, y = axis: height of liquid. C is steepest, B is the least.*

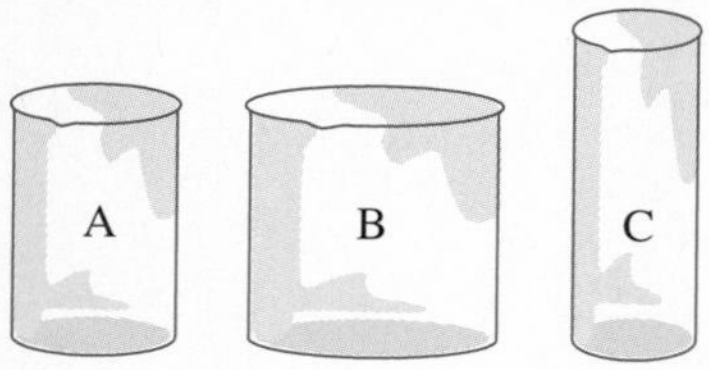

EF-10. The following table shows inputs and some of their corresponding outputs for a particular equation relating x and y. Fill in the rest of the table and explain how you got your answers. You can fill in the table without writing the equation.

x (input)	1	4	9	100	16	23	25	81	–25	$\frac{1}{4}$	$\frac{1}{9}$	0
y (output)	1	2		10						$\frac{1}{2}$		

Missing y values are 3 (for x = 9), 4 (for 16), ≈4.8 (for 23), 5 (for 25), 9 (for 81), 1/3 (for 1/9), 0 (for 0), and no y value for –25.

EF-11. Tony says that $(x + y)^2$ is the same as $x^2 + y^2$, but Patrice says Tony is wrong. Do you agree with Tony or with Patrice? Explain in at least *two* different ways (using numbers, algebra, diagrams, or other approaches) why the person you agree with is right. You have to convince the other person, who is very strong-willed.

EF-12. The cost of being connected to the Internet through the US-on-Line service is a base fee of $9.95 each month. The user is charged at the rate of $2.50 per hour of connect time.

a. How much would a total of 6 hours of connect time cost? *$24.95*

b. What would a total of 100 hours of connect time cost? *$259.95*

c. What would a total of x hours of connect time cost? Write your answer as an equation, with x representing the number of hours and y representing the total cost. If necessary, use a table to help develop your equation. *$y = 2.50x + 9.95$*

EF-13. A convenient way to show what a function machine does is to use **function notation**. For Carmichael's machine in PS-44 (in Chapter 1), we would write $f(x) = x^2 + 2x + 1$. The f is just the name or label of the function machine; it is not a variable. It could just as well be $C(x)$ or $\text{Car}(x) = x^2 + 2x + 1$ for Carmichael. In part c of PS-44, you actually calculated $f(-22.872)$. Use function notation to describe what each of the following machines does to x.

a. *$g(x) = 3x$*

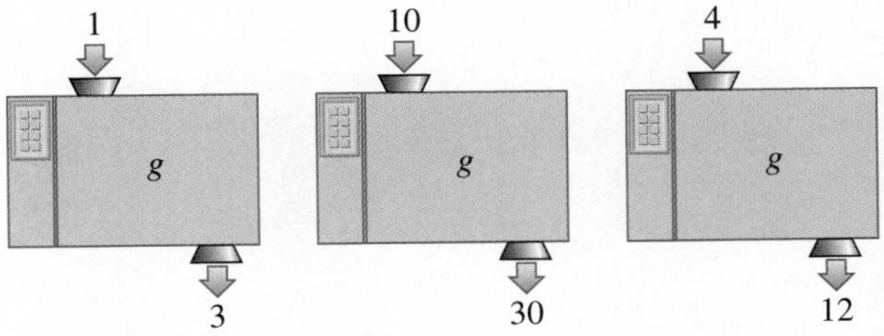

b. *$h(x) = x$*

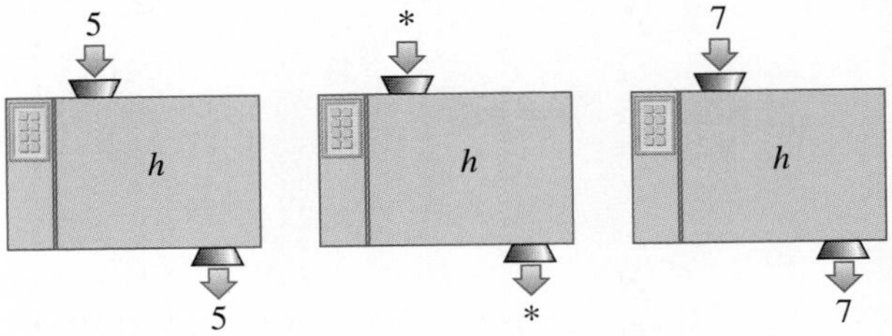

c. *$f(x) = x^2 + 1$*

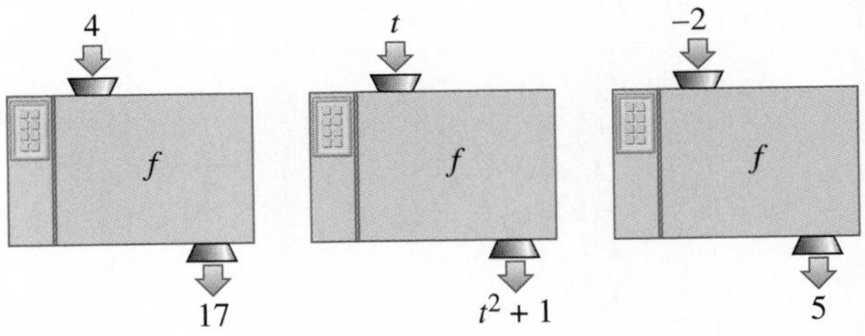

EF-14. Sketch a reasonable graph showing the relationship between the temperature outside and the time of day from 6:00 A.M. to 11:00 P.M. on an average summer day. Be sure to label your axes.

EF-15. The Downhill Slope There is a sign on Interstate 5 near Mt. Shasta that says "7% downgrade next four miles." What is the vertical change in feet for every 100 feet of horizontal change? What is the vertical change in feet for 1 mile (5280 feet) of horizontal change? *7 ft, 369.6 ft*

EF-16. If $f(x) = -4x^2 - 5x + 2$, what value(s) of x will make $f(x) = 0$? *$x \approx -1.57$ or $x \approx 0.32$*

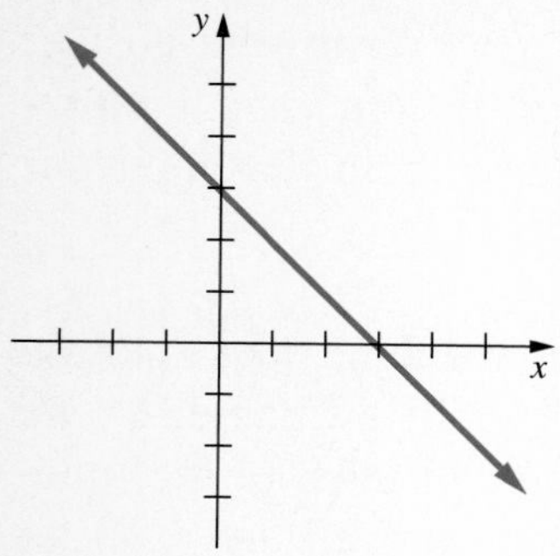

EF-17. For the graph shown here:

a. Determine the x- and y-intercepts. *(3,0) and (0,3)*

b. For a change of one unit horizontally what is the vertical change? For a horizontal change of two units what is the vertical change? What is the slope of the graph? *−1, −2, −1*

c. Write the equation of the graph. $y = -x + 3$

EF-18. Stacie says to Cory, "Reach into this standard deck of playing cards, and pull out any card at random. If it is the queen of hearts, I'll pay you $5.00." What is the probability that Cory gets Stacie's $5.00? What is the probability that Stacie keeps her $5.00? Justify your answers. *1/52, 51/52*

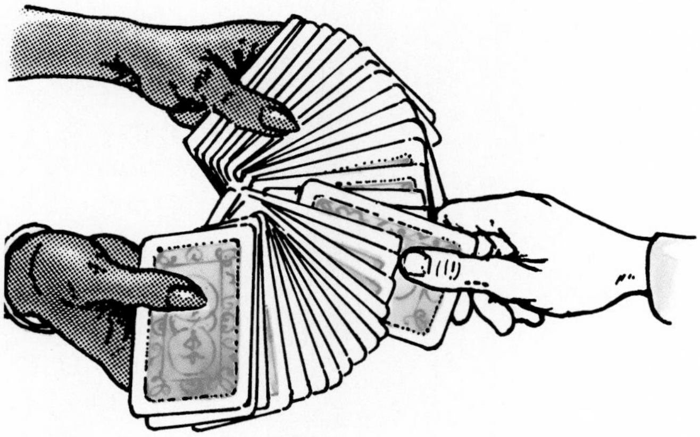

2.2 INVESTIGATING A FUNCTION WITH A GRAPHING CALCULATOR

Today we will investigate a function using the graphing calculator. Because we really want students to learn about using the graphing calculator, we chose to investigate a function that ***would not*** *be easy to do by hand.*

Be sure students read the introduction to the problem so that they see the point of this work. It is worth referring back to these goals since not all students can tie the ideas together at this point.

If students are not familiar with graphing calculators, you will need to give them a ***brief*** *introduction, but* ***don't*** *spend too much time on this. All they need to know is how to set window maximums and minimums and how to get the graph. It is more effective to go from group to group answering questions as they arise. This first function investigation with the calculator will lead the students to get an idea of what we are expecting. Later on there will be less guidance.*

We use the term ***function*** *here without definition. Students will develop a more formal definition in Chapter 5. For now* ***function*** *is just another way to talk about the equation for a graph or table.*

We want students' Tool Kits to be useful to them, and we recommend telling them they will be able to use their Tool Kit pages on tests. At the end of each chapter there is a Tool Kit Check. Periodically students should revise their

Tool Kits, getting rid of things they no longer need notes for and making their definitions more precise. Some instructors provide Tool Kit pages on yellow paper. Then the Tool Kits are the students' own personal "yellow pages."

EF-19.

> As you begin investigating functions, it is important that you understand what we expect when we ask you to sketch a graph. To **sketch a graph** means to show the approximate shape of the graph in the correct location with respect to your axes, to scale and label axes properly, and to clearly label all key points.

In the Data Walk and Sharpening Pencils Lab, we saw that we could write an equation to represent a collection of data points. When the points all lie on a line, we found that we could write the equation that produces the line. But not all data collections produce points that lie on a line. Consequently, we need to become familiar with the graphs of other functions to know what relationships they might represent. To begin this exploration, consider the following equation:

$$y = \sqrt{(x+9)} - 1$$

Use a graphing calculator to help you answer each of the following completely. Set the window so that your minimum x-value is -10 and the maximum value is 10. Set your y-values to range from -5 to 5.

a. Sketch the graph from your graphing calculator, using the guidelines in the box above.

b. Name the point where the graph starts. *(–9, –1)*

c. What is the output (the y-value) if you replace x with -5? with 16? with 91? with -13? Mark those points on your graph. *(–5, 1), (16, 4), (91, 9), (–13, no real number); no need to discuss complex numbers here—students just need to see that some input values are not possible.*

d. Recall that the y-intercept is where the graph crosses the y-axis. Similarly, the x-intercept is where the graph crosses the x-axis. What are the x- and y-intercepts of this graph? *(–8, 0), (0, 2)*

e. Does the graph ever cross the horizontal line $y = 15$? How about $y = 500$? How do you know? *yes, they will probably need to expand the window and check.*

f. What is the smallest possible value of y? What is the range for this function? $y \geq -1$

g. What is the smallest possible value of x? What is the domain for this function? $x \geq -9$

EF-20. After the Data Walk in PS-13, you wrote an equation to represent each of the three linear relationships. To refresh your memory, here are the three situations and the equations you probably wrote.

Situation 1, Cost of Milk: Milk costs $1.80 per gallon. If your integer is the number of gallons of milk you are going to buy, what is the cost?

x is the number of gallons of milk, y is the cost, and $y = 1.80x$ or $C(x) = 1.80x$.

Situation 2, Freeway Driving: Driving at a constant rate on the freeway, your car consumes 2 gallons of gas each hour. You start with 13 gallons. If your number is the number of hours you have been driving, how many gallons are left in your tank?

x is the number of hours you have been driving, y is the number of gallons left in the tank, and $y = 13 - 2x$ or $G(x) = 13 - 2x$.

Situation 3, Davis's Savings: For Davis's birthday his grandmother gave him $5.00, which he quickly put into his piggy bank. Each week, Davis added one quarter to the bank. If your number is the number of weeks that have passed since his birthday, what is the amount of money in his piggy bank?

x is the number of weeks since Davis's birthday, y is the amount of money in his piggy bank, and $y = 0.25x + 5.00$ or $M(x) = 0.25x + 5.00$.

a. Using your graphing calculator, graph the equation for the cost of milk. On your paper, make a sketch of the graph. What is different about the graph you originally drew after the Data Walk and the graph the calculator produced? Explain. Graph the other two equations and make sketches of what the calculator produces. ***Students should notice that calculator extends each graph beyond what they looked at.***

b. When graphed, an equation can produce points that we did not consider when we first considered the real situation. This is because the calculator simply uses values for x, pairs each value with its output y, and then plots all the points. The calculator does not know that x represents "the number of gallons of milk." For each of the three situations, what values are appropriate for x? Can x be a negative number? A fraction? These "appropriate values" for x are called the **domain** of the equation. ***Situation 1, 3, $x \geq 0$; Situation 2, $0 \leq x \leq 6.5$. Don't get into the discrete–vs.–continuous question here. If the question of fractions of gallons comes up, acknowledge that, "yes, the graph could be just points, but we'll return to it in more detail in Chapter 3."***

c. It also is important to realize that the graphing calculator may give extra y values. In Situation 1, can the y-values (the output values being the cost) be negative? Can they be fractions? For each situation what are the appropriate y-values? These "appropriate values" for y are known as the **range** of the equation. ***Situation 1, $y \geq 0$; 2, $0 \leq y \leq 13$; 3, $y \geq 5.00$, jumping in $0.25 increments, but they probably won't see that.***

The purpose of the next problem is to provide another way to understand domain and range. Domain and range can be visualized as the "flashlight shadows" of the graph of a function on the x- and y-axes. This problem is meant to be teacher-led, so watch your groups. As they finish EF-20 (and have already been thinking about domain and range), begin discussing this idea.

Imagine that it is possible to "squish" the graph onto the x-axis from both sides without pushing toward the y-axis at all. You might demonstrate this on the overhead projector, and use dotted vertical arrows from the endpoints to show the edges of the area in the shadow. The amount of the x-axis covered is the domain. Similarly, we can squish the graph toward the y-axis to find the range. Another method is to imagine we are "inside" the plane (in Flatland) shining a "flat" flashlight from an infinite distance beyond the graph toward the x-axis for the first graph. What would the shadow look like on the x-axis? Draw the shadow on the axis itself (colored pens are a good idea). You'll have to discuss the shadows cast by the points that lie beneath the axis, but students will catch on. Ask for a volunteer to draw the shadow cast on the y-axis by a horizontal flashlight.

Do EF-21 with the whole class, or make a transparency of the Resource Page on domains and ranges to do with the class and leave EF-21 for the groups to do. You will need to discuss the meaning of the open endpoints.

EF-21. Use a set of axes on which to represent the domain and another set on which to represent the range for each of the following functions. Each tick mark represents one unit.

a.

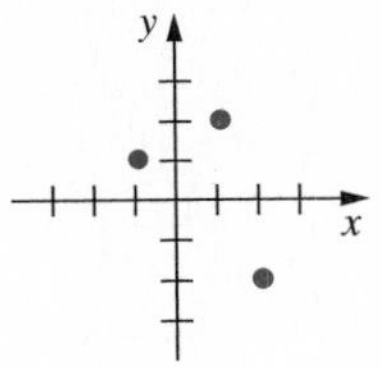

D = {–1, 1, 2},
R = {–2, 1, 2}

b.

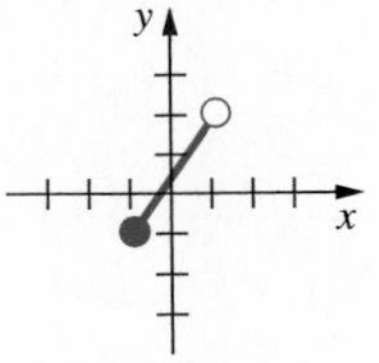

D = [–1, 1), R = [–1, 2)

c.

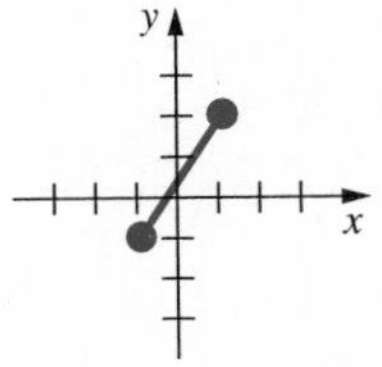

D = [1, 1], R = [–1, 2]

d.

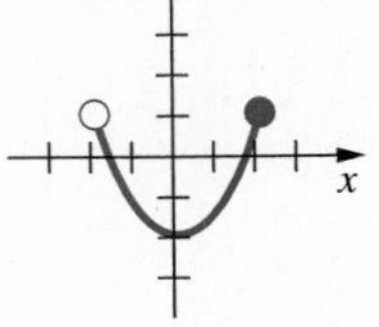

D = (–2, 2], R = [–2, 1]

We give the answers here in interval notation only to save space. We are not looking for students to use a particular notation for describing the domain and range. Verbal descriptions and/or shadow graphs are enough at this point, or inequalities would be great.

EF-22. Does the temperature outside depend on the time of day, or does the time of day depend on the temperature outside? This may seem like a silly question, but to sketch a graph that represents this relationship, we need to determine which axis will represent what quantity.

Solution:

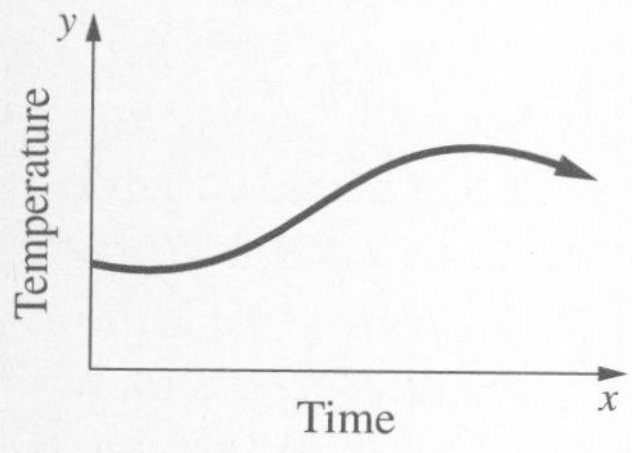

a. When you graph an equation such as $y = 3x - 5$, which variable—x or y—depends on the other? Which is not dependent, that is, which is **independent**? Explain. *y depends on x, x is independent.*

b. Which is dependent: temperature or time of day? Which is **independent**? *Dependent = temperature, independent = time.*

c. Sketch a graph, with appropriately named axes, showing the relationship between temperature outside and time of day. *Students may need to discuss the convention of using the horizontal axis for the independent variable.*

EF-23. For each of the linear situations from the Data Walk (repeated in EF-20), what is the independent variable? What is the dependent variable? *Situation 1, Independent: number of gallons of milk, dependent: cost; 2, ind: number of hours driving, dep: gallons of gas left; 3, ind: number of weeks since Davis's birthday, dep: amount of money in his bank.*

EF-24. This notion of independent and dependent variables is an important concept. The following definition would be a good thing to include in your Tool Kit along with domain and range.

> The set of possible values that the **independent variable** can take on (input values) has a special name. It is called the **domain** of the function. It consists of every number x can represent for your function.
>
> The **range** of a function is the set of possible values of the **dependent variable** (output values). It consists of every number y can represent for your function.

EXTENSION AND PRACTICE

EF-25. Examine each of the following graphs. Based on the shape of the graph and the labels of the axes, describe the relationship that each graph represents. State which is the independent variable and which is the dependent variable. The scaling is left off two of the graphs so you can focus your description on the general relationship.

a.

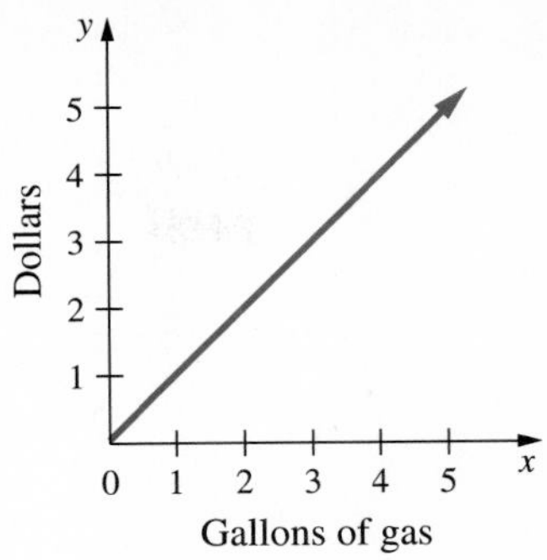

b.

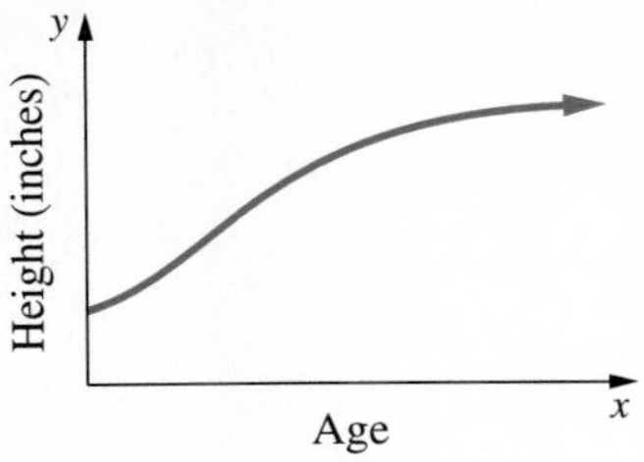

c.

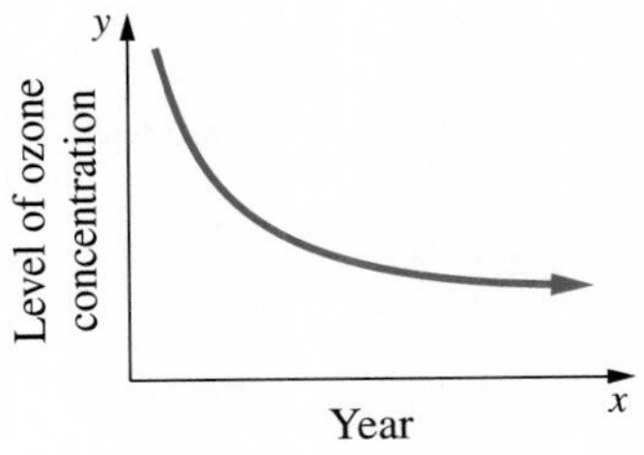

d.

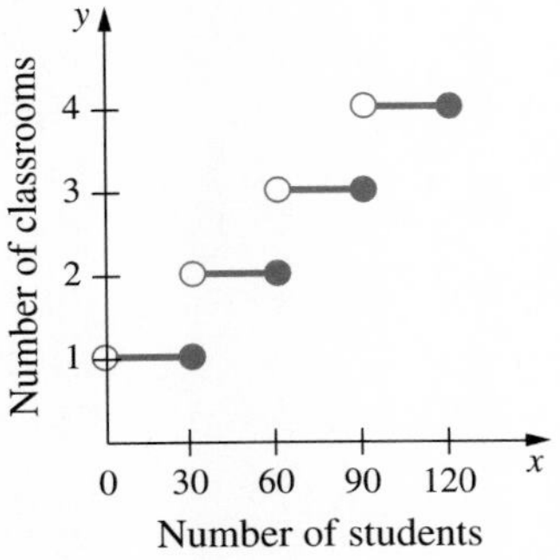

EF-26. For the situation described below, decide what is the independent variable and what is the dependent variable, and describe the domain and the range. Sketch a graph of the relationship.

As I left for work, I drove slowly through town at a constant rate. When I reached the freeway, it took me 15 seconds to get up to speed. I could maintain that speed for only 10 minutes when I came to a horrible traffic jam. I did not move for 45 minutes. *ind = time, dep = speed; D: $x \geq 0$, R: $0 \leq y \leq$ speed limit*

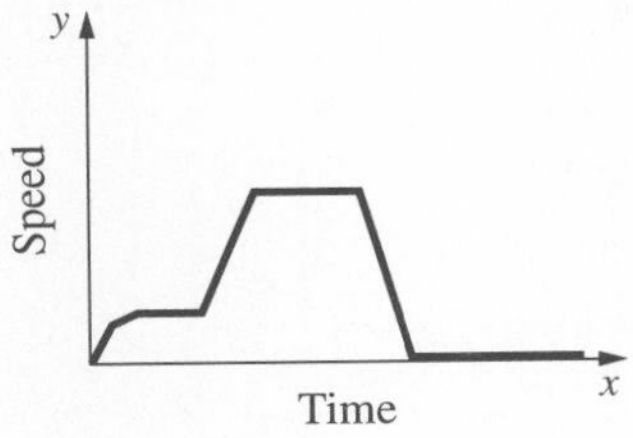

EF-27. Armand and Labe like to hike in the summertime, but they get very thirsty, so they always start out with 5 liters of water. Together they drink about 1 liter of water every 2 hours. The graph shown here represents the graphing calculator version of the graph for this situation.

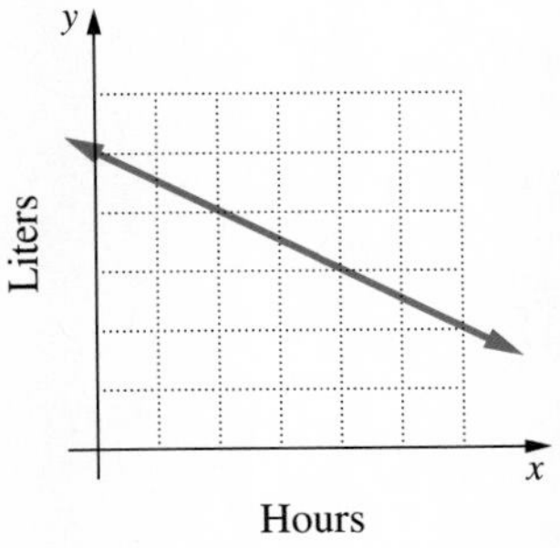

a. How much water do they drink per hour? What is the slope of the graph? What does the vertical change represent? What does the horizontal change represent? What is the y-intercept? *$\frac{1}{2}$ liter, $-\frac{1}{2}$, liters, hours, (0, 5)*

b. What is the equation of the line if the x- and y-axes are scaled by 1? *$y = -\frac{1}{2}x + 5$*

EF-28. What is the error in the following solution? Rework the problem correctly.

$$4(3-x)+2=5x+12$$
$$12-4x+2=5x+12$$
$$14-4x=5x+12$$
$$14=x+12$$
$$2=x$$

Should have added 4x to both sides in line 3; $x = 2/9$.

EF-29. Is it true that $\frac{x+2}{x+3}=\frac{2}{3}$? Explain your reasoning. *Only if $x = 0$*

EF-30. In Salem, California, 42 percent of the registered voters are Republicans and 44 percent are Democrats. The other 217 are Independents. How many registered voters are there in Salem? Write an equation that represents this problem. *1550, $0.14x = 217$*

EF-31. If $f(x) = x^2 - 4x - 5$, find the following:

a. $f(-2)$ *7* **b.** $f(10)$ *55* **c.** $f(0)$ *−5*

EF-32. For each of the following, state the domain and range. Each tick mark represents one unit.

a.

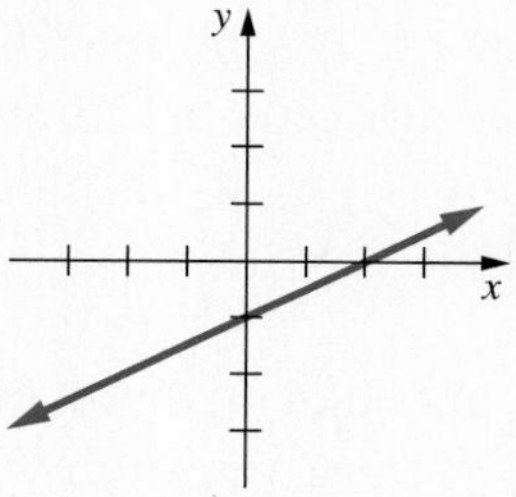

$\infty < x < \infty$; $-\infty < y < \infty$

b.

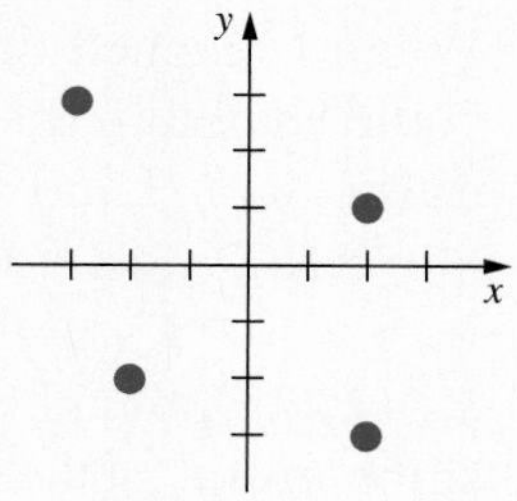

$x = -3, -2, 2$;
$y = 3, 1, -2, -3$

c.

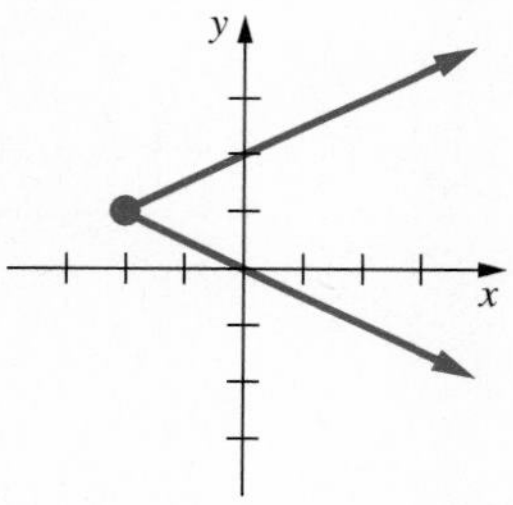

$x \geq -2$; $-\infty < y < \infty$

d.

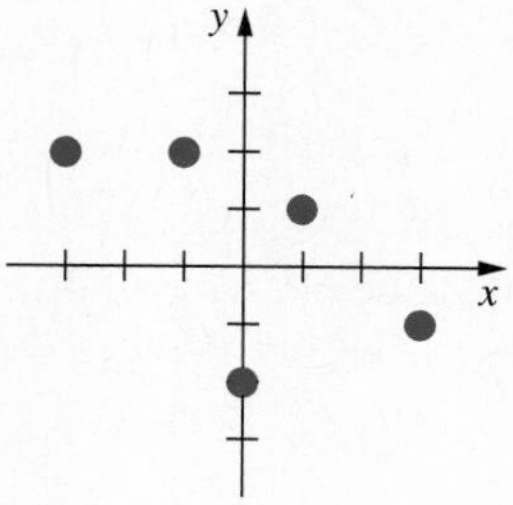

$x = -3, -1, 0, 1, 3$;
$y = -2, -1, 1, 2$

e.

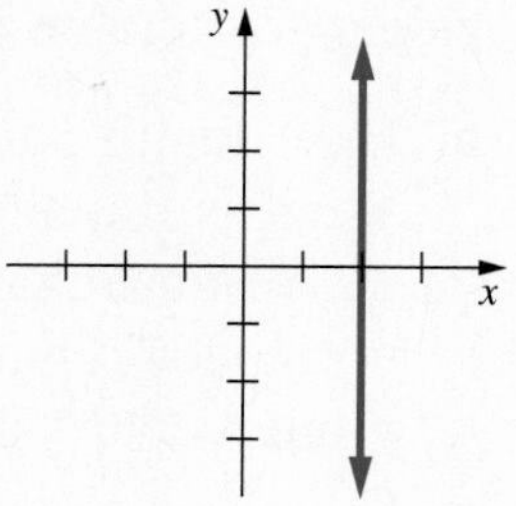

$x = 2$; $-\infty < y < \infty$

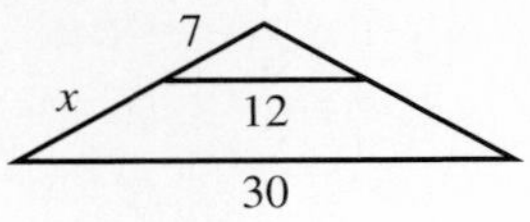

EF-33. Find x for this figure. The base and the other horizontal line are parallel, so the triangles are similar. *$x = 10.5$*

2.3 USING A GRAPHING CALCULATOR

In this section our goal is to have students become more familiar with the graphing calculator and the equation $y = mx + b$. The mechanics of using the calculator will require quick-thinking facilitation, so you need to become very familiar with how the calculator works. Take this hint from those who have gone before: Work all these problems ahead of time to be sure you know how to use your graphing calculator! In fact, it might not be a bad idea to work them out twice. If you are using an overhead graphing calculator, you will also want to work with it ahead of time, as it may be different from the regular one.

For this section, you will need to know how to set the viewing window, enter equations to be graphed, clear the screen, move back and forth between the graphing window and the text window, keep two graphs on the screen at once, use the trace and zoom features, and know what to do to fix mistakes. When students first use the calculators, they often do not notice that there is a standard window.

Many questions and problems will arise. The more experience you have had with the calculator the better. Handle these questions as students raise them in their groups: what to do if they do not see the graph (change the range or domain or both), what to do if the equation is not written as "y = " as the calculator requires (rewrite it using algebra), and what it means to see the complete graph.

Encourage group members to help each other so you can move quickly from group to group.

In EF-34 we ask for the "direction" of the line. Students generally understand this to mean up or down as the x-values increase from left to right, so most groups should handle this without explanation.

EF-34. Using a Graphing Calculator (EF-34 to EF-39) We have learned that equations of lines can be written as $y = mx + b$. For each of the following, predict where the graph will cross the y-axis and the steepness and direction of the line. Sketch your prediction. Then, check your accuracy with the graphing calculator and revise your graphs as needed. How does the coefficient of x relate to the steepness?

a. $y = 6x - 1$ **b.** $y = -6x + 1$

c. $y = \frac{1}{6}x + 4$ **d.** $y = x$

EF-35. Look back at the "cost of milk" equation you graphed in EF-20. When graphed, the equation is a line. Use a graphing calculator to graph the equation again, and draw it on graph paper.

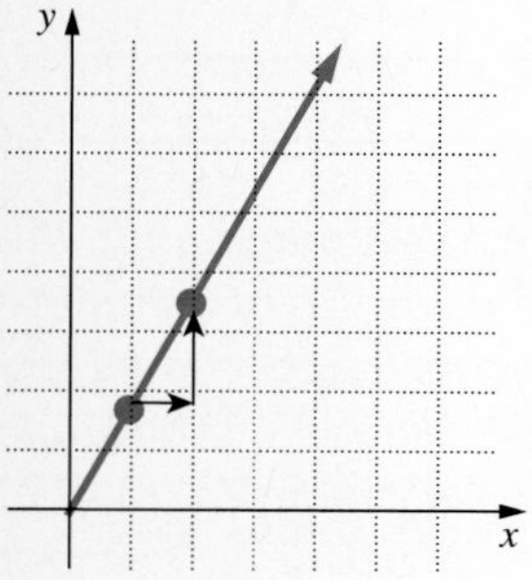

a. We have already interpreted the y-intercept [which is (0, 0)] as "Zero gallons of milk costs zero dollars." On your graph mark the points (1, 1.80) and (2, 3.60).

b. From the first point (1, 1.80), draw a horizontal segment one unit long to a point directly below the second point. Then, draw a vertical segment up to the second point. The triangle that is formed is called a **slope triangle**. What are the lengths of the horizontal and vertical sides of the triangle? *horizontal = 1, vertical = 1.8*

c. The slope of a line, commonly represented by the letter m, is the ratio

$$m = \frac{\textbf{vertical change}}{\textbf{horizontal change}}$$

On your graph of this line, what is the vertical change from the first point to the second? What is the corresponding horizontal change? Write the ratio. What do you notice? *1.80/1; hopefully students will notice that this is the number we already determined for the equation.*

d. Each member of your group should choose two *different* points on the line to create another slope triangle. Calculate the slope using your new triangles. What do you notice? *All slopes are the same.*

e. For $y = 1.80x$, write in words what the vertical change (dependent variable) of the slope represents. Now write in words what the horizontal change (independent variable) for this line represents. This is called **interpreting the slope**.

$$\frac{\textit{change in cost}}{\textit{change in no. of gallons}} \text{ or } \frac{\$}{\textit{gal}}.$$

f. The slope of 1.80 in this question is interpreted as cost per gallon of milk, or in other words, there is a \$1.80 change in price per one gallon of milk purchased. Write a statement in your Tool Kit explaining:

i. how to determine units of slope.

ii. how to interpret the slope.

(i) The units of slope are always the change in the units of the vertical axis over the change in the units of the horizontal axis. (ii) To interpret the slope, state in words what the vertical change over the horizontal change represents.

EF-36. Draw graphs of the "freeway driving" and "Davis's savings" equations from EF-20. Calculate each of the slopes using the vertical and horizontal sides of a slope triangle. Explain how and why this result for the slope shows up in the equation. *m = −2, m = 0.25*

EF-37. Not all graphs have slopes. What does the graph of $y = x^2 - 3x - 3$ look like? What do we call it? Why doesn't it make sense to talk about the slope of this graph? Make a sketch on paper. Use the trace and/or zoom buttons to approximate the x-intercept(s), y-intercepts, and lowest point (the vertex). ***a parabola; vertex (1.5, –5.25), x-intercepts (3.8, 0), (–0.8, 0), y-intercept (0, 3)***

EF-38. How would you set the window on the graphing calculator to graph $y = (x + 1)(x - 9)$ so that you can see the complete graph? By **complete**, we mean that we see everything that is important about the graph and that everything off the screen is predictable based on what we do see. Another way to think about completeness is to say that a graph is complete when further zooming or window changes will not add any new information. Sketch the graph and describe your window setting. ***$-5 \leq x \leq 15$, $-30 \leq y \leq 10$ is one possibility.***

*The next problem asks students for the vertex as well as the intercepts. We do not intend to stop here and define **vertex** or how to locate it. In their groups, students will generally decide it is the lowest point and make an estimate from their graphs. That's all we're looking for here. If they ask, just ask them, "What's the lowest point?"*

EF-39. Now that you have found the graph of $y = (x + 1)(x - 9)$, find the x-intercept(s), y-intercept, and the vertex. *(–1, 0), (9, 0), (0, –9), (4, –25)*

If students need more practice using the graphing calculator, there is a Resource Page with additional problems for practice. These problems are similar to the ones given here and can be used as an extra day in class or one at a time as you continue. The Graphing Calculator Resource Page is in the student text, so if your students have their own graphing calculators, you may want to assign those problems instead of the usual homework.

The students will not need to have graphing calculators for the rest of the problems in this section.

EXTENSION AND PRACTICE

Be sure your students get the idea of EF-40. We want them to realize that labeling and scaling graphs correctly is an important skill. A follow-up discussion in class will be worthwhile.

EF-40. Feliz completed the tables and drew the graphs shown on the following page but his instructor marked each one wrong. Be as specific as you can when you answer the following questions. For each graph do the following.

a. Identify as many mistakes as you can.

b. Describe in a sentence what to do to correct each mistake. ***Feliz needs to re-scale his axes and make complete graphs.***

c. Redraw the graphs so that Feliz knows what to do and can get full credit next time.

i. $y = 2x + 1$

x	y
1	3
2	5
3	7

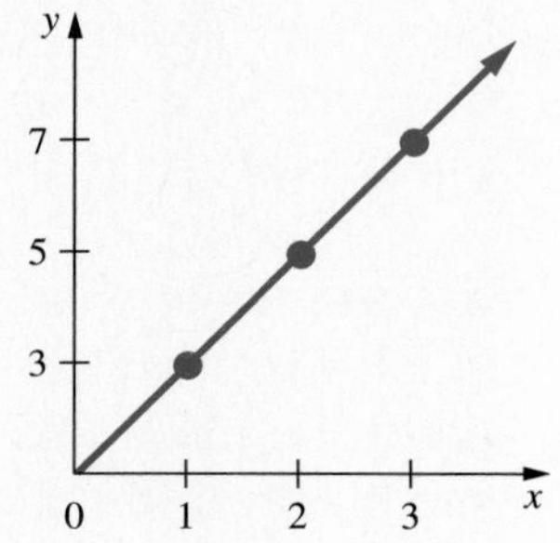

ii. $y = x^2$

x	y
0	0
1	1
2	4
3	9

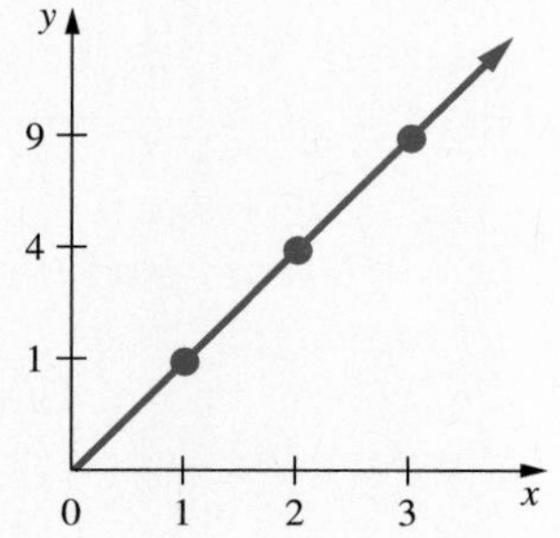

iii. $y = 3x + 50$

x	y
0	50
1	53
2	56

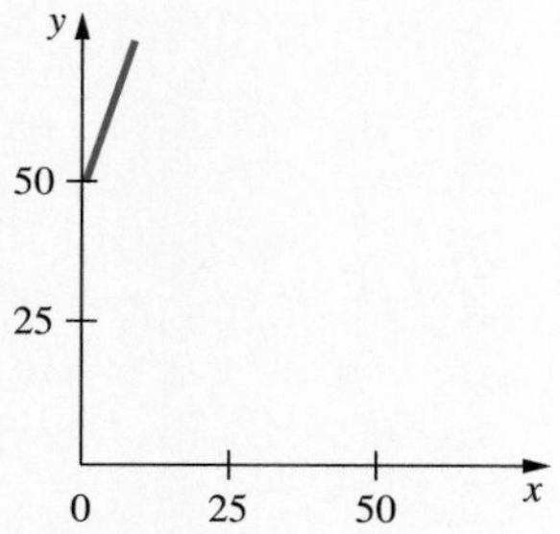

iv. $y = x^2 - 4$

x	y
0	−4
−1	−3
1	−3
2	0
3	5

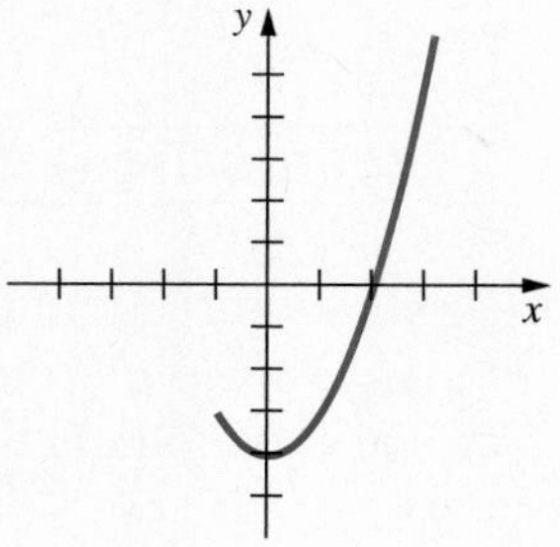

EF-41. Identify the independent variable and the dependent variable, describe the domain and the range, and write the equation representing the following situation. Sketch the graph and write the equation representing the graph. Interpret the slope and the y-intercept.

At the Olala Berry "Pick-Your-Own" Farm, the berry basket costs \$1.00. Then you can pick as many olala berries as you like, and when you are finished, the basket is weighed. The cost is \$0.75 per pound. *Ind = pounds, dep = cost; D, $x \geq 0$; R, $y \geq 1$; $y = 0.75x + 1$; m is \$/lb, y-intercept is 0 lb cost \$1.00.*

EF-42. Ashley wants her new little brother or sister to be born on her birthday. The doctor says that Ashley's mother will deliver the baby any day in August. Ashley's birthday is August 7. What is the probability that Ashley will share a birthday with her new sibling? *1/31*

EF-43. Solve the following for x. Show all of the steps you take or explain your answer.

a. $5(x + 3) = 4 - 3x$ *$x = -11/8$*

b. $8 + 2(3 + x) = 5(x - 2)$ *$x = 8$*

c. $8x^2 + x - 1 = 0$ *$x = (-1 \pm \sqrt{33})/16$*

d. $3(x^2 + 2x - 7) = 0$ *$x = -1 \pm 2\sqrt{2}$*

EF-44. If $f(x) = 3 - 7x$, what is

a. $f(9)$? *−60*   **b.** $f(0)$? *3*   **c.** $f(-4)$? *31*

EF-45. Shiho has overslept on the first day of classes and has to rush to class. She knows her class is in the B wing of Root Hall but can't remember which of the twelve rooms is the right one. If she chooses at random, what is the probability she chooses the correct one? *1/12*

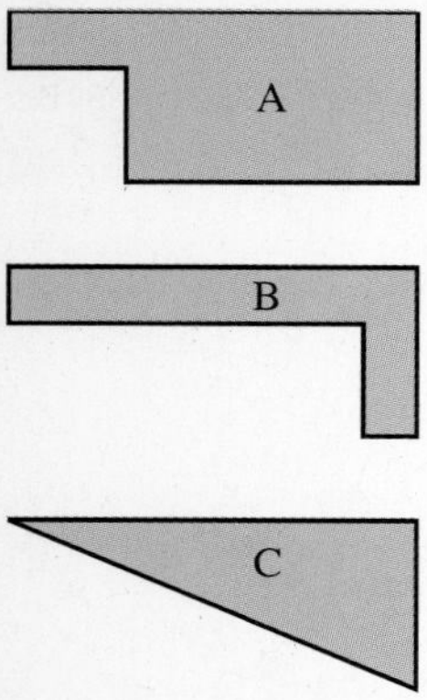

EF-46. The side views of three swimming pools are shown. Below are three graphs that show the relationship between the depth of the water (in the deep end, which is on the right side) over time as the pool is filled at a constant rate. Match each of the following pool profiles, A, B, and C, to the graph that could represent this relationship.

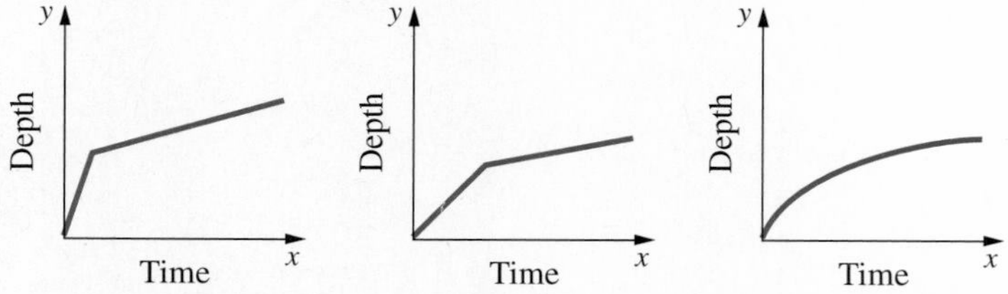

Graph 1 — *B*

Graph 2 — *A*

Graph 3 — *C*

EF-47. Kendall is creating a table of values for the function $K(x) = x^2$ using the following input values for x: −3, −2, −1, 0, 1, 2, 3. Amy, who is very lazy but clever, has to make a table using the same x values but for the function $A(x) = x^2 - 5$. Make Kendall's table. Then explain how Amy could use Kendall's results to make her table. *9, 4, 1, 0, 1, 4, 9. Just subtract 5 from each of Kendall's outputs.*

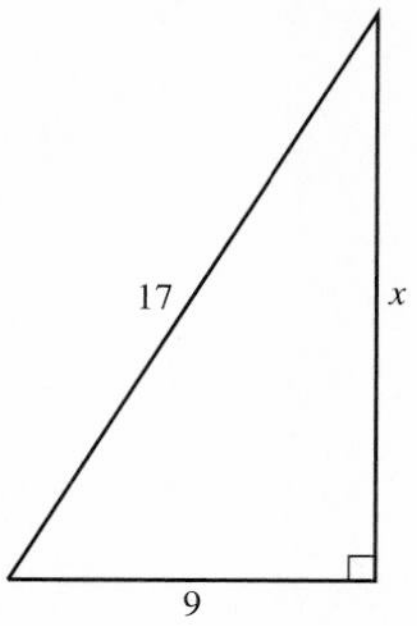

EF-48. Describe a situation or write a story problem whose solution could use the picture at the right. Then solve the problem. *14.42*

2.4 INTERSECTIONS OF GRAPHS

We use the graphing calculator again in this section to explore the intersections of graphs. The students should be getting more familiar with the graphing calculator, so they can focus on the math now. We expect students to mostly approximate here. You can hint at things to come by saying, "Wouldn't it be nice to have an exact answer? One we could count on?"

EF-49. Choosing a Phone Plan In most towns you have a choice of two cellular phone companies. The Fast Talkers company charges $10.00 per month plus 40 cents per minute; Relax and Chat charges $20.00 per month plus 15 cents per minute.

a. About how many minutes of phone calls do *you* make per month?

b. For each company, write an equation that represents the cost in a given month in terms of the minutes of phone calls. *Fast Talkers = 10 + 0.40x and Relax and Chat = 20 + 0.15x. These equations model the problem if x is a whole number.*

c. Graph each equation you wrote in part b on the graphing calculator. What are the independent and dependent variables? *ind = no. of minutes of calls, dep = cost*

d. Discuss how your two graphs relate to the solution of the problem. When are the costs for both companies the same? When is Fast Talkers a better choice? When is Relax and Chat a better choice? *Fast Talkers costs less when the no. of min < 40; they are equal at 40 min.*

e. Interpret the point of intersection of the two graphs. *At 40 min, both plans cost $26.*

f. How many minutes do you think the average student in your group spends on the phone each month?

g. How could you find out the answer to part e?

h. Carry out your plan from part g.

i. Decide which students in your group should use which phone company, and explain why.

EF-50. One of the situations of the Data Walk is repeated here along with the equation:

Driving at a constant rate on the freeway, your car consumes 2 gallons of gas each hour. You start with 13 gallons. If your number is the number of hours you have been driving, how many gallons are left in your tank?

x is the number of hours you have been driving, y is the number of gallons left in the tank; $y = 13 - 2x$.

Graph the equation on the graphing calculator. When will your car be out of gas? Sketch the graph and label the point that represents the answer to this question on the graph. *The car will be out of gas in 6.5 hr. (6.5, 0) is the x-intercept and is interpreted as "in 6.5 hours, there is zero gas."*

EF-51. Graph the equation $y = x(8 - x)$ on the graphing calculator. What are the x-intercepts? *(0, 0), (8, 0)*

EF-52. Graph the equation $y = -x^2 + 6x - 3$. What are the x-intercepts?

$\left(\dfrac{(-6 \pm 2\sqrt{6})}{-2}, 0\right)$ *or (0.55, 5.45)*

EF-53. Where do the graphs of $y = 2x - 6$ and $y = -4x + 6$ cross? Use your graphing calculator to find the point. Use your zoom key to get closer. *(2, –2)*

EF-54. Sketch the graphs of $y = 3x - 5$ and $y = 2x + 12$. Adjust the window and use your zoom-in feature to find out where they cross. *(17, 46)*

EXTENSION AND PRACTICE

EF-55. Find the error in the problem below. Identify the error and show how to do the problem correctly.

$$4.1x = 9.5x + 23.7$$
$$\underline{-4.1x = -4.1x}$$
$$5.4x = 23.7$$
$$\frac{5.4x}{5.4} = \frac{23.7}{5.4}$$
$$x = 4.39$$

Error: line 3, $x \approx -4.39$

EF-56. Factor or use the quadratic formula to solve the following quadratic equations:

a. $x^2 - 7x + 10 = 0$ *2, 5* **b.** $x^2 + x - 42 = 0$ *–7, 6*

c. $3x^2 + 15x = 0$ *–5, 0* **d.** $2x^2 - x = 3$ *–1, 3/2*

EF-57. Tracy is convinced that $x^3 + x^2$ must equal x^5. Your job is to convince her she is mistaken. Use at least two different ways to demonstrate to Tracy why $x^3 + x^2 \neq x^5$.

EF-58. Ivan overheard part of your discussion with Tracy and said, "I know why that can't be right. It's because $(x^3)^2 = x^5$." Now explain in at least two different ways to Ivan why he is also mistaken.

EF-59. There *is* a way to combine x^3 and x^2 that does give the answer x^5. What is it? Show why this is true.

In the next problem you should accept either variable as independent or dependent at this point as long as the student can justify his or her choice.

EF-60. An average transit bus holds 45 people. Sketch a graph showing the relationship between the number of transit buses and the number of people they can transport. Be sure to label the axes.

EF-61. State the domain and range of each graph. Each tick mark represents one unit.

a.

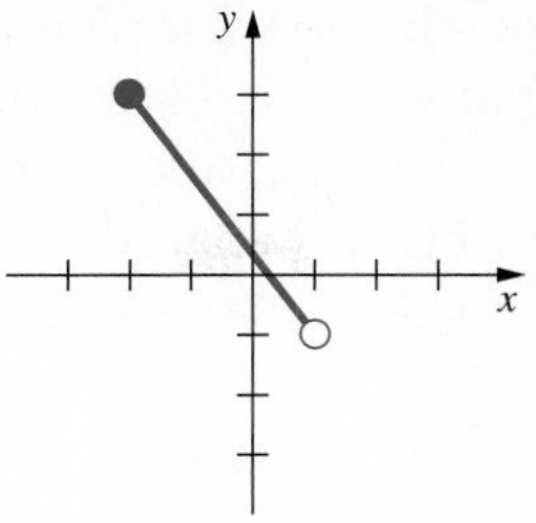

D: [–2, 1); R: (–1, 3]

b.

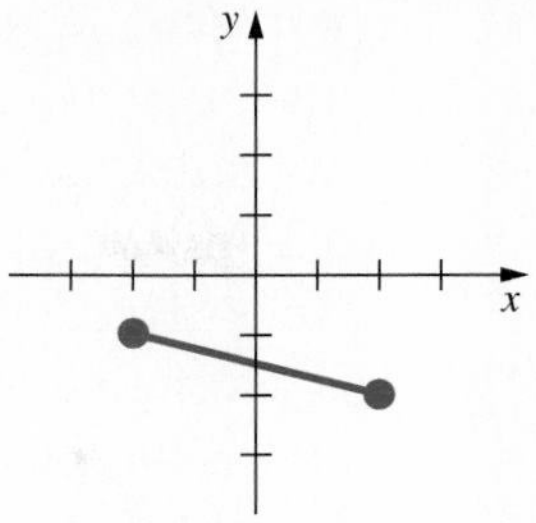

$-2 \le x \le 2$, $-2 \le y \le -1$

EF-62. Think about your solutions to the previous problem. If you know a specific domain and range, what do you know about the shape of the function? *very little*

EF-63. Ryan created a function machine that he called $R(x)$. He used several different input values and recorded the outputs shown below. What is the equation for $R(x)$? *$y = 5x - 1$*

x	0	1	2	3	4	5
$R(x)$	–1	4	9	14	19	24

EF-64. Maggie wants to ask Douglas out on a date, and she is trying to remember Douglas's phone number. The last four digits are easy: 1111. The first three digits are the hard part. She knows that the first three digits consist of a 6, a 7, and a 9, but she has no clue about the order. If she dials one combination at random, what is the probability she'll be talking to Douglas on her first try? ***The three digits can be arranged in six different ways. The probability is 1/6.***

2.5 SOLUTIONS, GRAPHS, AND THE GRAPHING CALCULATOR

We hope this section will drive home the connection between solutions of equations and their graphs. The problems are designed to make the algebraic method a more desirable procedure. We suggest that you tell the students they must show you their results to EF-65(c) and you must be satisfied with their response before they can continue.

EF-65. On the same set of axes, graph the equations $y = \frac{2}{3}x + 5$ and $2x + 3y = 24$ with your graphing calculator.

a. What did you need to do to the second equation before you could graph it? ***Solve for y.***

b. Where do the graphs cross? ***(2.25, 6.5)***

c. What should happen if you substituted the x and y values you found in part b into each of the two equations? Explain how you know. ***The point will make both equations true.***

EF-66. Dasam's stock in BoBearCo, a company that produces baby products, is worth \$87.50. But with the baby boomers and their children growing older and with families having fewer children, he noticed that the stock began to drop at a rate of \$0.25 each day. He has been watching another stock, BigCar, Inc., a company that refurbishes large American cars, and it is on the rise. When BoBearCo stock is \$87.50, BigCar's stock is \$42.65 and rising \$0.15 each day. At what point will the price of BigCar, Inc. exceed BoBearCo. *After 112 days, BoBearCo is worth \$59.50 while BigCar is worth \$59.45. On the 113th day, BoBearCo is worth \$59.25 and BigCar is worth \$59.60.*

EF-67. Where do the graphs of $y = x^2 - 6x + 5$ and $y = 2x + 5$ cross? Use your graphing calculator to find the point(s). *(0, 5), (8, 21)*

EF-68. Suppose we want to find where the lines $y = 3x + 15$ and $y = 3 - 3x$ cross, and we want to be more accurate than the graphing calculator or graph paper will let us be. We can use algebra to find this point of intersection.

a. If you remember how to do this, find the point of intersection using algebra and explain your method to your group members. If you don't remember, then do parts b through e below.

b. Since $y = 3x + 15$ and $y = 3 - 3x$, what must be true about $3x + 15$ and $3 - 3x$? *They are equal.*

c. Write an equation which contains no y's, and solve it for x. $x = -2$

d. Use the x-value you found in part c to find the corresponding y-value. $y = 9$

e. Write the coordinates of the point where the two lines cross. *(−2, 9)*

EF-69. There are several ways to find the intercepts of the graph of an equation or function. A sketch of the graph $5x + 3y = 14$ is shown here.

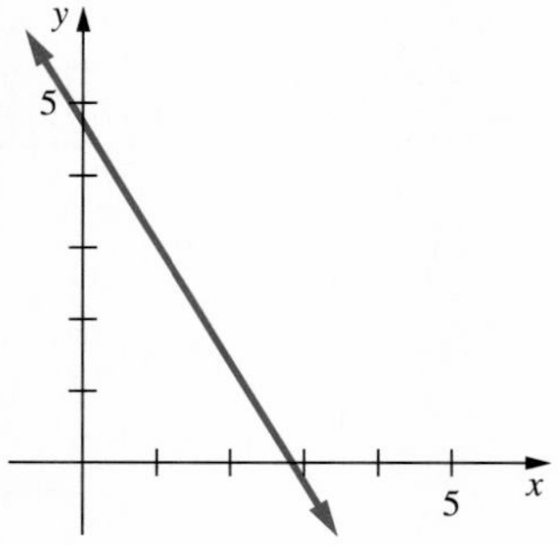

a. From the graph, estimate the y-intercept. Write it as a coordinate. *(0, 4.7)*

b. As you have probably guessed, there are more accurate methods. Another way to find the y-intercept is to write the equation in the form $y = mx + b$. Write the equation in this form. How can you use the equation to determine the y-intercept? How close was your estimate from the graph?

$$y = -\frac{5}{3}x + \frac{14}{3}$$

Another way to find the y-intercept algebraically is use the fact that, at the y-intercept, the x-coordinate is zero. We can calculate the y-intercept by substituting $x = 0$ into the original equation:

$$5(0) + 3y = 14$$
$$0 + 3y = 14$$
$$y = \frac{14}{3}$$

With this method we also find the y-intercept to be $\left(0, \frac{14}{3}\right)$.

c. From the graph, estimate the x-intercept. Write it as a coordinate. *(2.8, 0)*

d. What is true about the y-coordinate of the x-intercept? *It is zero.*

e. Use this information to find the x-intercept. *(14/5, 0)*

f. Be sure to check your Tool Kit to see if you have an entry on how to find the x- and y-intercepts of a graph. If you don't have an entry, do it now!

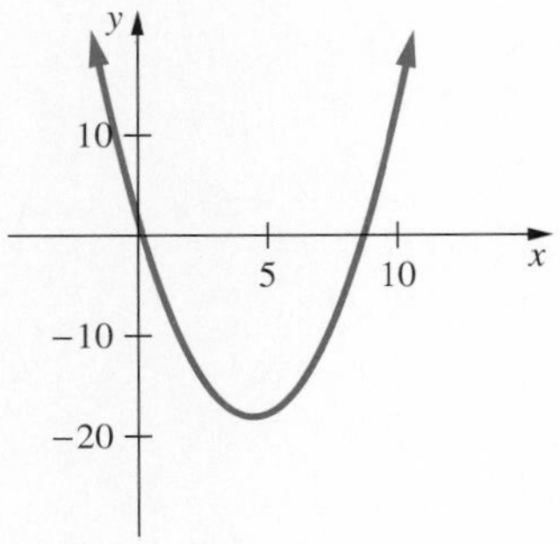

EF-70. A sketch of the graph of $y = x^2 - 9x + 2$ is shown here.

a. From the graph, estimate the y-intercept and write its coordinates. *(0, 2)*

b. From the graph, estimate the x-intercepts and write their coordinates. *(0.2, 0), (8.8, 0)*

c. Describe an algebraic method to improve on your estimate. Use your method, and compare your results with your estimates.

EXTENSION AND PRACTICE

EF-71. Find where the following pairs of lines intersect:

a. $y = 5x - 2$
$y = 3x + 18$
(10, 48)

b. $y = x - 4$
$2x + 3y = -17$
(–1, –5)

EF-72. Where does the graph of each equation cross the y-axis?

a. $y = 3x + 6$ *(0, 6)*

b. $x = 5y - 10$ *(0, 2)*

c. $y = x^2$ *(0, 0)*

d. $y = 2x^2 - 4$ *(0, –4)*

e. $y = (x - 5)^2$ *(0, 25)*

EF-73. What value of y allows you to find the x-intercept? For each equation in EF-72, find where its graph intersects the x-axis. *(–2, 0), (–10, 0), (0, 0), ($\pm\sqrt{2}$, 0), (5, 0)*

EF-74. If $f(x) = -x^2 + 2x^3$, find:

a. $f(-1)$ *–3*

b. $f(-2)$ *–20*

c. $f(2)$ *12*

d. $f(1)$ *1*

e. $f(0)$ *0*

f. Sketch a graph of this function.

g. What value(s) of x will make $f(x) = 0$? *x = 0, 1/2*

EF-75. Match each of the following four graphs with the correct story. For the graph that does not fit, write a story that does.

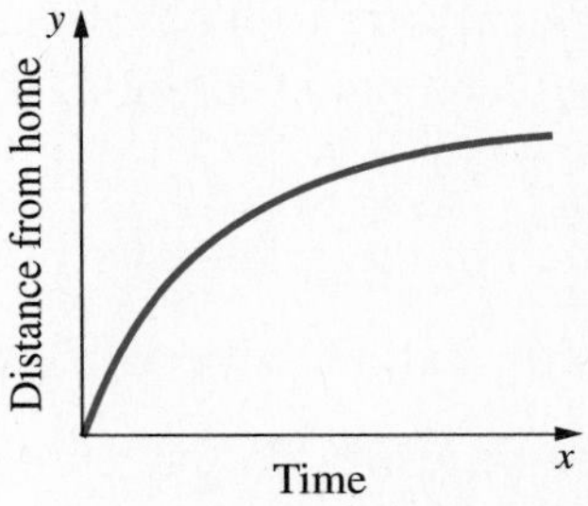

Graph i

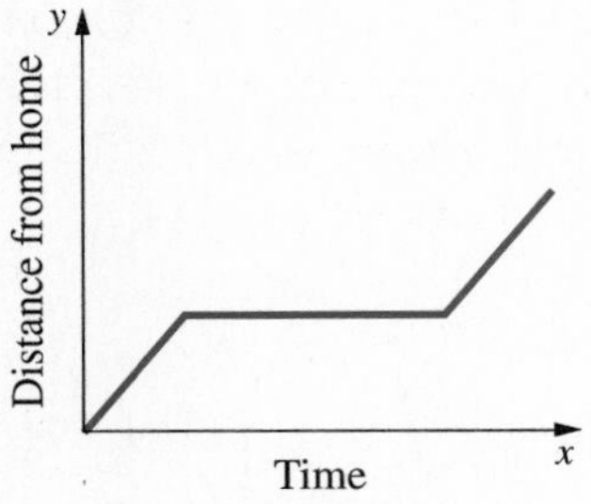

Graph ii

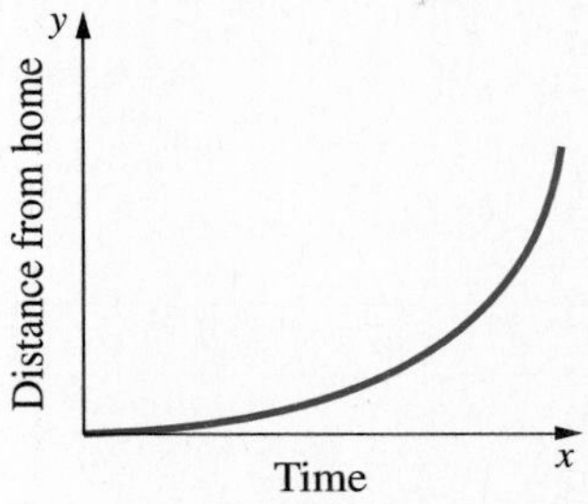

Graph iii

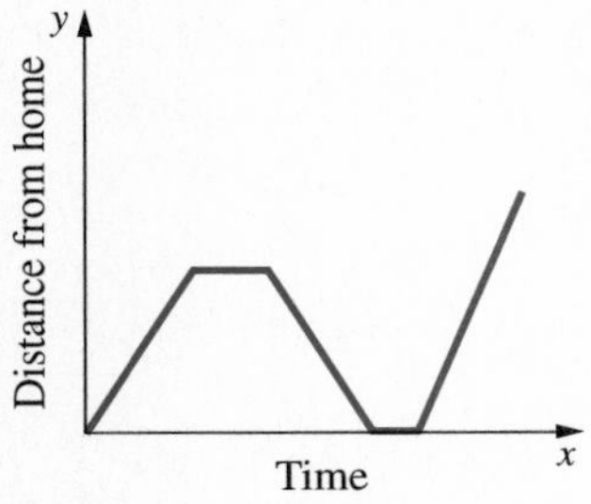

Graph iv

a. I had almost reached my class when I realized I had left my backpack at home, so I had to go back and find it. *iv*

b. I was riding my bike to class enjoying the beautiful fall day when I got a flat tire. I fixed it and made it to class on time. *ii*

c. I was strolling to class taking my time, because I thought it was only 7:30. After several minutes my watch *still* said 7:30, so I knew I was in trouble! I had to run the rest of the way to class. *iii*

EF-76. Ayla and Sean are about to paint the side of their house facing their neighbor Brutus. Brutus tells them that since the wall they will paint on their house is only 6 feet away from his wall, they might as well paint his wall as well. Both walls are 10 feet high, and Ayla and Sean's ladder is

10 feet long. If they place the ladder exactly halfway between the two buildings, and let it lean against one side and then the other (as if it were hinged on the ground), will they be able to reach the tops of both walls? Explain your answer. *Yes: The ladder reaches ≈ 9.5 ft up each wall.*

EF-77. From each of the following pairs of domain and range graphs, sketch two different graphs of functions.

a.

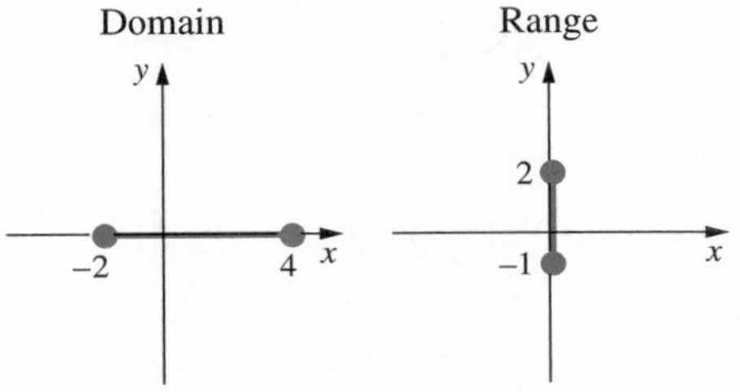

b.

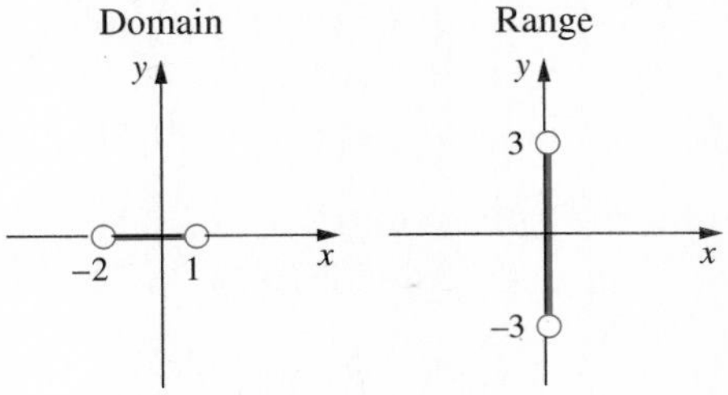

2.6 SUMMARY

Summarizing is different from just doing another bunch of problems. There are three or four problems at the end of most chapters that are designed to help you to organize and articulate what you have been learning.

The summary problem is not just another problem to be done. Its parts are selected to represent the big ideas of the chapter and their interrelationships. Your job is not just to do the problem, but to write out and explain what you are doing as you go, partly for your own information and future use and partly as a way of convincing your instructor and others that you really do understand.

Your instructor may use these summaries as assessments, either in place of tests or along with them. Your instructor may consider the summaries as important as tests because the explanations you write can represent very well what you know. Summaries also allow you to give thoughtful responses as opposed to the quick responses required in some testing situations.

EF-78. Summary In this chapter you have (1) solved linear and quadratic equations, (2) found mathematical models to fit real data, (3) drawn graphs of lines and parabolas, (4) investigated functions, and (5) been introduced to independent and dependent variables and the concepts of domain and range. Your responses to the following situations should convince your fellow group members and your instructor that you can apply what you have learned.

a. The graphs $y = 5x - 4$ and $-2x + 3y = 14$ cross at the point (2, 6).

i. Find the slope, y-intercept, and x-intercept of each equation.

ii. Solve the system of equations algebraically. Show your work.

iii. Explain how the point (2, 6) relates to the graph.

b. Look back to your first function investigation, $y = \sqrt{(9+x)} - 1$ from EF-19. What are the possible x- values for the equation? What are the possible y-values for the equation? What did y depend on? Explain. *$x \geq -9$, $y \geq -1$; x*

c. A teacher asked 10 students to keep track of the time they spent on math assignments outside of class. The table lists their time in hours for one week and their percentage score in the course to date.

i. Use a whole sheet of graph paper to make a large scatterplot of the data, and draw in a line of best fit.

ii. Find the equation of your line of best fit. *One possibility: $y = 0.8x + 20$*

iii. Describe the independent and dependent variables. *ind = work time, dep = percentage score*

iv. What does the slope represent? Explain. *percentage gain per hour*

v. Explain what the y-intercept represents on this graph. *potential score for no work outside of class*

Study Time (hours)	Score
3.4	62
7.8	87
6.5	85
4.6	70
5.5	72
2.5	46
5.0	74
6.0	81
4.2	68
7.1	90

Use your model to predict how many hours a student would have to study in order to have a percentage of 95. Explain how you arrived at your prediction.

d. Imagine a situation different from the one given in part c that also would be linear. Describe it, and explain the domain and range associated with it.

EF-79. Growth over Time—Problem No. 1 This growth-over-time problem will be repeated in Chapters 5 and 7. So you will be doing the same problem three times, but each time you should have something new to add. After completing it for the third time, you will be asked to reflect on your three responses, compare them, and discuss what you have learned.

On a separate sheet of paper (you will be handing this problem in separately or putting it into your portfolio) explain everything that you know about the following two equations:

$$y = x^2 - 4 \text{ and } y = \sqrt{(x+4)}.$$

EXTENSION AND PRACTICE

EF-80. Self-Evaluation There are a lot of ideas covered in this chapter. Some ideas should have been a review while others were probably new. Answer the following questions completely:

a. Write down at least three major ideas of the chapter, and find one or more problems that illustrate each idea. Be sure to include a description of the original problem and a completely worked-out solution.

b. If you had to choose a favorite problem from the chapter, what would it be? Why?

c. Find a problem you still cannot solve or one you are worried that you might not be able to solve on a test. Write out the question and as much of the solution as you can until you get to the hard part. Then explain what it is that keeps you from solving the problem. Be clear and precise.

d. In this chapter you worked in your group daily. What role did you play in the group? (Discuss ideas similar to these: Were you a leader, taking charge? Did you keep your group on task? Did you ask questions? Did you just listen and copy down answers? In what ways are you a good group member? How could you do better? Did your group work well? Why?)

e. What did you learn in this chapter? Explain how your learning relates to your response in part d.

EF-81. Tool Kit Check Now would be a good time to review and revise your Tool Kit. Consider including examples and explanations that will help you understand and remember each of the following. Remember, you want your Tool Kit to be useful, so just include what *you* need. Also be sure you are updating and revising your definitions because we will continue to clarify details and add new examples. Exchange your Tool Kit with your group members, and read each other's to check for clarity and accuracy.

- domain and range
- slope
- independent and dependent variables
- x- and y-intercepts
- function notation

EF-82. Graph these two lines on the same set of axes: $y = 2x$ and $y = -\frac{1}{2}x + 6$.

a. Find the x- and y-intercepts for each equation. *$y = 2x$: (0, 0), $y = -\frac{1}{2}x + 6$: (0, 6), (12, 0).*

b. Shade in the region bounded by the two lines and the x-axis.

c. What is the range and domain of the region? *$0 \le x \le 12$, $0 \le y \le 4.8$*

d. Find the area of this region, accurate to the nearest tenth. *28.8 sq. units*

e. What is the angle between the two lines? There is a word that describes the relationship between the two lines. What is it? *90°, perpendicular*

f. What do you notice about the slope of the lines? *They are reciprocals and opposites.*

EF-83. In the last problem, did you need to find the point where the two lines intersected?

a. Explain how to find the point of intersection for the two lines in the previous problem. *We hope they will say, "Use algebra and solve," but they may count instead.*

b. Is using the graph an accurate way to find the point of intersection? Explain. *Answers will vary.*

c. What method do you think is the most accurate? Explain.

EF-84. Nissos and Chelita were arguing over a math problem. Nissos was trying to explain to Chelita how to find the x-intercepts of the function $y = x^2 - 10x + 21$. "Okay," Chelita said, "I know how to find x-intercepts. You make the y equal to zero and solve for x. But how do I solve *this* equation?"

$$x^2 - 10x + 21 = 0$$

Nissos responded, "If you can't factor it, you can use the quadratic formula." Finish the problem for Chelita. Explain your answer completely. *(3, 0), (7, 0)*

EF-85. Solve each of the following equations for x.

a. $1234x + 23{,}456 = 987{,}654$ *$x \approx 781.36$*

b. $-3x^2 + 2x + 8 = 0$ *$x = -4/3$ or 2*

c. $5x^2 - 6x + 1 = 0$ *$x = 1$ or 1/5*

EF-86. A circle has an area of 45 cm^2 (square centimeters). Find its circumference (perimeter). Remember that $C = 2\pi r$ and $A = \pi r^2$.

a. List all the subproblems you will need to complete to solve this problem. *First use one equation to find the radius, then use the other to get the circumference.*

b. Solve the original problem. Give your answer exactly. (This will involve π and a square root.) Then use your calculator to approximate your answer to two decimal places. *Students will probably get $2\pi\sqrt{45\pi} \approx 23.78$ cm*

EF-87. Find the domain and the range for each of the following functions. Each tick mark represents one unit.

a.

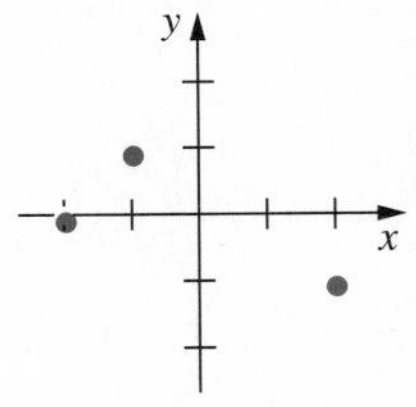

D = {−2, −1, 2}, R = {−1, 0, 1}

b.

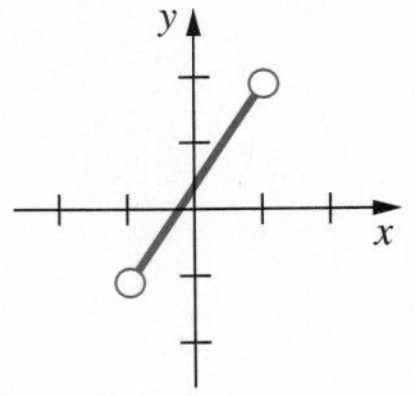

D = (−1, 1), R = (−1, 2)

c.

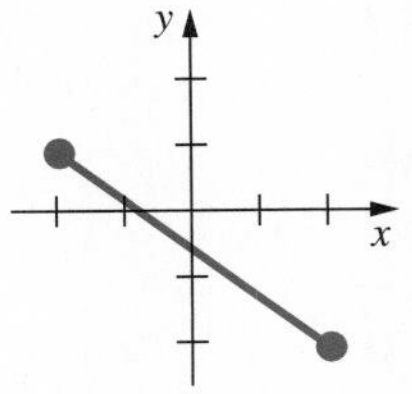

D = [−2, 2], R = [−2, 1]

d.

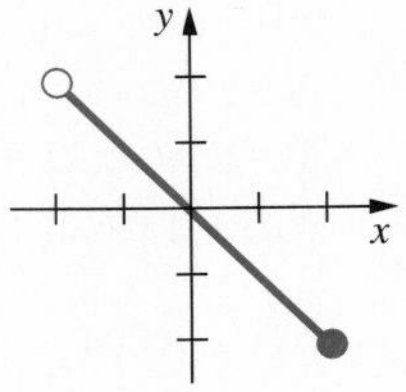

D = (−2, 2], R = [−2, 2)

EF-88. I noticed that my outdoor faucet was dripping even after I turned it off. So I decided to see how much water was being wasted. I put a glass under the faucet at 7:33 P.M. and removed it at 9:19 P.M. When I poured the water out of the glass into a measuring cup, the water measured almost exactly $1\frac{1}{3}$ cups. About how much water leaks from that faucet each day?

What subproblems did you do to solve this? *18.11 cups/day*

EF-89. Write an equation that represents the relationship between the number of days the faucet in EF-88 has been dripping and how much water has been wasted. *$y = 18.11x$*

a. What are the independent and dependent variables? *ind = time (days), dep = cups of water*

b. What are the domain and range? *domain: 0 to a large number; range: 0 to some large number*

c. What should the graph look like? Justify your answer. *straight line, positive slope*

Problem EF-90 introduces the idea of a Mathematics Portfolio. If you are planning to have students keep portfolios, this problem may provide a useful introduction to your portfolio assignment.

If the teaching methods and approach of this text are new to you and you have not used portfolios as a form of assessment in the past, we recommend not introducing their use the first time through. Portfolio assessment adds one more new method, which may be one too many. Those who have tried it recommend using the materials for a quarter or semester first and waiting until the second time around to introduce portfolios as a method of assessment. Moreover, using the materials for a quarter or semester first will allow better anticipation of what you would like to see in the portfolios.

EF-90. What Is a Portfolio? If you heard someone talking about an artist showing his portfolio of work, or an architect displaying her portfolio of designs, you would probably know exactly what each of those portfolios are. Those portfolios represent the artist's and architect's best work. It is their way to demonstrate to prospective clients that they are skilled and know their trade or that they are talented and creative in their fields.

A landscape architect might include photographs of work in progress so that he could show the stages of the job he performs. This would give the client an idea of how hard and meticulously the landscape architect works.

Sometimes this portfolio also shows changes in the person's work. For example, if a product designer developed a new technique, she might include pictures of how her products have improved and developed over time.

Keeping all these examples in mind, what would you put in your Mathematics Portfolio? You want to make sure your best work goes into it, but this does not mean just your neatest or prettiest work. Your portfolio must show the person who reads it not only that you know some mathematics, but that you are competent—in fact good—at mathematics! You must not assume, however, that the reader knows a lot of mathematics. Sometimes the reader will be your mathematics instructor, who does know a lot of mathematics. But you may also be showing your portfolio to other students and other instructors, some from mathematics who are not familiar with this course and some from other disciplines. So you will need to show examples of the steps you went through to produce a finished piece.

Possibly, you would include examples of work that show how you have improved or have learned new mathematics. Samples of an initial attempt followed by several revisions are important in demonstrating your ability to persevere, follow through, and grow in your understanding.

One of the reasons an algebra course is required, in addition to providing needed preparation in mathematics, is to determine which students are willing and able to keep going when the going gets tough. Are you a student who is willing to persevere in working on an idea or problem that is difficult or that does not make sense at first? Are you able to try a variety of approaches? Are you willing to revise? These are all skills that can be demonstrated in a portfolio.

So in this course, one of your goals is to produce a Mathematics Portfolio that will be an honest reflection of the mathematics you know, have learned, and are still considering. It should be something you would be proud to show a future university admissions panel or an employer.

CHAPTER THREE

THE BOUNCING BALL AND OTHER SEQUENCES

IN THIS CHAPTER YOU WILL LEARN TO RECOGNIZE AND DEVELOP PATTERNS, PARTICULARLY IN ARITHMETIC AND GEOMETRIC SEQUENCES, AND UNDERSTAND THE MEANING OF THOSE SEQUENCES BY REPRESENTING THEM NUMERICALLY, GRAPHICALLY, AND ALGEBRAICALLY. THROUGHOUT THIS CHAPTER YOU WILL:

- continue to learn new mathematics by working in groups and using problem-solving strategies including guess and check, looking for patterns, making data tables, drawing diagrams and graphs, and solving subproblems;
- use patterns to make conjectures and to write algebraic representations;
- become familiar with numeric patterns and graphs of functions that are multiplicative or geometric, as compared with functions that are additive or arithmetic;
- determine relationships between discrete functions (such as arithmetic sequences) and continuous functions (such as straight lines);
- continue to use previously learned algebraic skills in new problem-solving contexts, particularly in solving linear and quadratic equations, solving systems of equations, and using exponents.

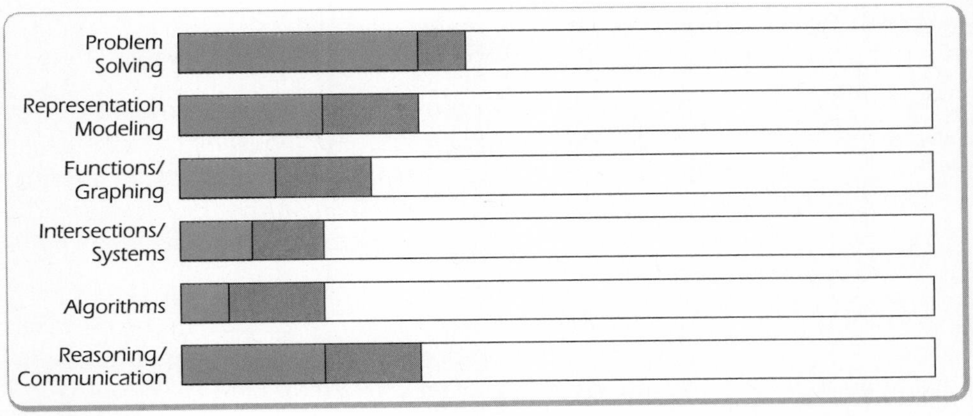

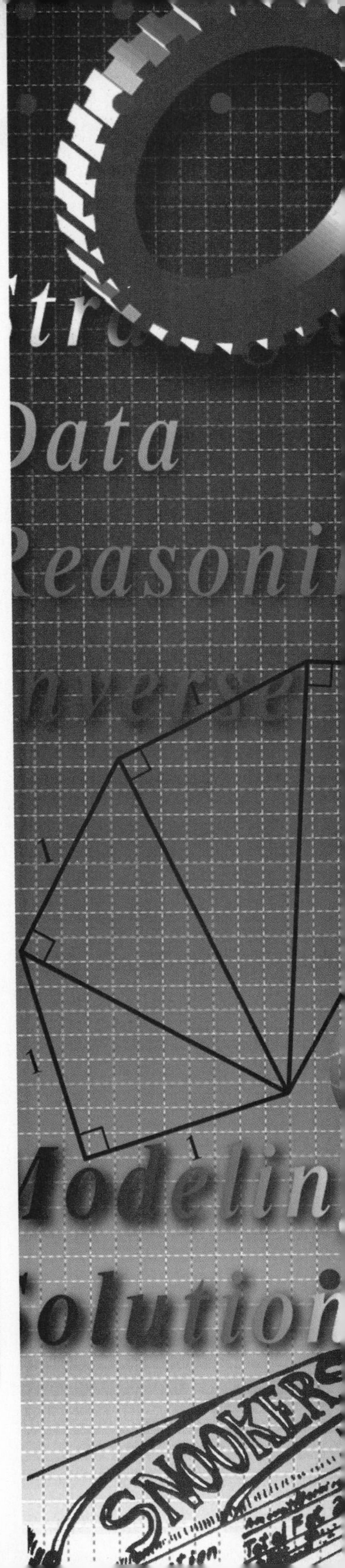

MATERIALS		CHAPTER CONTENTS	PROBLEM SETS
	3.1	Introduction to Sequences	BB-1 – BB-15
BB-16 Resource Page (two copies per student)	3.2	Graphing Sequences	BB-16 – BB-32
	3.3	Classifying Sequences, Function Notation, and Arithmetic Sequences	BB-33 – BB-51
Superballs, meter sticks, graph paper	3.4	The Bouncing Ball Lab	BB-52 – BB-68
		Extra Problems	BB-69 – BB-73
	3.5	Geometric Sequences and the nth Term	BB-74 – BB-90
	3.6	Multipliers, Applications, and Chapter Summary	BB-91 – BB-104
		Extra Problems	BB-105 – BB-107

Although this chapter on patterns develops arithmetic and geometric sequences, its main purpose is to review basic algebraic skills in a new context and to prepare for work with exponential functions in Chapter 4. Early in the chapter we expect students to guess and check, but as the chapter progresses we want them to move to the algebraic and graphical representations so that we can review skills such as solving for an indicated variable and solving two equations with two unknowns.

Throughout the chapter we use $n = 0$ as our initial value. Although most traditional approaches start sequences with $n = 1$, we have found that students become greatly confused when they need to apply the sequences to real situations. In addition, using the initial value $n = 0$ provides a direct correspondence with the y-intercept on the graph. When the students start to work with a sequence, you should describe the sequence as follows: "$t(0) =$ the initial value, $t(1) =$ the first value **after** *the initial value, $t(2) =$ the second value* **after** *the initial value," and so on. To demonstrate a need for this approach, we begin the chapter with an application problem in which the students can use their problem-solving skills to develop an equation with an initial value at $n = 0$.*

3.1 INTRODUCTION TO SEQUENCES

Problem BB-1 gives students an example of an arithmetic sequence. They should be able to solve this problem using the problem-solving skills developed earlier. Notice that we are starting with an initial value at zero months. After the students have completed BB-1, spend a few minutes discussing the problem, in particular, part e. Remind them of this problem as they work through other sequences.

Problems BB-1 through BB-3 introduce sequences on a numeric level, whereas Problems BB-16 through BB-20 present sequences in a graphical context. If your class meets in longer time blocks, these two groups of problems, which we cover in separate sections, 3.1 and 3.2, could be done together. Problem BB-16 requires two copies of the Sequence Graph Resource Page.

BB-1. Lona has received a stamp collection from her grandmother. The collection is in a leather book and currently has 120 stamps. Lona joins a stamp club that sends her 12 new stamps each month. The stamp book can contain a maximum of 500 stamps.

a. Copy and complete the following table:

Months (n)	**Total Stamps** $t(n)$
0	120
1	132
2	144
3	
4	
5	

b. How many stamps will she have after one year? *264*

c. When will the book be filled? *after 32 months*

d. Write an equation to represent the total number of stamps Lona has in her collection after n months. Let the total be represented by $t(n)$. $t(n) = 12n + 120$

e. Solve your equation for n when $t(n) = 500$. Explain why Lona will not exactly fill her book with no stamps remaining. $n = 31.67$

The next problem develops a way of looking at sequences recursively. The goal here is to give the students a means to distinguish arithmetic and geometric sequences by examining the patterns generated by terms.

BB-2. Samantha was looking at one of the function machines in Chapter 2, and decided that she could create her own machine by recycling her *output values* through it. She called it her "output recycling machine" and tried it out by dropping in the initial value of 8. She recorded each output before recycling it as a new input.

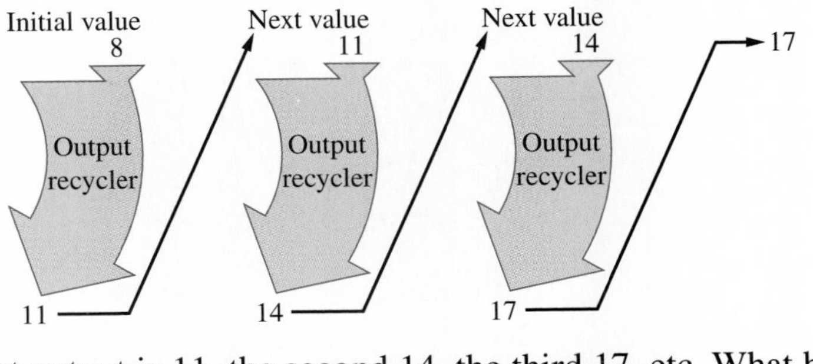

a. The first output is 11, the second 14, the third 17, etc. What happens to each output when it is recycled ? *3 is added.*

b. In a new experiment Samantha uses the initial value –3 and the recycling rule "multiply by –2." Make a list including the initial value and the first four outputs of the recycling machine. *–3, 6, –12, 24, –48*

c. What sequence will she generate if she uses an initial value of 3 and the recycling rule *square*? Make a list as you did in part b. *3, 9, 81, 6561, 43,046,721 . . .*

BB-3. Samantha has been busy building new output recycling machines and has created several sequences based on simple rules. Her instructor also has been busy thinking up sequences using his own devious methods. The following list of sequences, parts a through i, is a mixture of Samantha's sequences and her instructor's. Recall that these machines recycle outputs to create new outputs. In your groups, follow directions 1 through 4 for each sequence a through i.

1. For the given sequence, make a table like the one shown. The sequence in part d is given as an example. We denote the initial value, or **term**, in the sequence as $t(0)$, the first term *after* the initial value as $t(1)$; the second term *after* the initial value $t(2)$, and so on.

Position n	Term $t(n)$
0	2
1	3.5
2	5
3	6.5
4	8

2. Extend each table to show the next three terms for each sequence.
3. Use words or algebraic symbols to describe a rule for finding the next term.
4. Decide whether the sequence could be produced by a recycling machine with a single simple rule for repeatedly adding or repeatedly multiplying by the same number in order to go from one number to the next.

a. 0, 2, 4, 6, 8, . . .
10, 12, 14; add 2; yes

b. 1, 2, 4, 8, . . .
16, 32, 64; multiply by 2; yes

c. 7, 5, 3, 1, . . .
−1, −3, −5; subtract 2 or add −2; yes

d. 2, 3.5, 5, 6.5, . . .
8, 9.5, 11; add 1.5; yes

e. 1, 1, 2, 3, 5, . . .
8, 13, 21; add previous 2 terms; no

f. 27, 9, 3, 1, . . .
1/3, 1/9, 1/27; divide by 3 or multiply by 1/3; yes

g. 40, 20, 10, . . .
5, 2.5, 1.25; divide by 2 or multiply by 1/2; yes

h. −4, −1, 2, 5, . . .
8, 11, 14; add 3; yes

i. 3, 6, 12, . . .
24, 48, 96; multiply by 2; yes

Be sure to save your tables and other work from this problem for BB-16.

EXTENSION AND PRACTICE

BB-4. Samantha thought about another sequence recycling machine. "If I can square, I can unsquare," she reasoned. "I wonder what will happen if I start with 625 and use the rule *unsquare*." Samantha had just started her machine when the phone rang. When she returned, her machine was generating the 4275th result after the initial value.

a. What is another word for *unsquare*? *square root*

b. What was the 4725th number Samantha found? (*Note*: This can be found fairly easily using a scientific or graphing calculator. You may need to consult the directions of the one you use.) *1: if students keep hitting the "$\sqrt{\ }$" button, they will get one after 30 to 40 iterations.*

BB-5. Consider the following sequence of rectangles:

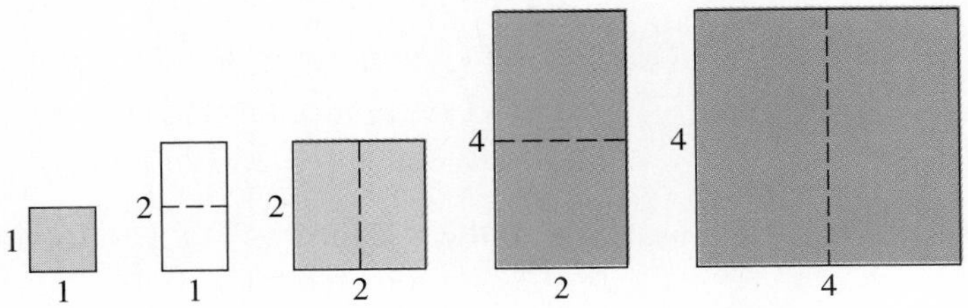

a. On a sheet of graph paper, draw a picture of the next two rectangles in the sequence of rectangles above.

b. Describe how each is formed from its preceding rectangle.

c. Write the areas of these rectangles as a sequence. *1, 2, 4, 8, 16, 32, 64, . . .*

BB-6. Rewrite each expression in parts a through f in a simpler form. (Examples 1 and 2 are reminders about how to do this. After doing parts a through f, add basic exponent rules to your Tool Kit.

Examples: 1. $\frac{7^5}{7^3} = \frac{7\cdot7\cdot7\cdot7\cdot7}{7\cdot7\cdot7} = 7\cdot7 = 7^2$

2. $(5^3)^4 = (5^3)(5^3)(5^3)(5^3) = (5\cdot5\cdot5)(5\cdot5\cdot5)(5\cdot5\cdot5)(5\cdot5\cdot5) = 5^{12}$

a. $\frac{5^{723}}{5^{721}}$ *$5^2 = 25$*

b. $\frac{3^{300}}{3^{249}}$ *3^{51}*

c. $\frac{3\cdot4^{1001}}{7\cdot4^{997}}$ *$\frac{3\cdot(4^4)}{7}$*

d. $\frac{(6^{54})^{11}}{(6^{49})^{10}}$ *6^{104}*

e. $\frac{x^p}{x^q}$ *x^{p-q}*

f. $\frac{(x^p)^k}{(x^q)^m}$ *x^{pk-qm}*

BB-7. Write the equation of a line with:

a. a slope of –2 and a *y*-intercept of (0, 7). *$y = -2x + 7$*

b. a slope of $\frac{-3}{2}$ and an *x*-intercept of (4, 0). *$y = \frac{-3}{2}x + 6$*

BB-8. Consider the tables below. See if you can find some patterns within each table:

Table 1

n	$t(n)$
0	2
1	3
2	4
3	5

Table 2

n	$s(n)$
0	0
1	3
2	6
3	9

Table 3

n	$p(n)$
0	−2
1	2
2	6
3	10

a. How is each sequence above being produced? *add 2, multiply by 3, multiply by 4 and subtract 2*

b. Extend each table to $n = 5$. Write an equation that shows (or describe in words) the relationships between $n = 5$ and $t(5)$, between $n = 5$ and $s(5)$, and between $n = 5$ and $p(5)$. *$t(5) = 5 + 2$; $s(5) = 3 \cdot 5$; $p(5) = 4 \cdot 5 - 2$*

c. Without extending the tables any farther, find $t(25)$, $s(25)$, and $p(25)$. *27, 75, 98*

d. Write a rule that will find the nth term without going through all of the previous terms. For example, Table 1 would have the equation $t(n) = n + 2$. *$t(n) = n + 2$; $s(n) = 3n$; $p(n) = 4n - 2$*

For BB-9, students will need to draw the triangle and use the Pythagorean theorem to find the lengths of the sides. These suggestions for getting started are in the students' answers.

BB-9. A triangle has vertices of $A(3, 2)$, $B(-2, 0)$, and $C(-1, 4)$. What kind of triangle is it? Be sure to consider all possible triangle types. Include sufficient evidence to support your conclusion. Sufficient evidence includes knowing the lengths of the sides. Calculate the length of each side. *scalene triangle*

BB-10. Factor each expression. Then simplify the rational expression.

a. $\dfrac{x^2 - 4}{x^2 + 4x + 4}$ **b.** $\dfrac{2x^2 - 5x - 3}{4x^2 + 4x + 1}$

$\dfrac{x - 2}{x + 2}$ $\dfrac{x - 3}{2x + 1}$

BB-11. Find the x-intercepts for the graph of $y - x^2 = 6x$. *(0, 0), (−6, 0)*

BB-12. Jonathan pitches a baseball to Carianne, who hits a home run. Sketch a graph showing the height of the ball from the moment it leaves Jonathan's hand. What is the independent variable? the dependent variable? the domain? the range?

Solution: Some students may make the path trace back over itself. If you see that, remind the student that the x-axis represents time.

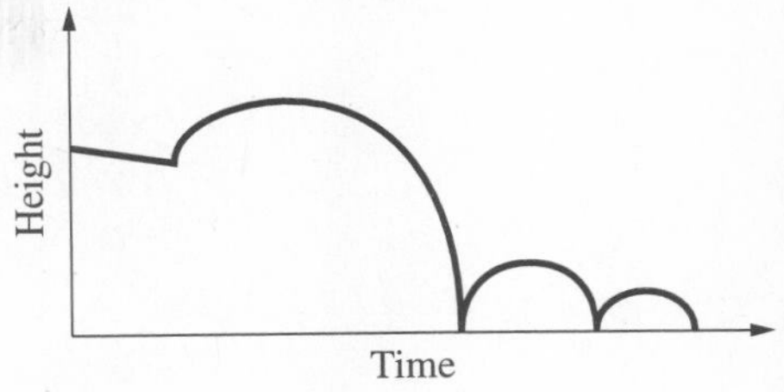

BB-13. Solve the following equation for x: $-4x^2 + 5x + 7 = 0$. *$x = (-5 \pm \sqrt{137})/-8$*

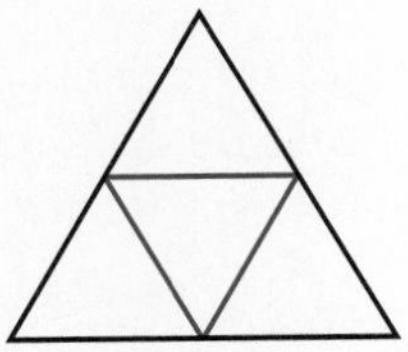

BB-14. A dartboard is in the shape of an equilateral triangle with a smaller equilateral triangle in the center made by joining the midpoints of the three edges. A dart hits the board at random. What is the probability that:

a. the dart hits the center triangle. *1/4*

b. the dart misses the center triangle. *3/4*

BB-15. For each step a through e, explain what mathematical operation or relationship to use to get the next result. It might be helpful to read Appendix A for hints on solving equations with fractions in them.

a. $\frac{x}{4} + \frac{2}{3} = \frac{5}{2}$

Multiply by a common denominator.

b. $12\left(\frac{x}{4} + \frac{2}{3}\right) = 12\left(\frac{5}{2}\right)$

Use distributive property.

c. $12\left(\frac{x}{4}\right) + 12\left(\frac{2}{3}\right) = 12\left(\frac{5}{2}\right)$

Complete multiplication.

d. $3x + 8 = 30$

Subtract 8.

e. $3x = 22$

Divide by 3.

$x = \frac{22}{3}$

3.2 GRAPHING SEQUENCES

For these problems, students will be graphing sequences. It is important to note that the graphs are discrete, so ask students whether they should connect the points. Since the domain represents the number of times an input passes through the recycling machine, the domain can only consist of non-negative integers. Be sure to tune in to student discussions as they debate the nature of their graphs.

Here it is important to state that the initial value occurs when $n = 0$, not when $n = 1$. Doing so avoids a great amount of confusion later in this chapter and in Chapter 4. It also provides a natural correspondence between the initial value and the y-intercept. For example, in part a of BB-3 you can state that 0 is the initial value and 2 is the first number after the initial value or the first application of the rule. We refer to this rule as a ***recycling rule****.*

Each student will need two copies of the Resource Page containing sets of axes for BB-16. Using these established sets of axes, students will be able to complete these problems much more efficiently, allowing them to get to the main point.

BB-16. You'll need at least two copies of the sequence graph Resource Page for this problem. You may want to divide this task among the members of your group. Be sure that each group member has his or her own set of graphs. For each designated sequence in BB-3, draw a separate graph using the Resource Page provided (at the end of the text). Use n, the *position* of the term after the initial value, as the independent variable. Use the term itself, $t(n)$, as the dependent variable. Notice that the domain for each sequence is whole numbers. *Carefully consider whether or not to connect the points on your graphs. Discuss this with the other members of your group.*

a. Use the tables you made to plot the sequences given in parts a, b, d, f, h, and i of BB-3.

b. Write the initial value and the recycling rule on each graph.

BB-17. Discuss the similarities among the graphs of BB-16 with your group. Write down your observations in relation to the following questions:

a. Which graphs look similar? *a, d, h: linear; b, f, i: curves*

b. Which graphs have similar recycling rules?

c. What is the significance of the initial value in each graph? *y-intercept*

The following problem should make some of the patterns in the sequences more visible. We are not yet seeking a rule but rather a graphic and numeric understanding of sequences. Class discussion should lead to a connection between the ideas of common difference and the rise in slope for the sequences whose graphs are lines. At this point the connection may not be well articulated, but it should be developed over several days and not given all at once in a definition.

We also are looking for a difference in the way the graphs of exponential, or geometric, sequences increase. Some students should be able to point out and define the difference in the rise of parts a and d compared with the rest. In the example below, be sure to ask students why the points cannot be on the same line.

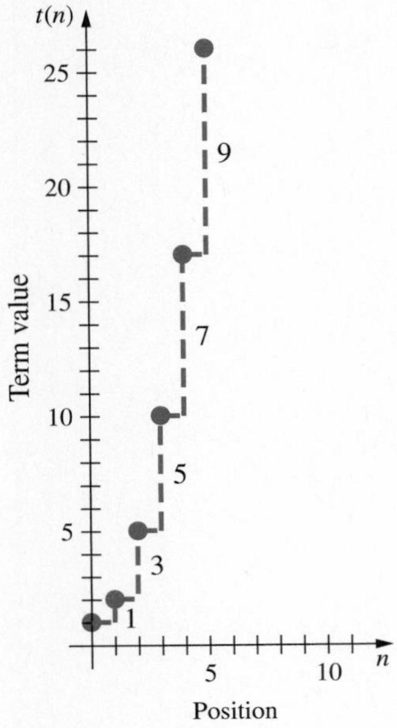

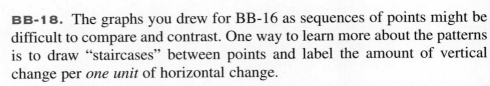

BB-18. The graphs you drew for BB-16 as sequences of points might be difficult to compare and contrast. One way to learn more about the patterns is to draw "staircases" between points and label the amount of vertical change per *one unit* of horizontal change.

Draw a "staircase" for each graph, and write down at least two observations of the vertical and horizontal relationships. For example, the sequence 1, 2, 5, 10, 17, . . . gives the table:

Position n	0	1	2	3	4
Term $t(n)$	1	2	5	10	17

which in turn gives the graph shown.

BB-19. Look at the sequence given in part c of BB-3. Plot the sequence. How is this sequence similar to the sequences in parts a and d? Show that sequence c is similar by writing its recycling rule using addition. How is sequence c different?

BB-20. Use a similar argument to show that sequence f can be grouped with the sequences in parts b and i. Find another sequence in BB-3 that belongs with this group.

EXTENSION AND PRACTICE

BB-21. Systems of Equations (A Reminder from Elementary Algebra) Solve each system of equations algebraically. If you remember how to do these, go ahead and solve. If you are not sure how to start, read the examples in the following boxes first. This may be something to enter into your Tool Kit.

a. $y = 3x + 1$
$x + 2y = -5$
−1, −2

b. $4x + 2y = 14$
$x - 2y = 1$
3, 1

The first subproblem in solving a system of equations is to **eliminate one variable.** One way to do this is **by substitution**.

Consider this system:

$$8y - 3x = 10$$
$$5x + 10y = -5$$

Look for the equation that is easiest to solve for either x or y. Let's try solving the second equation for x:

$$5x + 10y = -5$$
$$5x = -5 - 10y$$
$$x = -1 - 2y$$

Now replace the x in the other equation with $(-1 - 2y)$:

$$8y - 3(-1 - 2y) = 10$$
$$8y + 3 + 6y = 10$$
$$14y + 3 = 10$$
$$14y = 7$$
$$y = 0.5$$

Find x by substituting 0.5 for y:

$$x = -1 - 2(0.5)$$
$$x = -2$$

So the solution is $(-2, 0.5)$.

Some people prefer to **eliminate one variable by adding the two equations**:

First, rewrite the equations so we can see the x's and y's lined up. This makes it easier to decide what to multiply by to make the coefficients of either the x's or the y's opposites of each other.

$$-3x + 8y = 10$$
$$5x + 10y = -5$$

We could multiply the top equation by 5 and the bottom equation by 3 to get:

$$-3x + 8y = 10 \quad \xrightarrow{\text{by } 5} \quad -15x + 40y = 50$$
$$5x + 10y = -5 \quad \xrightarrow{\text{by } 3} \quad 15x + 30y = -15$$

Adding:

$$70y = 35$$
$$y = 0.5$$

Go back now and substitute 0.5 for y:

$$8(0.5) - 3x = 10$$
$$4 - 3x = 10$$
$$-3x = 6$$
$$x = -2$$

Again we find the solution: $(-2, 0.5)$. Be sure to make note of this method in your Tool Kit. Remember, in Chapter 2 you solved these same kinds of problems using a graphical approach.

BB-22. How can you solve a system of equations by graphing? What are some advantages of choosing to use an algebraic approach as opposed to solving a system by graphing? Explain.

BB-23. One week Thomas bought five cans of chicken noodle soup and three cans of tuna and spent \$8.12. The next week the prices were the same, and his total bill was \$7.52 for six cans of chicken noodle soup and two cans of tuna. Write two equations and solve them to figure out the price of a can of chicken noodle soup and the price of a can of tuna. *soup, \$0.79; tuna, \$1.39*

BB-24. Determine whether the points $A(3, 5)$, $B(-2, 6)$, and $C(-5,7)$ are on the same line. Justify your conclusion. *no: see graph, or using algebra:* $m_{AB} = -1/5$; $m_{BC} = -1/3$; $m_{AC} = -1/4$

BB-25. Sketch the graphs of $y = x^2$, $y = 2x^2$, and $y = \frac{1}{2}x^2$ on the same set of axes. Describe the differences between the graphs.
Width of graph changes.

BB-26. Allene wants to examine these equations on her graphing calculator. Rewrite each of the equations as "$y =$ " so that she can enter them into the calculator.

a. $5 - (y - 2) = 3x$ $\quad y = -3x + 7$ $\qquad$ **b.** $5(x + y) = -2$ $\quad y = -x - \frac{2}{5}$

For BB-27, students may represent domains and ranges in a variety of ways — in words, interval notation, inequality notation, or even with number-line graphs. It is not necessary to restrict them to a particular method, so long as they communicate effectively. Students will gravitate to a form they are most comfortable with, and for now, they should be encouraged to use the method that makes the most sense to them.

BB-27. Find the domain and range for each of the following functions. Each tick mark represents one unit.

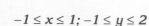

a.

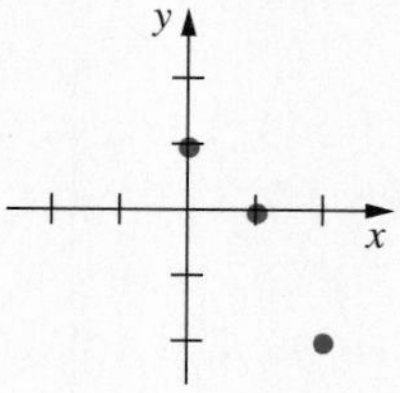

x: 0, 1, 2; y: −2, 0, 1

b.

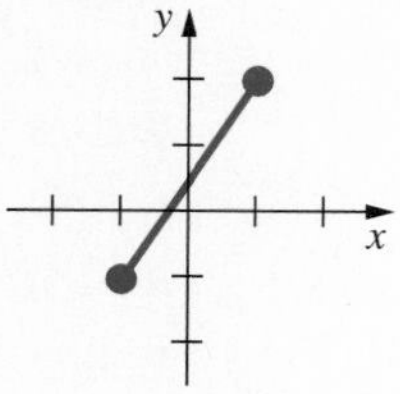

$-1 \le x \le 1; -1 \le y \le 2$

c.

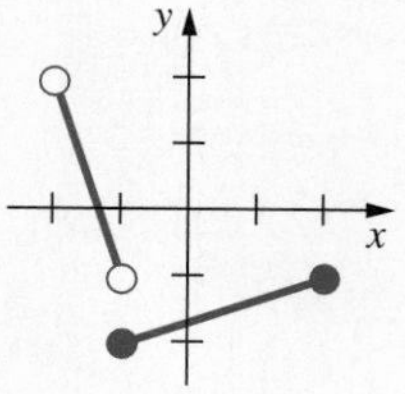

$-2 < x \le 2; -2 \le y < 2$

d.

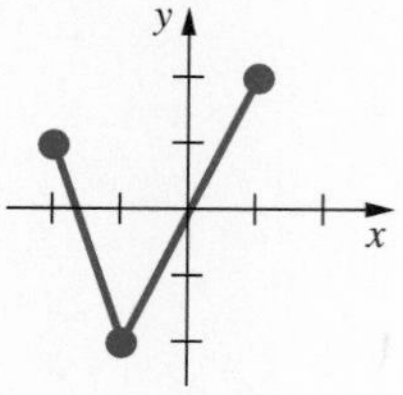

$-2 \le x \le 1; -2 \le y \le 2$

BB-28. Simplify each of the following. Add to your Tool Kit any basic exponent rules that you don't already have.

a. $\dfrac{3^5 \cdot 9^4}{27^4}$ 3

b. $\dfrac{(b^2)^3}{(b^3b^4)}$ *1/b*

c. $\dfrac{(b^n)^2}{b^n \cdot b^{n+1}}$ *1/b*

BB-29. If the two cans shown here hold the same amount of tuna, find h. Reminder: $V = \pi r^2 h$. *4*

BB-30. Solve the following equations for x and/or y. Show all work.

a. $5x - 7 = -3(5 - x)$ $x = -4$

b. $\dfrac{x}{12} = \dfrac{14}{9}$ $x = 56/3$

c. $y = 4x - 1$
$y = -2x + 5$ *(1, 3)*

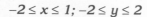

d. $2x^2 + 9x - 5 = 0$ $x = 1/2, -5$

BB-31. The theme for the spring dance this year is "Under the Sea." Marty is in charge of supplying the punch. He decides to make a tropical punch with guava juice and mango juice. The ratio of guava to mango must be 3:5, and he needs 15 gallons total. How many gallons of each juice must he use? *guava, $5\frac{5}{8}$ gal; mango, $9\frac{3}{8}$ gal*

BB-32. Toss three coins in the air. Make a list of all the possible outcomes, and find the probability that:

a. all of them land heads up, *1/8*

b. two of them land tails up. *3/8 for exactly 2 tails; students might argue for the answer 1/2*

3.3 CLASSIFYING SEQUENCES, FUNCTION NOTATION, AND ARITHMETIC SEQUENCES

Have students write their examples from parts a and c in BB-33 on the board to see how many common characteristics they can come up with. Discussion of the examples students put on the board also gives you a chance to assess their understanding so far.

BB-33. Some types of sequences are given names based on their common characteristics.

a. Sequences such as those in parts a, c, d, and h of BB-3 are called **arithmetic sequences**. Describe what these sequences have in common, and give two more examples. Add this concept to your Tool Kit.

b. What do the "staircases" of the graphs of the arithmetic sequences have in common? *The stairs are equally spaced and can be found by adding the same number each time.*

c. Sequences such as those in parts b, f, g, and i of BB-3 are called **geometric sequences**. Describe what these sequences have in common and give two more examples. Add this concept to your Tool Kit.

d. What do the "staircases" of the graphs of the geometric sequences have in common? *The stairs are spaced in increments that increase by multiplying by the same number.*

e. The sequence in part e of BB-3 is special and is named after the mathematician who discovered, explored, and wrote about it. His name was Fibonacci. What is different about the staircase of the graph of this sequence? Explain why this sequence is neither an arithmetic sequence nor a geometric sequence. *The staircase turns out to be almost the same as the sequence.*

BB-34. Notation for Writing Sequences. When we write a sequence such as 3, 7, 11, 15, . . . , we need to have a way of describing the individual terms. The most common way to do this is to think of the sequence as a function whose domain is the numbers 0, 1, 2, 3, . . . , n. For each input n, an expression for $t(n)$ is used to generate the term $t(n)$ (also called the output). So a sequence is a function of the position number n. Since the initial term of our sequence is 3, $t(0) = 3$. We could write this sequence as $t(0) = 3$, $t(1) = 7$, $t(2) = 11$, and so on.

a. Find $t(3)$, $t(4)$, and $t(5)$. A table, like the ones you made in BB-3 might help. *t(3) = 15, t(4) = 19, t(5) = 23*

b. Find $t(25)$. Try to use the pattern among the $t(n)$ values to figure out $t(25)$. Focus on just the relationships between 3 and $t(3)$, 4 and $t(4)$, and 5 and $t(5)$, so you don't have to make a huge table. *t(25) = 103*

c. Find an expression for $t(n)$. *t(n) = 4n + 3*

d. Find $t(100)$. *t(100) = 403*

e. Put information about function notation for sequences in your Tool Kit.

BB-35. A vertical or horizontal table of values is very useful when analyzing a sequence to determine its pattern. Consider the sequence 7, 10, 13, . . . shown in this table:

Position n	0	1	2	3	4	5	6
Term $t(n)$	7	10	13	16			

a. Fill in the rest of the table. *19, 22, 25*

b. What is the initial value and the rule for the sequence? What kind of sequence is it? *7, add 3, arithmetic*

c. The number added in an arithmetic sequence is frequently referred to as the **common difference**. Why might this term be used? Add this concept to your Tool Kit. *It is the difference between the terms.*

d. How many 3's must be added to the initial term in order to get the term for $n = 4$? *4*

e. What is an algebraic expression for the value of the nth term after the initial term? *t(n) = 3n + 7*

f. Graph this function and find the vertical change along the line for each unit of horizontal change. What is the slope of the line that would go through the points on the graph? *Slope is 3.*

BB-36. Look at the sequence given in BB-35:

a. Is it possible for the value 2.4 to show up in the position n (that is, could we place a 2.4 in the top row)? Could –6 show up in position n? Explain why or why not.

b. Could $t(n) = 30$ show up in the $t(n)$ row? Explain your thinking. *No, $7\frac{2}{3}$ is not in the domain of this function.*

BB-37. What kinds of sequences have graphs whose points lie on a line? How is the common difference for the sequence related to the graph of the line? *Arithmetic; common difference is the slope.*

BB-38. Using what you have learned in BB-35 through BB-37, find the slope of the line containing the points given by each sequence listed below. Write an expression for the nth term in each case.

a. 5, 8, 11, 14, . . . *$t(n) = 3n + 5$* **b.** 3, 9, 15, . . . *$t(n) = 6n + 3$*

c. 26, 21, 16, . . . *$t(n) = -5n + 26$* **d.** 7, 8.5, 10, . . . *$t(n) = 1.5n + 7$*

BB-39. Find an equation for the function $t(n)$ where the initial value is a and we repeatedly add the value d (d stands for common difference). Add this function to your Tool Kit. Because it uses the variables a and d instead of specific numbers, this function represents *all* arithmetic sequences. *$t(n) = dn + a$, or $t(n) = a + dn$*

BB-40. Find the initial value and the common difference for a sequence where $t(5) = -3$ and $t(50) = 357$. One (not the only) procedure for doing this is as follows:

a. Use the equation you found in BB-39, and substitute –3 for $t(n)$ and 5 for n.

b. Then use the equation again and substitute 357 for $t(n)$ and 50 for n.

c. Now you should have two equations with a and d as variables. Solve the two equations for a and d. If necessary, refer back to BB-21, and review the procedures for solving systems of equations with two variables. *$t(0) = -43$; $d = 8$*

BB-41 can be turned in by students as an assessment of their understanding so far. If you use it as an assessment, encourage them to refer to their Tool Kits.

BB-41. Write a paragraph about the relationship between an arithmetic sequence and its graph. Be sure to explain the relationships among the common difference, slope, y-intercept, and the initial value of the sequence. *Students should provide information that includes the fact that the slope is the common difference and the y-intercept is the initial value.*

EXTENSION AND PRACTICE

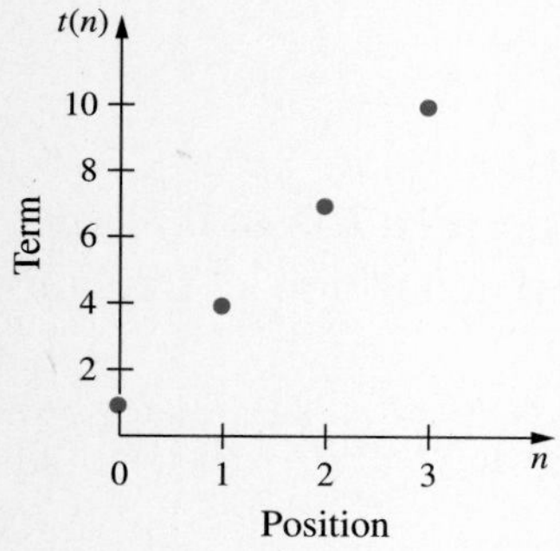

BB-42. For the sequence 1, 4, 7, 10, 13, 16, . . . , we know that $t(2) = 7$ and $t(5) = 16$.

a. What is $t(8)$? *25*

b. What is $t(14)$? *43*

c. The terms, $t(0)$, $t(1)$, $t(2)$, $t(3)$, of this sequence are graphed for you. What is the equation of the line that passes through these points? Write your equation in the form $t(n) =$. *$t(n) = 3n + 1$*

d. What is the significance of the initial value in terms of the graph? *the y-intercept*

e. What does the slope of the line represent in terms of the sequence? *the common difference*

BB-43. Solve the following system for m and b: *$m = 4$, $b = -5$*

$$15 = 5m + b$$
$$7 = 3m + b$$

BB-44. The equations given in BB-43 could have been created by substituting the x and y values for the points (5, 15) and (3, 7) into the equation $y = mx + b$. To get the first equation we replace x with 5 and y with 15. We get the second equation by replacing x with 3 and y with 7. By solving the system to find the slope m and the y-intercept b, you can find the equation of the line through the two points. What is the equation of the line determined by these two points? *$y = 4x - 5$*

BB-45. Using what you learned in BB-43 and BB-44, find the equation of the line through the points (2, 3) and (5, –6). The first equation you need is $3 = 2m + b$. *$y = -3x + 9$*

BB-46. Given that n represents the length of the bottom edge of the L-shaped figures below, and that $t(n)$ is the total number of dots on the L-shape, what type of numbers are the $t(n)$ values? Find $t(46)$. What is $t(n)$?
odd numbers; 93; $2n + 1$

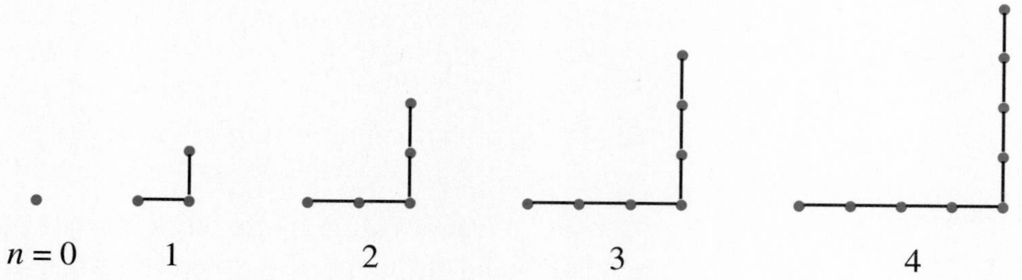

The following is a prelude to the Bouncing Ball Lab.

BB-47. Many games depend on the players' response to the bounce of a ball. For this reason manufacturers have to make balls so that their bounce conforms with certain standards. Some published standards for different balls are:

tennis ball: rebounds 106 to 116 cm (centimeters) when dropped from 200 cm

handball: rebounds 62 to 65 inches when dropped from 100 inches at 68°F

lacrosse: rebounds 45 to 49 inches when dropped from 72 inches on a wooden floor

squash: rebounds 28 to 31 inches when dropped from 100 inches onto a steel plate at 70°F

a. Only squash and handball authorities recognize the importance of both temperature and surface in setting their standard. What effect could surface and temperature have on the bounce of a ball?

b. Which ball bounces proportionately higher than the others? *The range of rebound ratios for lacrosse balls (0.625 to 0.681) is slightly higher than the range for handballs (0.62 to 0.65).*

BB-48. The following data points show both actual scores and adjusted scores for a recent algebra quiz. Graph these data points, and write the equation used by the instructor to adjust the scores. $y = 3x + 10$

Actual Score	27	15	20	18
Adjusted Score	91	55	70	?

a. What is the adjusted score for 18? *64*

b. What *score* would have been adjusted to 82? *24*

c. What does the vertical change of this line represent? What does the horizontal change represent? What does the slope of this line represent? *Vertical: change in adjusted score; horizontal: change in actual score. Slope tells us that a 3:1 ratio was used to adjust the points.*

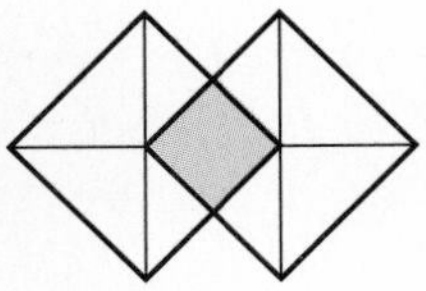

BB-49. Two congruent overlapping squares are shown. If a point inside the figure is chosen at random, what is the probability that it will be in the shaded region where the figures overlap? *1/7*

Solution:

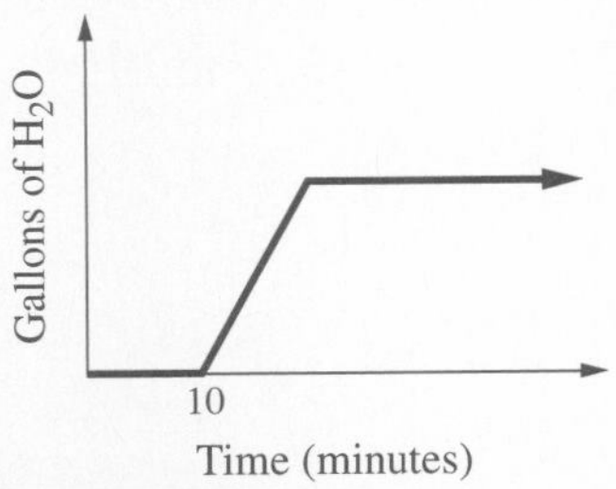

BB-50. Burt Balding takes a shower in a 40-gallon tub. The drain in his tub can drain up to 10 gallons of water per minute when the drain is clear. Unfortunately, Burt's rapid hair loss slows the drainage considerably! For each minute that Burt showers, he loses a set amount of hair and the drain becomes more and more clogged. Suppose the water is flowing into the tub at a rate of 5 gallons per minute, the water in the tub starts to back up after 10 minutes, and Burt likes to take *long* showers. Make a reasonable graph of this situation showing how the gallons of water in the tub and the showering time (in minutes) are related. Label the axes, important points, and any other critical items.

BB-51. As a candle burns, the mass of the candle steadily decreases. Suppose two candles are lit at the same time. The first candle is 8 inches tall and burns at a rate of 1 inch each hour. The second candle is 15 inches tall and burns at a rate of 2.25 inches per hour.

a. On the same set of axes, graph the height of the candles over time.

b. Find and interpret the y-intercepts, the x-intercepts, and the slope of each graph. ***y-intercept: first, (0, 8); second, (0, 15); both represent the original height of the candles. x-intercept: first, (8, 0); second, (6.6, 0); time each candle burns completely. slopes: −1, −2.25, the rates the candles burn.***

c. Do the graphs intersect? If so, where? Interpret the point of intersection. If not, what does that tell you? ***Intersect at (5.6, 2.4); the candles are at the same height, 2.4 inches tall, after 5 hr 36 min have passed.***

3.4 THE BOUNCING BALL LAB

The Bouncing Ball Lab, BB-52 through BB-57, is the theme problem for this chapter. Students will complete most of the lab now but will be left with two questions that they will return to later in the chapter. As a minimum, you want to complete all the data gathering in one class session.

One way to speed up data gathering is to set up the experiment and have one group demonstrate for the whole class. Note that predictions come before data gathering as part of the scientific process. Students need to be well organized and efficient, and they should ***start right at the beginning of class****, or this investigation may take more than one day.*

In case you find it necessary to use two class sessions for BB-52 through BB-57, there are extra problems BB-69 through BB-73 that you can assign. The intention of this investigation is to cause students to bring up questions about sequences. They will need to learn more about sequences to answer the last two questions about the bouncing ball in BB-57.

For this investigation, each group will need:

- *a Superball*
- *meter sticks or a 2.5-meter measuring device. (A length of computer paper marked in 10-cm increments* ***and*** *taped to the wall, or a 2.5-meter pole will work. You could ask the groups to assign the job of marking the computer paper to one of their group members as homework.)*
- *graph paper*

Each group should assign the jobs of ball dropper, data recorder, and at least two spotters. Remind spotters to sight the height of the ***bottom*** *of the ball.*

BB-52. Designate one person in your group as recorder, one as ball dropper, and at least two as spotters. Make a table like the one shown. Choose a starting height and record it in your table. *Predict* how high the ball will bounce (rebound) when it is dropped from the starting height. Now drop the ball from the chosen height and record the rebound height of the ball. *Make at least three more trials from this starting height*, and record each resulting rebound height in the second column. It will be easier to compare heights if you measure from the ground to the *bottom* of the ball in all cases.

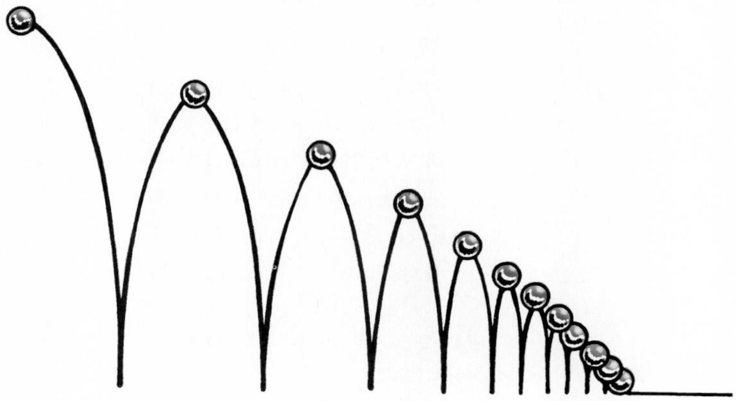

a. What do you notice about the rebound heights? Once you have recorded your observation, average the last three heights and record the average r in the third column. Why should you average them? (Note: You will not use the fourth column of the table until BB-54.) *Rebound heights are all about the same; averaging gives one sound data point for the rebound height.*

b. Select another four starting heights, and repeat the procedure for each one (so you will have at least 15 entries in your table).

Remember to measure the heights to the bottom of the ball.

Starting height, s	Three rebound heights	Average of three rebound heights, r	Rebound ratio, r/s

BB-53. Referring to the data collected in BB-52, identify the independent variable as x and the dependent one as y, and graph your results carefully on graph paper. What do you notice? Draw a line of best fit. Write the equation for your line.

BB-54. For each trial starting height, calculate the ratio of the average rebound height, r, to the starting height, s, or r/s, and record it in the fourth column on your table. What do you notice? ***The ratio r/s will be close to constant if they measure to the bottom of the rebounding ball.***

a. How is this result related to the slope of your line of best fit? *Slope is r/s.*

b. Explain why it makes sense for the y-intercept of your graph to be zero. *A ball dropped from 0 cm rebounds 0 cm.*

c. Physicists call a relationship such as this one—between the height of the rebound and the height from which the ball was dropped—**a direct variation**. Why do you think they use this description? Think of another direct variation relationship, describe it, and write an equation for it. ***Called a direct variation because the dependent variable varies directly with the independent one; double the independent variable and the corresponding dependent variable doubles as well.***

BB-55. Use the information you have gathered to predict how high the Superball will rebound the first time when it is dropped from 2 meters.

a. Now that you have made a prediction, drop the ball from a height of 2 meters. How does the predicted rebound height compare with the actual rebound height? Why might they be very close? What might explain any discrepancy?

b. Suppose that the Superball is dropped and you notice that its rebound height is 60 centimeters. From what height was the ball dropped?

c. Suppose the Superball were dropped from a window in the Empire State Building, 200 meters above the ground. What would you predict the rebound height to be after the first bounce? Is your prediction reasonable considering the nature of a Superball? Explain your thinking on this issue. ***Rebound will be less than the ratio indicates, for a variety of reasons, such as friction and wind resistance.***

d. How high will the ball rebound after the second bounce? after the third?

e. What factors might affect the accuracy of your predictions?

The goal of the next problem is to draw students intuitively to the idea that as the ball repeatedly bounces, the rebound ratio stays the same, causing the rebound height to decrease in a geometric pattern. This is a key idea leading to exponential decay, yet they probably won't see the algebra behind the curve—that will come in a later lesson. For now, we simply want them to see that our ball's rebound height is dropping by a relatively constant percentage each bounce.

BB-56. Now let's consider a slightly different problem: drop the Superball from 2 meters, and allow it to bounce six times.

a. Measure the rebound height for each bounce. You may want to discuss just how to gather this data with your group before beginning. If you get stuck, consider letting the ball bounce, dropping it again from the height of the bounce, and repeating this until you get enough bounces to give you an accurate result.

b. When you have all six rebound heights, graph the relation between the bounce number (x) and the rebound height for each bounce (y).

c. Compare your graph in part b to the graph in BB-53. How does it differ? ***This graph should be discrete and curved, decreasing exponentially.***

d. Use your answer to part c to explain as simply as possible why your graph took on the shape it did.

For part e, we want students to use their rebound ratio to figure out the results. When the graphs are complete, have some groups show their graphs on the overhead projector so you can discuss differences, such as: Should the point (0, 10) be included? Should the graph be points (discrete), or should a continuous curve be drawn? How would it be drawn if it were continuous? Why? You may have to point out that the graph is not a picture of the path of the bouncing ball.

e. Elaine now drops the *same* Superball from a height of 10 feet and lets it bounce. She then turns to you and says she can estimate the rebound heights and graph the height of each successive bounce without having to gather any data. Explain how she can do it, and then sketch her graph.

f. Should the points on her graph be connected? Explain your thinking. ***The real issue here is whether the bounce number can be other than integer values. If necessary, ask questions that help students to see the difference.***

g. When the points on a graph are connected, and it makes *sense* to connect them, we say the graph is **continuous**. If it is not continuous, and is just a sequence of separate points, we say the graph is **discrete**. Look back at the graphs you drew for this lab. Should the graphs be continuous? Explain your thinking on this. Add the definitions of discrete and continuous to your Tool Kit. ***Elaine's graph is discrete; the linear graph is continuous.***

BB-57. The saga of the Bouncing Ball will continue, so be sure your data are accurate, neatly organized, and handy because you will be referring to them later in this chapter to answer the following questions:

a. What equation could represent your graph in BB-56?

b. How long will it take for the ball to stop bouncing?

EXTENSION AND PRACTICE

BB-58. We often call $t(n)$ the ***nth term*** of the sequence. Actually, it is the nth term after the initial term. For example, $t(7)$ is term 7 (the seventh term after the initial term) of the sequence. For each sequence determine what number n could be if 447 is the nth term. Is it always possible for 447 to be a term?

a. $t(n) = 5n - 3$ *yes, 90th term*

b. $t(n) = 24 - 5n$ *no*

c. $t(n) = -6 + 3(n - 1)$ *yes, 152nd term*

d. $t(n) = 14 - 3n$ *no*

e. $t(n) = -8 - 7(n - 1)$ *no, $-64 = n$ is not in the domain.*

In BB-59 we ask students to choose one of the sequences in BB-58 for which they answered "no" and to give a sound argument as to why they answered no. This is a good place to push the use of ***algebra****! Guessing and checking, you can argue, is not a good justification because the student just might not have guessed the correct number yet. Using algebra is the most efficient way to prove that a number is or is not a term.*

BB-59. Choose one of the sequences in BB-58 for which 447 was *not* a term. Write an explanation clear enough for an elementary algebra student to understand how you were able to determine that 447 was not a term of the sequence. *When solving for n, you will find a value for n that is not a positive integer value.*

BB-60. Seven years ago Patrick found some old baseball cards in a box in the garage. During the first year he had the collection he added a certain number of cards and decided to add that same number each year. He had 52 cards in the collection after three years and now has 108 cards. How many cards were in the original box? *10*

a. Patrick plans to keep the collection for a long time. How many cards will the collection contain 10 years from now? *248*

b. Write an equation that determines the number of cards in the collection after n years. Explain what each number in your equation stands for. *$t(n) = 14n + 10$*

BB-61. Dr. Sanchez asked her class to simplify $x + 0.6x$.

Terry says, $x + 0.6x = 1.6x$. Jo says, $x + 0.6x = 0.7x$.

Whose response is correct? Justify your conclusion. *Terry*

BB-62. Recall sequence g from BB-3. The sequence was 40, 20, 10, ...

a. What kind of sequence is 40, 20, 10, . . . ? Why? *geometric, each term multiplied by 1/2*

b. Plot the sequence up to $n = 6$.

c. Since the sequence is decreasing, will the values ever become negative? Explain. *No, the sequence approaches zero. Half of a positive number is still positive.*

BB-63. Solve the system by graphing each line and estimating their point of intersection, then solve the system algebraically to check.

$$x + y = 5$$
$$y = \frac{1}{3}x + 1 \quad (3, 2)$$

BB-64. At Food-R-Us they have a special on bread and soup. Scott bought four cans of soup and three loaves of bread for $4.36. Greg bought eight cans of soup and one loaf of bread for $3.32.

a. Write equations for both Scott's and Greg's purchases.

b. Solve the system to find the price for a can of soup and the price for one loaf of bread. *The soup cost $0.28 and the bread cost $1.08.*

BB-65. Solve each of the following equations.

a. $\frac{m}{6} = \frac{15}{18}$ $\quad m = 5$ **b.** $\frac{\pi}{7} = \frac{a}{4}$ $\quad a = \frac{4\pi}{7}$

BB-66. Find the x- and y-intercepts for the graph of $y = x^2 + 4x - 17$. *(0, −17), $(-2 \pm \sqrt{21}, 0)$ or (2.58, 0), (−6.58, 0)*

BB-67 is a preview for Chapter 6 on linear systems.

BB-67. Plot the points (3, −1), (3, 2) and (3, 4). Draw the line through these points. The equation of this line is $x = 3$. Why do you think the equation is defined this way?

a. Plot the points (5, −1), (1, −1), and (−3, −1). What will be the equation of the line through these points? Explain. $y = -1$

b. Choose any three points on the y-axis. What will be the equation of the line that goes through those points? $x = 0$

BB-68 previews the concept of multipliers; be sure to assign it.

BB-68. How did you answer BB-61? Did you agree with Terry? If so, do these problems. If not, go back and reconsider. Remember: $x + 0.6x = 1x + 0.6x$; which is the same as $1.0x + 0.6x$.

a. $y + 0.03y$ *1.03y* **b.** $z - 0.2z$ *0.8z* **c.** $x + 0.002x$ *1.002x*

EXTRA PROBLEMS

These are included in case you use an extra day.

BB-69. The sequence 5, 10, 15, . . . has 5 as the initial term, then 10 as the first term after that, then 15, and so on. Suppose this sequence is part of another arithmetic sequence where 5 is the initial term, 10 is the *second* term after that, and 15 is the *fourth* term after the initial term . List the five terms at the beginning of the new sequence. *5, 7.5, 10, 12.5, 15, . . .*

BB-70. Examine the graph below which shows the cost of renting in-line skates from two different shops: Bif's Big Wheels and In-Line with Inez.

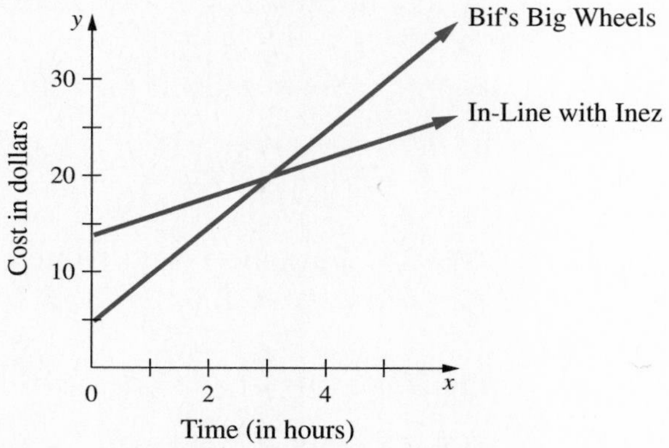

a. What are the y-intercepts for each of the lines? Interpret these points. *(0, 5) and (0, ≈14); these would be the base cost.*

b. Estimate the point of intersection for these lines. Interpret this point. *≈(3, 20), the cost is the same for this amount of time.*

c. What is the best deal? Explain. *depends on how long the renter plans to use the skates*

BB-71. Write the range and domain for each of the following graphs. Each tick mark represents one unit. *all domains: real numbers*

a.

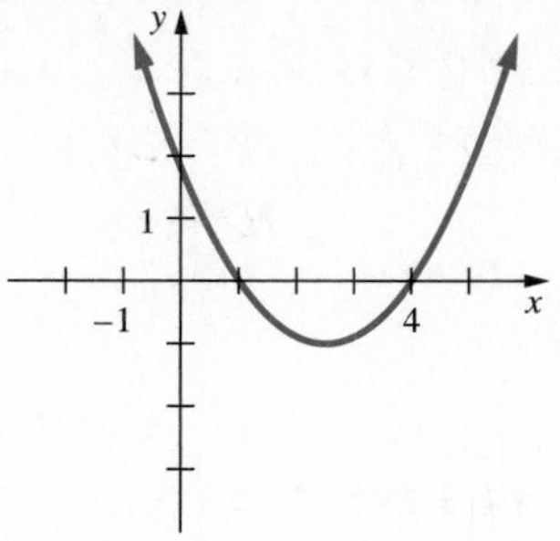

$y \geq -1$

b.

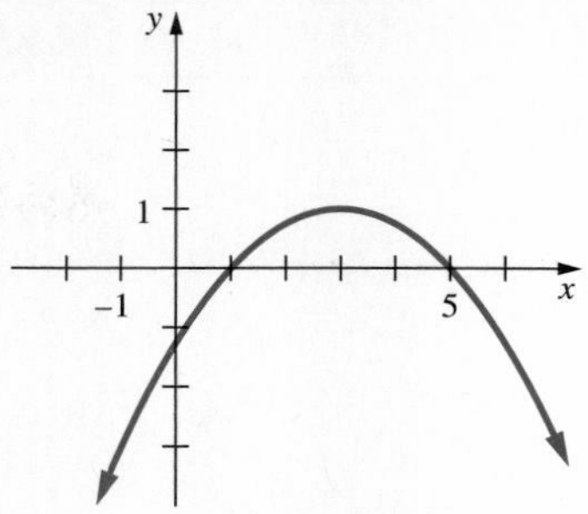

$y \leq 1$

c.

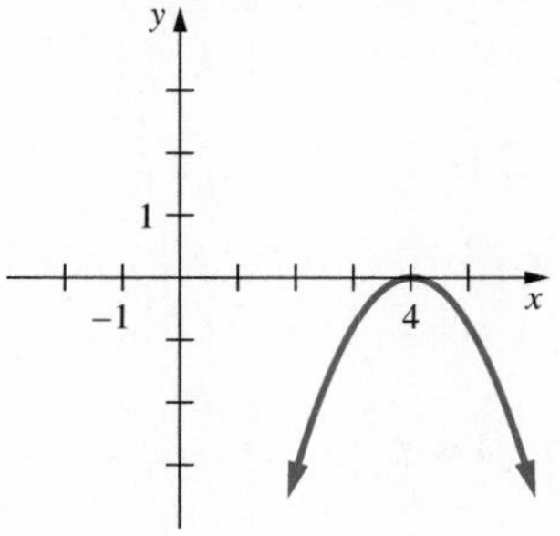

$y \leq 0$

BB-72. For each step, parts a through d, explain what mathematical operation or relationship to use to get the next result.

a. $4(7 - 2x) = -5(2x - 3)$
Distribute.

b. $28 - 8x = -10x + 15$
Add 10x.

c. $28 + 2x = 15$
Subtract 28.

d. $2x = -13$
Divide by 2.

$$x = \frac{-13}{2}$$

BB-73. Examine each of the following sequences. For those that are arithmetic sequences, find the expression for the nth term, $t(n)$. For those that are geometric, find the multiplier used to get from one term to another.

a. 1, 4, 7, 10, 13, . . .
arithmetic; $t(n) = 3n + 1$

b. 0, 5, 12, 21, 32, . . .
neither

c. 2, 4, 8, 16, 32, . . .
geometric; $r = 2$

d. 5, 12, 19, 26, . . .
arithmetic; $t(n) = 7n + 5$

e. $x, x + 1, x + 2, x + 3, \ldots$
arithmetic; $t(n) = x + n$

f. 3, 12, 48, 192, . . .
geometric; $r = 4$

3.5 GEOMETRIC SEQUENCES AND THE NTH TERM

BB-74. Consider the sequence 3, 6, 12, 24, ...

a. What kind of sequence is it? How is it generated? ***geometric; multiply by 2.***

b. Pairing the values in the sequence with their position in the sequence, we get the following table:

n	0	1	2	3	4
$t(n)$	3	6	12	24	

Make your own table, and extend it to include the next five terms of the sequence.

c. How many times do we multiply the *initial value* of 3 by 2 in order to get the result 48 in the sequence? ***4***

d. Is there a shortcut for representing repeated multiplication? What is it? ***yes, using exponents***

e. When $n = 6$, $t(6)$ can be obtained by multiplying $3 \cdot 2 \cdot 2 \cdot 2 \cdot 2 \cdot 2 \cdot 2$. Rewrite this expression using exponents. $3(2)^6$

f. What will be the representation for $t(100)$, the 100th term after the initial value? $3(2)^{100}$

g. How could you represent any value $t(n)$ in the sequence? $3(2)^n$

BB-75. Consider this table:

n	0	1	2	3
$t(n)$	1	3	9	27

a. What is the rule to move from one $t(n)$ value to the next? ***Multiply by 3.***

b. If the table showed only the two $t(n)$ values $t(5) = 243$ and $t(6) = 729$, how could you find the multiplier? ***Divide.***

c. The multiplier is sometimes called the **common ratio**. Based on your answer to part b, why do you think this is so? Add this term to your Tool Kit.

BB-76. Write a function $t(n)$ with an initial value a and multiplier r. Why do you think we chose r ? Refer to BB-75. $t(n) = ar^n$

BB-77. A tank contains 8000 liters of water. Each day, one-half of the water in the tank is removed. Make a table showing the first 4 days of water loss. Then determine how much water will be in the tank after:

a. the sixth day? *$8000\left(\frac{1}{2}\right)^6 = 125$ liters*

b. the 12th day? *$8000\left(\frac{1}{2}\right)^{12} \approx 1.95$ liters*

c. the *n*th day? *$8000\left(\frac{1}{2}\right)^n$ liters*

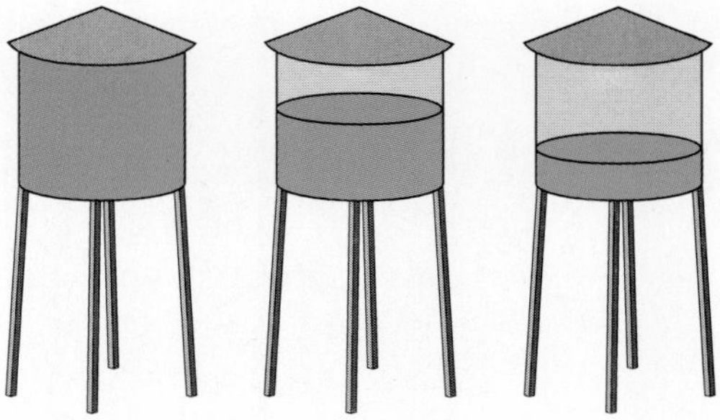

Students may not generate the general expression in part c of BB-78, but future multiplier problems will enable them to do this. At this point they will probably rely on "number crunching" to accomplish the solution. Invite them to set up their data numerically and look for patterns if they get stuck. This problem will help to generate the need for using multipliers. Plan on coming back to this problem later to reconsider part c. Part d can lead to a discussion of the value of mathematical modeling in fields such as public health and others.

BB-78. Flu Epidemic Hits Local Area! The evening news reports that the annual flu season has begun with 100 cases and a 30 percent increase in cases projected by the end of the week.

a. If this rate continues, how many cases will be reported four weeks from now? *286*

b. Make a data table for weeks 0 through 4.

c. How many cases will be reported after 7 weeks? *$100(1.30)^7 \approx 627$ cases*

d. Write a general expression that enables you to determine the number of cases reported after *n* weeks. *$100(1.30)^n$*

e. For how long do you think this pattern is likely to continue? *This should be discussed.*

In the next problem we return to the Bouncing Ball Lab. It's better if students use their own data from their investigation, but in case they don't have it, we have provided some. Now that the nth terms of both arithmetic and geometric sequences have been examined, students are asked to generalize their observations of the ball. BB-79 requires that they ponder whether the ball will actually stop bouncing and that they consider the sum of the series. Give students time to carefully think through and discuss this problem, perhaps by playing devil's advocate. Ask them about the typical scenario: "If I walk halfway to the wall, then stop, then walk halfway to the wall and stop, and so forth, when will I get to the wall?" Also, you may want to program a computer or have students program their graphing calculator to analyze the total distance traveled by the ball.

BB-79. The Return of the Bouncing Ball! Cheryl collected the data shown below when she conducted the Bouncing Ball Lab. If you have your own data handy, use yours; otherwise, use her data.

	Height, s (meters)	Rebound heights, r (meters)	Rebound ratio, r/s
Start	2	1.61	0.805
1st bounce	1.61	1.23	0.7643
2nd bounce	1.23	1.00	0.813
3nd bounce	1.00	0.81	0.81
4th bounce	0.81	0.67	0.827
5th bounce	0.67	0.52	0.776
6th bounce	0.52		

a. Determine an equation that could represent either your data or Cheryl's data.

b. Draw a graph based on the equation you wrote in part a.

c. How high will the ball be after the seventh bounce? ***If rebound ratio is about 0.8 with initial height of 2 m, result is approximately 0.4 m.***

Part d of this problem could generate some good discussion among the students. They know from experience that the ball eventually will stop bouncing. They may decide on the point where the rebound is so small that the ball is no longer bouncing.

d. How long will it take for the ball to stop bouncing?

The following problem is a preview to parent graphs in Chapter 5. If students do not have their own graphing calculators, have them make a quick sketch of the graph in class, and then draw complete graphs at home.

BB-80. Sketch $y = x^2$, $y = -1x^2$, $y = -3x^2$, and $y = -0.25x^2$ on the same set of axes. What is the effect of a negative coefficient on the last three graphs compared to the first? ***It flips the parabola downward.***

EXTENSION AND PRACTICE

BB-81. Look back at the data given in BB-47 regarding the rebound ratio for an approved tennis ball. Select a rebound ratio from within the stated range for tennis balls. Suppose you dropped your tennis ball from a height of 10 feet:

a. How high would it bounce on the first bounce? *5.3 to 5.8 ft*

b. How high would it bounce on the 10th bounce? *0.0175 to 0.0431 ft*

c. How high would it bounce on the nth bounce? *$10(0.53)^n$ to $10(0.58)^n$ ft*

BB-82. For parts c and d, show how you arrived at your answers. Give answers to the nearest one-hundredth (two decimal places). You may need to guess and then check with your calculator.

a. $x^2 = 49$ *$x = 7, -7$*

b. $x^3 = 27$ *$x = 3$*

c. $x^3 = 30$ *$x = 3.11$*

d. $x^6 = 93$ *$x = 2.13, -2.13$*

BB-83. For parts a through d, find the $t(n)$ values of the sequence for $n = 0$ to $n = 4$.

a. $t(n) = 8 + 7(n - 1)$
1, 8, 15, 22, 29

b. $t(n) = -5$
−5, −5, −5, −5, −5

c. $t(n) = 2^{n-4}$
1/16, 1/8, 1/4, 1/2, 1

d. $t(n) = (-2)^n$
1, −2, 4, −8, 16

e. Describe the domain for each sequence. *whole numbers*

f. What is the range for each sequence? *the result or sequence values*

g. Allene claims she had to graph each sequence to find the domain and range. Jon disagrees, saying he never had to use a graph. How do you suppose Jon knows the domain and range without graphing? *Domain is always whole numbers, range is always the sequence itself.*

BB-84. A sequence is defined as $t(n) = 4n - 3$.

a. List the first four terms. *−3, 1, 5, 9*

b. What type of sequence is this? Explain. Could you have answered this question without finding the first four terms? Explain. *arithmetic; yes, 4n means each time n increases by one the result increases by 4 so there is a common difference.*

c. 321 is in this sequence. What term number is it? Explain how you found this answer. *term 81; let $4n - 3 = 321$, and solve for n.*

The following problems introduce the idea of obtaining a multiplier from a percentage increase. Exponential functions often represent a constant percent rate of increase (or decrease).

BB-85. ClothTime Problem What a deal! ClothTime is having a sale: 20 percent off. Beth decides to buy 14 shirts. When the clerk rings up her purchases, Beth sees that the clerk has added the 5 percent sales tax first, before taking the discount. She wonders whether she got a good deal at the store's expense or should complain to Ralph Nader about being ripped off. Your mission is to find out which is better: receive the discount first and then be taxed, or tax first, and then discount. Before you begin calculating, make a guess. *Using a convenient number such as $100 for the initial cost is a popular strategy. The surprising answer is that both situations are the same, with a final amount of $84.*

BB-86. In BB-85, suppose that the total Beth paid for the shirts is x dollars.

a. What is the tax amount (in terms of x)? *0.05x*

b. What is the total cost in terms of x? *1.05x*

c. The coefficient in part b is what we call a **multiplier**. It is greater than 1 because the tax made the cost increase. What would be the multiplier for a tax rate of 7 percent? *1.07*

d. Convert each of these increases to a multiplier:

3 percent	8.2 percent	2.08 percent
1.03	*1.082*	*1.0208*

e. Record the term **multiplier** in your Tool Kit along with an understandable definition or explanation.

Students will probably struggle on part c of BB-87. Refer students to the suggestion in their answer section that they consider approaching it as an area problem. This can lead to a discussion of a quadratic expression as an expression of area.

BB-87. Rectangular Numbers Listing the number of dots in each "rectangle" in Figures 0 to 4 gives the sequence: 0, 2, 6, 12, 20 . . . . The numbers in this sequence are called rectangular numbers.

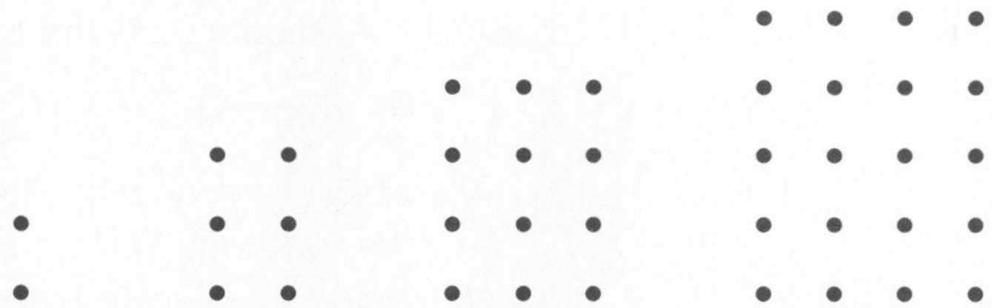

Fig. 0 Fig. 1 Fig. 2 Fig. 3 Fig. 4

a. What is the next rectangular number in the sequence? *30*

b. What is the 10th rectangular number? *110*

c. What is the nth rectangular number? *n(n + 1)*

d. What is the 100th rectangular number? *10,100*

BB-88 is probably too difficult for students to do individually, but it could be used for a group assessment.

BB-88. In 1992, Charlie received the family heirloom marble collection, consisting of 1239 marbles. The original marble collection was started by Charlie's great grandfather back in 1898. Each year Charlie's great grandfather had added the same number of marbles to his collection. When he passed them on to his son, he insisted that each future generation add the same number of marbles per year to the collection. When Charlie's father received the collection in 1959, there were 810 marbles.

a. How many marbles are added to the collection each year? *13*

b. How many years has the collection been maintained? *94 years*

c. Use the information you found in part b to figure out how many marbles were in the original collection when Charlie's great grandfather began it. *17*

d. Write a generalized expression describing the growth of the marble collection since it was started by Charlie's great grandfather. *$t(n) = 17 + 13n$*

e. When will Charlie have more than 2000 marbles? *in 59 years*

BB-89. In a geometric sequence $t(0)$ is 3 and $t(2)$ is 12. Find $t(1)$ and $t(3)$. Write a sentence to explain how you could start with the first term, 3, and generate the other terms of this sequence. *$t(1) = 6$; $t(3) = 24$; $12/3 = 4$, which is split into two intervals giving the relationship $r^2 = 4$, making the common ratio 2; $t(n) = 3(2^n)$.*

BB-90. The figures shown here are pieces of graphs that were left in place when a computer glitch erased the grids they were on. Match each figure with one of the following tables, and explain why you made the choices you did.

a.

g(x)

b.

f(x)

c.

h(x)

x	$g(x)$
8	56
9	50
10	45
11	41
12	38
13	35

x	$f(x)$
10	22
11	26
12	30
13	34
14	38
15	42

x	$h(x)$
9	26
10	34
11	40
12	44
13	45
14	45

3.6 MULTIPLIERS, APPLICATIONS, AND CHAPTER SUMMARY

BB-91 will help to clarify the idea of a multiplier. It is important for the students to think about the percentage that remains and not the amount that is removed or added.

BB-91. Multipliers and Applications Karen works for Macy's and receives a 20 percent employee discount on any purchases that she makes. Today Macy's is having their end-of-the-year clearance sale in which any clearance item will be marked 30 percent off. When Karen includes her employee discount with the sale discount, what is the total discount she will receive? Does it matter which discount she takes first?

Using graph paper, outline two 10-by-10 grids as shown here.

Case1: 20% discount first
30% discount next

Case2: 30% discount first
20% discount next

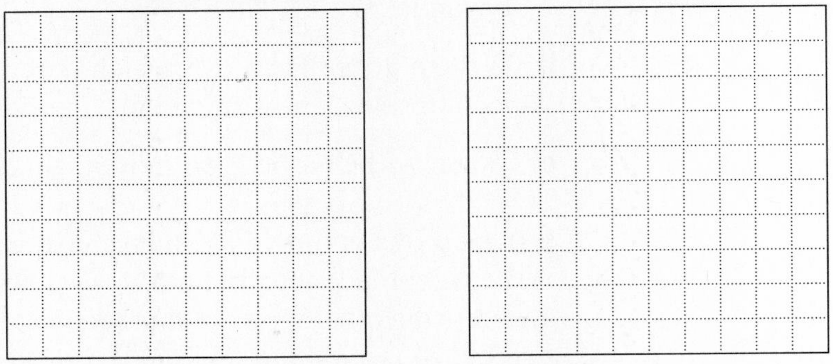

a. Each grid has 100 squares. Let's start by considering Case 1: Take the first discount (20 percent) by horizontally shading the appropriate number of squares in the top rows of Case 1's grid.

i. How many squares remain after the first discount? What percent is this? *80, 80%*

ii. Now take the second discount of the remaining squares in Case 1's grid. (*Note*: you no longer have 100 squares to start.) Shade in this second discount along the *vertical* rows on the left. How many unshaded squares now remain in Case 1? What *percent* of the original 100 is this? *56, 56%*

b. Next let's consider Case 2: take its first discount (30 percent) by shading in 30 percent of Case 2's squares, *horizontally* along the top rows.

i. How many squares remain after the first discount in Case 2? What *percent* is this? *70, 70%*

ii. Using the remaining squares in the grid, take Case 2's second discount. Remember you no longer have 100 squares to start. Shade the second discount along the vertical rows on the left. How many squares remain in Case 2? What *percent* of the original 100 is this? *56, 56%*

c. Now suppose you had simply multiplied the percentage left after the employee discount by the percentage left after the sale discount. What do you get? *56%*

d. What if you had done it the other way around (that is, you multiplied the percentage left after the sale discount by the percentage left after the employee discount)? *56%—the same as part c*

e. Based on the work you just completed, what is the total discount Karen will receive? Does it matter which discount she takes first? Explain why this is true. *44%*

BB-92. Starting with an amount x in each case, find a simplified form for each increase or decrease. For example: A 30% increase is represented by the equation: $x + 0.3x = 1.3x$.

a. a 5 percent increase *1.05x*

b. a 12 percent discount *0.88x*

c. an 8.25 percent tax *1.0825x*

d. a reduction of 22.5 percent *0.775x*

BB-93 is the chapter summary. Students will benefit from having some class time to start working on this in their groups. Creating examples to fit a verbal description is not something students have had much experience with, and it is often more difficult for them than you might anticipate.

BB-93. Summary Assignment You have looked at a variety of sequences throughout this chapter. You have graphed them, and analyzed their "staircases" and other characteristics such as slope and y-intercept. You have also come across several situations in which these various sequences arise. Imagine that there is an Algebraic Review Board. Your responses to the items below should be complete and clear enough to convince them that you understand these ideas.

Write an example for each:

a. an increasing arithmetic sequence

b. a decreasing arithmetic sequence

c. an increasing geometric sequence

d. a decreasing geometric sequence

For each example for parts a through d, include the following:

1. the sequence

2. a graph of the sequence

3. an equation to represent the sequence

4. a description of what the various parts (initial term, common difference or multiplier, and so on) of the sequence represent

5. a description of how the parts of the graph (intercepts, slope, and so forth) relate to its sequence

BB-94. Growth over Time—Problem No. 2 The following problem is another growth-over-time problem. Draw the graph of the function now, and finish the rest outside of class. Do the best you can with it and keep it in your portfolio to refer to later. This one will appear again at the end of Chapter 6 and Chapter 8.

Explain *everything* that you now know about

$$f(x) = 2^x - 3$$

Do this on a separate piece of paper so you can hand it in or keep it separately.

EXTENSION AND PRACTICE

BB-95. Remember the flu epidemic in BB-78? Imagine that the epidemic has now become a statewide crisis. However, you have also developed a deeper understanding of how a multiplier can help you solve a problem such as this (see your Tool Kit). If the epidemic in Los Angeles started with 1250 reported cases and the number increased 27 percent per week, how many cases would be reported in L.A. after 4 weeks of the epidemic? *3252*

a. Write the general expression from which you can determine the number of cases in any week of the L.A. flu season. *$1250(1.27^n)$*

b. If Health Services provided 8000 doses of a special medication to fight the virus, when will the medication be used up? (Assume one dose per person affected.) *Guess and check to get 8 weeks.*

BB-96. Just after John finished his math assignment, his nephew colored all over the paper. All that remained of a sequence was:

5, 12, 19,   390, 397

He knew that 5 was the initial term and 397 the last, but he did not remember anything else about the sequence.

a. What is the 40th term? *285*

b. How many terms after the initial value is the last value in John's sequence? *56*

BB-97. A multiplier can be used for a percent decrease as well as for an increase. Suppose there is a 20 percent discount. Again let x be the original cost.

a. What is the amount of the discount (in terms of x)? *0.20x*

b. What is the new price in terms of x? *0.80x*

c. The coefficient in part b is a multiplier. It is less than 1 because the discount made the cost decrease. What would be the multiplier for a sale rate of 15 percent off? *0.85*

BB-98. Find the multiplier that is determined by each percent decrease:

a. 3 percent decrease *0.97*

b. 25 percent decrease *0.75*

c. 7.5 percent decrease *0.925*

BB-99. Use the idea of a multiplier to look back at the ClothTime problem (BB-85); explain why both answers were the same. Go to your Tool Kit and expand on your definition of *multiplier* based on your work on the last two problems.

BB-100. If an 8-ounce Snookers Bar has 27 grams of fat, how many grams of fat are in a 12-ounce bar? *40.5*

BB-101. Consider the following tables:

n	$t(n)$	x	$f(x)$
0	100	0	50
1	50	1	
2		2	100
3		3	

a. Copy and complete each table to make the sequences arithmetic. *0, –50; 75, 125*

b. Copy and complete each table to make the sequences geometric. (*Hint*: recall that multiplier is also a common ratio.) *25, 12.5;* $50\sqrt{2}$, $100\sqrt{2}$

c. Write the rule for each sequence in parts a and b. $t(n) = 100 - 50n$, $f(x) = 50 + 25x$; $t(n) = 100(0.5^n)$, $f(x) = 50(\sqrt{2})^x$

BB-102. Triangular Numbers Listing the number of dots in each "triangle" in Figures 0 to 4 gives the sequence: 0, 1, 3, 6, 10 . . . . The numbers in this sequence are called **triangular numbers**. What previous problem would be related to this sequence? Find it. It will be very useful in figuring out the answers for this problem.

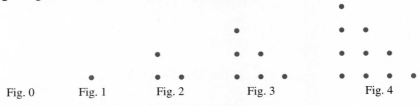

Fig. 0 Fig. 1 Fig. 2 Fig. 3 Fig. 4

a. What is the 15th triangular number? *120*

b. What is the *n*th triangular number? $n(n + 1)/2$

c. What is the 200th triangular number? *20,100*

BB-103. *Self-Evaluation* You did a self-evaluation of your skills at the end of Chapter 2:

a. Here are a few more skills and areas for you to check. Are you confident that you can:

i. solve a system of equations, and

ii. solve any quadratic equation that has real solutions?

b. Are there any problems you still can't do? If you are not confident in these skills, ask for help from your group or your instructor. You may need to get some review help and do some extra practice outside of class. Find a sample problem for each type that you cannot do. Then, by working with your group members, instructor, or someone else who can help, write out a complete solution to the problems. Enter this sample into your Tool Kit.

c. Reflect on your experiences in the course up to now to answer the following:

i. How has working in groups affected your understanding of the material? What has your role in your most recent group been? Do you prefer groups over working individually? Explain.

ii. One of the goals of this course is understanding. How have problem-solving strategies of guess and check and looking for patterns affected your understanding of the material?

BB-104. *Tool Kit Check* Now would be a good time to review your Tool Kit and make sure you have included any new ideas you have learned in this chapter. Include examples and explanations that will help you understand and remember what each of the following means:

- arithmetic sequence
- geometric sequence
- position number
- term of a sequence
- domain of a sequence
- range of a sequence
- multiplier
- common difference
- common ratio
- function notation, $t(n)$
- addition method of solving
- substitution method of solving
- slope
- y-intercept
- x-intercept
- quadratic expression
- quadratic equation
- system of equations

EXTRA PROBLEMS

These problems extend or expand on concepts previously covered and so are generally more difficult. Most of these problems can be assigned individually; however, we have noted where groups are advisable.

*BB-105 and BB-106 do not easily allow for t(0). We do **not** expect students to come up with the equations. They can solve by graphing or comparing tables of values. You might want to challenge your more advanced students to figure out the two equations. This could serve well as a take-home group problem.*

BB-105. You decide to hire two high school students to completely landscape your yard, rather than do it yourself. They will work an hour each afternoon until the job is completed. (You assume this will take no more than three weeks.) You give them a choice about which payment plan they prefer. Plan A pays $11.50 per afternoon, while Plan B pays 2 cents for one day's work, 4 cents for two days' work, 8 cents for three days' work, 16 cents for four days' work, and so on. Each student chooses a different plan. On which *day* will they be paid approximately the same? Support your thinking with data charts and graphs. If you need some guidance, complete the subproblems below: *In 14 days they will earn close to the same amount. Plan A: $c(n) = 11.50n$; Plan B: $c(n) = 0.01(2)^n$*

a. Make a table representing data from each case. Your table should include values for 5, 10, 15, and 20 days.

b. Plot and graph the data and write an equation. According to the graph, when will the plans be approximately the same?

c. Which plan grows faster between zero days and the day the plans pay the same? Which grows faster after that point? *Plan A grows faster from 0 to 11 days, while Plan B grows faster after 11 days. Exponential functions with $x > 0$ grow faster on a global scale than linear functions.*

BB-106. A rookie basketball player was recently drafted by the Phoenix Suns. His salary will be $673,500 for the first year, with an increase of 20 percent each year of his five-year contract.

a. How much will he make in the fifth year of his contract? *$1,396,569.60*

b. A teammate is currently in the fourth year of a five-year contract. His current salary is $875,900, and this represents an annual raise of 15 percent for the life of his contract. What was his starting salary? *$575,918.47*

c. Write the expression that describes the terms of the contract for each player. When you think you have the equations, be sure to check your answers to see if they work. $s(n) = 673{,}500(1.20)^{n-1}$; $s(n) = 575{,}918(1.15)^{n-1}$

d. What was the total amount of a five-year contract for each of these players? *$5,011,917; $3,883,058*

BB-107 presents an opportunity to investigate the functions $t(n) = \sqrt{(n+1)}$ for the hypotenuse length and $t(n) = \sqrt{n}/2$ for the area of the last triangle. We suggest giving this as a group problem, where the groups break up into two pairs: one pair investigates the hypotenuse and the other pair investigates the area. They can then share their findings with each other.

BB-107. The following spiral is made by using a sequence of right triangles, each with a leg that measures one unit, and the second leg being the hypotenuse of the triangle before it. Although only four right triangles are shown, we could continue this spiral forever. As you work on this problem, *think back to previous function investigations that you have conducted. Be thorough and provide relevant data, graphs, and conclusions.* Here you will investigate two functions. Each function is described as follows:

a. Consider a function for which the input is the number of right triangles in the figure and the output is the length of the *hypotenuse* of the last triangle.

b. Now, consider the function for which the input is the number of right triangles in the figure, and the output is the *area* of the last triangle formed.

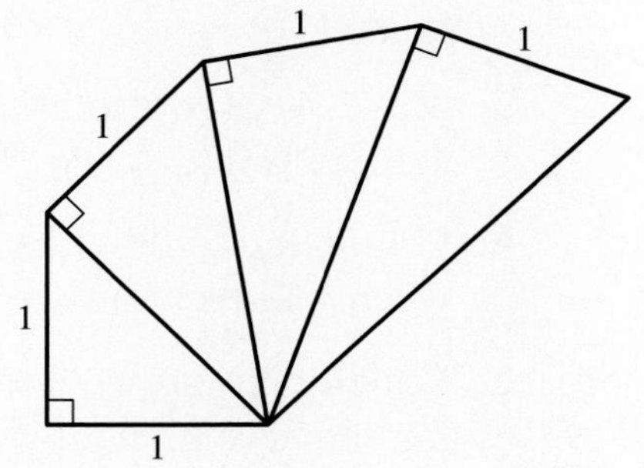

CHAPTER FOUR

FAST CARS AND DEPRECIATION

Exponential Functions

IN THIS CHAPTER YOU WILL HAVE THE OPPORTUNITY TO:

- see a relationship between geometric sequences and exponential functions that is similar to the relationship between arithmetic sequences and linear functions;
- use exponential functions to represent situations modeling growth and decay;
- develop your ability to write equations and interpret the meaning of fractional and negative exponents;
- continue to become familiar with the graphing calculator;
- continue to practice the use of basic algebraic skills in solving some not-so-basic problems.

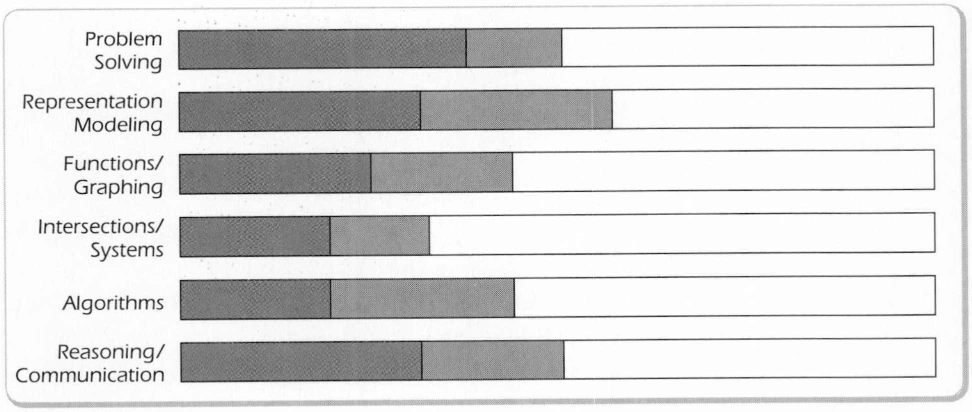

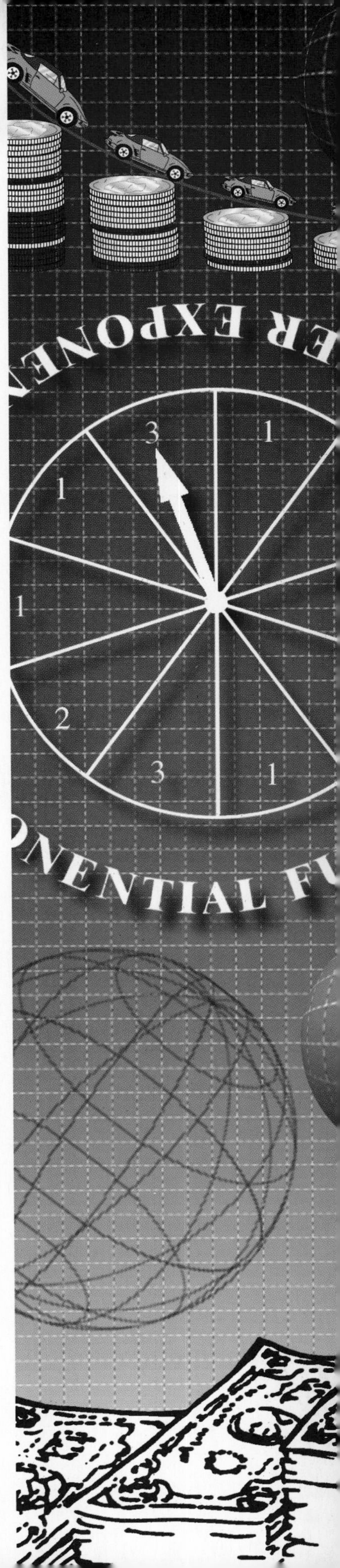

MATERIALS	CHAPTER CONTENTS		PROBLEM SETS
Graph paper	4.1	Multipliers and Compound Interest	FX-1 – FX-20
Per group: 100 pennies for the Penny Lab, graph paper, graphing calculator, one overhead transparency, overhead transparency pens	4.2	Half-Life and Other Applications	FX-21 – FX-35
4-color index cards, outdoor axes, graphing calculators	4.3	Negative Exponents: The Function Walk	FX-36 – FX-52
Graphing calculators	4.4	Roots and Fractional Exponents	FX-53 – FX-72
Graphing calculators	4.5	Applications of Exponential Functions	FX-73 – FX-89
Graphing calculators	4.6	Fast Cars—The Theme Problem	FX-90 – FX-106
Graphing calculators	4.7	Exponential Investigation and Summary	FX-107 – FX-122
		Extra Problems	FX-123 – FX-127

In this chapter, we examine exponential functions as an extension of geometric sequences. Exponential functions can be used to represent situations that involve growth or decay, appreciation or depreciation. They are often very useful as mathematical models in science, business, banking, and economics.

4.1 MULTIPLIERS AND COMPOUND INTEREST

FX-1 is the theme problem for this chapter. While it might be possible for you to solve it today, it will be easier to solve it later in the chapter when it reappears. For now read it and go on to FX-2.

FX-1. Fast Cars As soon as you drive a new car off the dealer's lot, the car is worth less than what you paid for it. This situation is called **depreciation**. Chances are you will sell it for less than the price you paid for it. Most cars depreciate, some more than others (that is, at different rates). On the other hand, some classic older cars actually increase in value. This situation is called **appreciation**. Let's suppose you have a choice between buying a 1993 Mazda Miata for $17,000 that depreciates at 22 percent a year, and a used 1990 Honda CRX for $12,000 that depreciates at 18 percent a year. In how many years will their values be the same? Should you instead buy a 1967 Ford Mustang for $4,000 that appreciates at 10 percent per year? Which car will have the greatest value in 4 years? in 5 years?

Although FX-2 is presented as an application, it is one of the key problems leading to the concept of an exponential function.

FX-2. Bud's uncle, Nathan, offers to pay him an 8 percent annual interest rate to encourage him to save some money for his education. Unlike a bank, Uncle Nathan will pay **simple interest**. (Banks generally pay compound interest, which you will encounter in FX-3, but we want to start by looking at simple interest.) Suppose Bud deposits $100 with his uncle, and

Nathan pays interest on the $100 at the end of each year. The initial deposit is made in year zero.

Simple Interest

Year	Amount of Money (in dollars)
0	100.00 (initial value)
1	108.00
2	116.00
3	124.00
4	132.00

a. By what percent had Bud's account increased at the end of the fourth year? *32%*

b. Continue this table for the next 8 years.

c. Draw a graph to show the relationship between years (independent variable) and amount of money in the account. Label each axis, and state whether the graph represents an arithmetic sequence, a geometric sequence, or neither. *linear or arithmetic*

FX-3. As you may know, banks and credit unions do not pay interest in this simple way. They normally pay **compound interest**. That means that interest is paid not only on the amount invested (called the **principal** or **initial value**) but *also* on the accumulated interest. In other words, you get interest on your interest. To see this clearly, look at the following table for calculating compound interest. We are using the same interest rate, 8 percent per year, and the principal is still $100.00.

Compound Interest

Year	Amount of Money (in dollars)
0	100.00 (initial value)
1	108.00
2	116.64
3	125.97
4	*136.05*

a. Explain to other members of your group why the amount in the account at the end of the second year would be $116.64.

b. Copy the table and enter the appropriate value for the fourth year. Be sure to round off appropriately—you can't have fractions of a cent.

c. The simple interest that Bud's uncle paid increased his account by 32 percent over the four-year period. Based on the compound interest table, by what percent would the account increase over a four-year period? *36.05%*

FX-4.

a. Extend the table of compound interest given in FX-3 for another eight years.

b. Draw the graph for the table of compound interest. Label each axis *and* state whether the graph represents an arithmetic (linear) sequence, a geometric sequence, or neither. *geometric*

c. Are the graphs for FX-2 and FX-4 **discrete** (points only) or **continuous** (connected)? Explain. *Actually, they are step functions whose trends are linear and exponential, but "discrete" will do as a rough description at this point.*

FX-5. Look at the compound interest table you just made for part a of FX-4.

a. Explain how to get the amount in *any* line from the amount in the line above it. *Multiply by 1.08.*

b. Represent the amount of money in your account after two, three, and four years using powers of 1.08. Include these in your table in a new column. $100(1.08^2)$, $100(1.08^3)$, $100(1.08^4)$

c. Suppose you invested $1000 in this credit union and left it there for 20 years. How much money would be in your account then? $1000(1.08^{20}) \approx 4660.96$

FX-6 relates directly to the previous ideas of compound interest. The students may need help to interpret the problem as "2% per quarter," but it's OK for now if they don't get that answer. They still will get practice at compounding, which is the major point.

FX-6. Suppose you have a young cousin, and yesterday was her sixth birthday. She got $100 from your grandfather, and you want to convince her to deposit it into a savings account at the credit union. In order to convince your cousin, you need to explain to her how much money she will have by her 18th birthday if the credit union is offering an 8 percent annual interest rate compounded *quarterly*.

a. If 8 percent is the annual interest rate, what is the quarterly interest rate? What is the multiplier each quarter? *2%, 1.02*

b. Write an expression that represents the amount of money in the account after 10 quarters. $100(1.02^{10})$

c. Write an expression that represents the amount of money in the account after x quarters. $100(1.02^x)$

d. Write an expression that represents the amount of money in the account on your cousin's 18th birthday *and* find the value of that expression. *$100(1.02)^{48} \approx \$258.71$*

FX-7. What if you could earn 12 percent per year (3 percent per quarter) and you started with \$356? Using the expression you wrote in FX-6 as a model, write an equation for this new function, where x represents the number of quarters and y represents the amount of money you end up with. *$y = 356(1.03^x)$*

FX-8. The concert has been sold out for weeks, and as the date for the concert draws closer, the price of the tickets increases. The cost of a pair of concert tickets was \$150 yesterday, and today it is \$162. Assuming that the cost continues to increase at this rate, answer the following questions:

a. What is the daily rate of increase? What is the multiplier? *8%, 1.08*

b. What will be the cost one week from now (the day before the concert)? *\$277.64*

In this next part, students may want to use the multiplier 0.92, which is, of course, not correct. They could use the equation $a(1.08)^{14} = 162$, or they could use the expression $162(1.08)^{-14}$. If you think that this issue will sidetrack your class at this point, you may not want to assign part c at this time. We will deal with it later.

c. What was the cost two weeks ago? *\$55.15. Students who get an answer of \$50.41 used a multiplier of 0.92.*

EXTENSION AND PRACTICE

FX-9. In each of the following equations decide what the number x represents.

a. $2^3 = 2^x$ *$x = 3$*

b. $x^3 = 5^3$ *$x = 5$*

c. $3^4 = 3^{2x}$ *$x = 2$*

d. $2^7 = 2^{2x+1}$ *$x = 3$*

We provide a variety of explanations for the result of using zero as an exponent because different examples click with different students. With enough variety, every student should find a way to understand this critical concept. Students should work on these in groups and discuss them as they work.

FX-10. Consider the following pattern:

$$2^4 = 16, \qquad 2^3 = 8, \qquad 2^2 = 4, \qquad 2^1 = 2$$

What should 2^0 equal? *half of 2 is 1*

FX-11. Mindy, an only child, is constructing a family tree and notices a pattern. She sees that she has two first-generation predecessors (her natural parents), four in the second-generation (her grandparents), eight in the third-generation, and so on. Create a table to show the number of natural parents, grandparents, great grandparents, and so forth, as a function of the generation number.

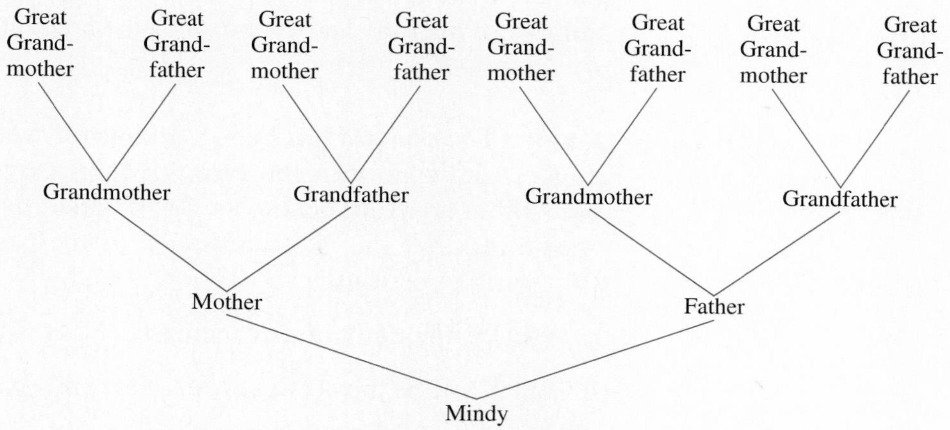

a. If Mindy's grandparents are the second generation and her parents are the first generation, which generation is she? *zeroth*

b. She had $8 = 2^3$ great grandparents, $4 = 2^2$ grandparents, and $2 = 2^1$ parents. The number of Mindys is the initial value, so how many Mindys are there? What power of 2 represents the number of Mindys? *just one, 2^0*

You may want to use FX-12 as a minilab.

FX-12. This problem involves folding a rectangular sheet of scratch paper. Make a table to record the number of folds and number of rectangular regions.

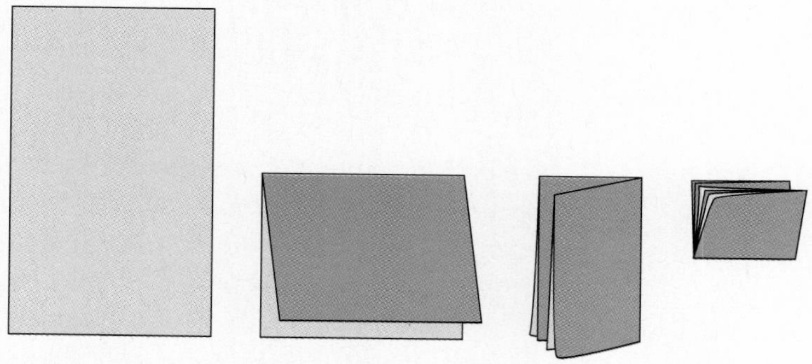

a. Fold the paper in half. How many times has the paper been folded? How many rectangular regions are there? *1, 2*

b. Fold the paper in half again. With two folds, how many regions are there? *2, 4*

c. With three folds, how many regions are there? *3, 8*

d. How many regions are there with four folds? with five folds? Describe the pattern or rule for this relationship.

e. Let x represent the number of folds and y represent the number of regions. Write an equation to represent the rule. $y = 2^x$

f. Explain how your equation accounts for how many regions there were with zero folds (in other words, before you folded the paper at all).

FX-13. Zero Power In FX-10, FX-11, and FX-12 you found the value of 2^0 by looking at patterns.

a. What does your calculator give for the value of 2^0? What value(s) do you get for 3^0, 4^0, 10^0, and 101^0?

b. Experiment with lots of different numbers raised to the zero power. Try $(-2)^0$, $\left(\frac{3}{5}\right)^0$, and at least three others. Write down your results.

c. Try more negative numbers raised to the zero power.

d. Try fractions and decimals raised to the zero power.

e. Write a general rule about numbers raised to the zero power. Put this general rule in your Tool Kit.

FX-14. Solve $2x^2 - 3x - 7 = 0$. Give your solutions both in radical form and as decimal approximations. Remember, you can refer to your Tool Kit for a review of how to use the quadratic formula. *$(3 \pm \sqrt{65})/4$, or 2.77 and −1.27*

FX-15. Write each of the following in its smallest base. (For examples, use prime factoring to see that 16 can be written using the base 2 and the exponent 4, $2 \cdot 2 \cdot 2 \cdot 2 = 2^4$, and that $27^2 = (3^3)^2 = 3^6$.)

a. 64 2^6

b. 8^3 2^9

c. 25^x 5^{2x}

d. 16^{x+1} 2^{4x+4}

e. $\frac{16}{81}$ $\left(\frac{2}{3}\right)^4$

f. 81^2 3^8

FX-16. Solve the following systems of equations. If you need a review of how to solve systems, refer to your Tool Kit or to BB-21 in Chapter 3.

a. $3x - 2y = 14$ and $-2x + 2y = -10$ *(4, −1)*

b. $y = 5x + 3$ and $-2x - 4y = 10$ *(−1, −2)*

c. Which system is most efficiently solved by substitution? Explain. *b*

d. Which system is most efficiently solved by elimination? Explain. *a*

FX-17. Give the coordinates of the x- and y-intercepts for the graphs of each of the following functions:

a. $f(x) = x^2 + 6x - 72$
x: (6, 0) and (−12, 0);
y: (0, −72)

b. $g(x) = -5x + 4$
x: (4/5, 0); y: (0, 4)

FX-18. Sketch a graph showing the relationship between a person's height and his or her age.

a. What are the independent and dependent variables?

b. What are the domain and range for this relationship?

c. Does your graph have an x-intercept? If so, interpret this point.

d. Does your graph have a y-intercept? If so, interpret this point.

FX-19. Graph $y = x^2 + 3$ and $y = (x + 3)^2$. What are the similarities and differences between the graphs? How would these graphs compare with the graph of $y = x^2$? *Both have the same shape as $y = x^2$; one is shifted up 3 units, while the other is shifted left 3 units.*

FX-20 is a preview for FX-21, the Penny Lab. Be sure to assign it.

FX-20. Suppose you flip a fair coin.

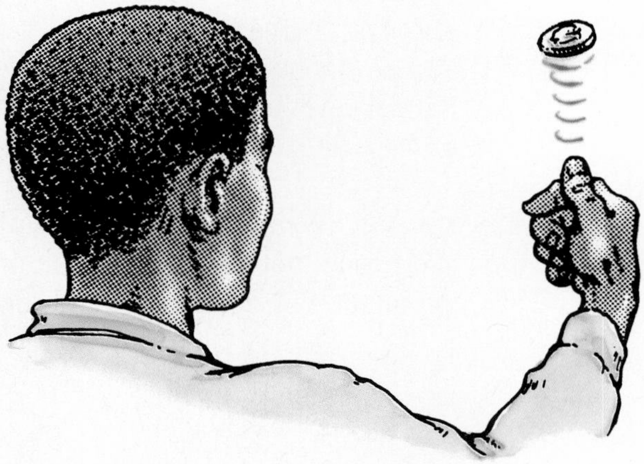

a. What is the probability of getting heads? *1/2*

b. What is the probability of getting tails? *1/2*

4.2 HALF-LIFE AND OTHER APPLICATIONS

Each group will need two rolls of pennies for the following experiment. If pennies are not available, use two-sided counters, or as a last resort have each group make 100 little scraps of paper with "H" on one side and "T" on the other. The students should work in groups to investigate this application of exponential decay and half-life. They will be preparing and submitting a lab report for this investigation. A suggested format for this report is included in the description of the problem, but you may prefer your own format or the general lab report outline in the resource folder for Chapter 2.

To compare their graphs, give each group a transparency with the axes already drawn and scaled (so that all the transparencies can be aligned when stacked), and ask them to plot their data on it. Different-colored pens for writing on overhead transparencies are also useful. Then one by one each group can stack its transparency on the overhead so the class can see how closely the graphs align.

FX-21. The Penny Lab With your group, conduct the following experiment:

- *Trial 0*: Start with 100 pennies. This is your initial value.
- *Trial 1*: Dump the pennies in a pile on your desk. Remove any pennies with tails side up. Record the number of pennies *left*, the heads.
- *Trial 2*: Gather the remaining pennies, shake them up, and dump them on your desk. Remove any pennies that have the tails side up, and record the number of pennies left.
- *Trial 3 to Trial* ?: Continue this process until the last penny is removed.

Answer the following questions:

a. Would the results of this experiment have been significantly different if you had removed the heads pennies each time? ***no general difference***

b. Would the results have been significantly different if you had alternated heads and tails after each trial? ***no general difference***

c. How would starting with 200 pennies have affected your results? ***Everything shifts over one trial.***

d. How does FX-20, about probability, relate to this investigation?

e. Decide what your dependent and independent variables are, clearly label them, and draw a graph of your data. ***Students should come up with discrete graphs.***

f. Select someone in your group to plot your data on an overhead transparency provided by your instructor.

g. Is it possible that some group conducting this experiment might never remove their last penny? Explain.

h. What are the domain, range, x- and y-intercepts of your graph?

FX-22. When a radioactive isotope undergoes decay, it does so at a fixed rate. The time it takes for half of it to decay is called the **half-life**. For example, if a substance has a half-life of ten years and you start with 100 grams of it, after ten years you will have only 50 grams. In another ten years, half of that 50 grams will decay, leaving you with 25 grams, and so on.

Years	Amount Left
0	100
10	50
20	*25*
30	*12.5*
40	*6.25*

a. Make a table of values like the one shown above, and fill in the blanks.

b. Extend the table for values up to 60 years. *(50, 3.125); (60, 1.5625)*

c. Sketch a graph to represent this situation. *continuous graph*

d. How much will be left after 25 years? *≈17.7*

e. When will this isotope disappear completely? *never, but . . .*

f. Explain why you think most states will not allow radioactive materials to be dumped inside their borders.

Part f of FX-23 is best done with a graphing calculator.

FX-23. A video loses 50 percent of its value every year that it is in a video store. The initial value of the video is $60.

a. What is the multiplier? *0.50*

b. What is the value of the video after one year? *$30*

c. What will the value be after four years? *$3.75*

d. Write a function $V(t) = ?$ to represent the value of the video in t years. $60(0.5^t)$

e. When does the video have no value? *never*

f. Sketch a graph of this function. Be sure to scale and label your axes.

EXTENSION AND PRACTICE

FX-24. Two very unlucky gamblers sat down in a casino with $10,000 each. Both lost every bet they made. The first gambler bet half of his current total at each stage, while the second bet one-fourth of her current total at each stage. They had to quit betting after each of them was down under $1. How many bets did each person get to place? *The first gambler bet 14 times, and the second gambler bet 33 times.*

If any students get stuck on parts c and d of FX-25, ask them how much $1 worth of groceries would cost one year from now.

FX-25. Find the **annual multiplier** for each of the following:

a. A yearly increase of 5 percent due to inflation. *1.05*

b. An annual decrease of 4 percent on the value of a television set. *0.96*

c. A monthly increase of 3 percent in the cost of groceries. $1.03^{12} \approx 1.426$

d. A monthly decrease of 2 percent on the value of a bicycle. $0.98^{12} \approx 0.785$

FX-26. Judy does not believe that x^0 can possibly have any meaning, and if it does, she can't remember it anyway. Kelly volunteers to convince her and starts by showing her the following pattern:

$$\frac{x^7}{x^2} = x^5, \quad \frac{x^7}{x^3} = x^4, \quad \frac{x^7}{x^4} = x^3, \quad \frac{x^7}{x^5} = x^2$$

a. Explain how continuing this pattern will help Judy understand x^0.

b. Kelly helps Judy see the pattern by asking her to simplify each of these fractions:

$$\frac{x^{103}}{x^{98}} \quad \text{and} \quad \frac{x^{1052}}{x^{1049}}$$

Simplify these fractions for Kelly. x^5, x^3

c. How would you represent $\frac{x^A}{x^B}$? x^{A-B}

d. Kelly then says that if $\frac{6}{6} = 1$ and $\frac{x^3}{x^3} = 1$, then $\frac{x^B}{x^B} = \cdots$

Use the idea of subtracting exponents in part c and Kelly's pattern in part d to write an explanation that shows Judy why x^0 should equal 1.

FX-27. Katya wrote the following as a solution to a problem but got stuck at the end. Explain the mistake she made, and then correct and solve the problem.

$$3 + 2(x-5) = 5(x-2)$$
$$5(x-5) = 5x - 10$$
$$5x - 25 = 5x - 10$$
$$5x - 5x = 15$$

No solution!

The second step should be $3 + 2x - 10 = 5x - 10$; $x = 1$.

FX-28. What are the coordinates of the y-intercepts for each of the graphs below?

a. $y = 2^x$ *(0, 1)*

b. $y = 3^x$ *(0, 1)*

c. $y = 101^x$ *(0, 1)*

d. $y = \left(\frac{3}{5}\right)^x$ *(0, 1)*

e. State in a sentence any conclusion you can make about the y-intercept for the graph of $y = a^x$.

If students ask what happens when a < 0 (though they probably won't), ask them what they think. We'll investigate the family of functions later.

FX-29. Graph $y = 0^x$ and $y = x^0$ on the same set of axes. Use values of x such that $-4 \le x \le 4$. (This is just a short way to say, "use numbers from –4 through 4.")

a. Where do these graphs intersect? *They do not.*

b. What do you think 0^0 equals? What does the calculator give for 0^0? Explain why the calculator gives this result. *undefined; error*

FX-30. Antoine was stuck trying to solve $25^x = 125^4$. Budge suggested changing 25 to 5^2 and 125 to 5^3 and showed Antoine how to solve this problem. Complete Budge's explanation. *$x = 6$*

FX-31. Solve each of the following for x. For parts c and d use the idea from FX-30 so that you don't have to guess and check.

a. $2^{x+3} = 2^{2x}$ *$x = 3$* **b.** $3^{2x+1} = 3^3$ *$x = 1$*

c. $9^{40} = 3^x$ *$x = 80$* **d.** $8^{70} = 2^x$ *$x = 210$*

FX-32. Consider the function $y - 4 = 4(x - 3)$.

a. What is the slope of this line? *4*

b. What is the y-intercept? *–8*

c. What is the x-intercept? *2*

d. Make a quick sketch of this function.

FX-33. Consider the function $f(x) = \dfrac{6}{x-1}$. Find the value of x that will make $f(x) = 5$. *11/5 or 2.2*

FX-34. Solve the following system of equations:

$$5x - 4y = 7$$
$$2y + 6x = 22$$

(3, 2)

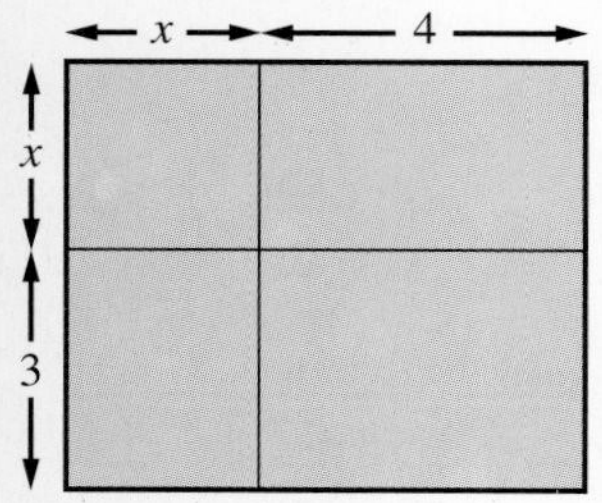

FX-35. What is the area of the largest rectangle shown at left?

a. Express the area as the product of the base times the height.

b. Now write the area as the sum of the areas of the four smaller rectangles.

4.3 NEGATIVE EXPONENTS: THE FUNCTION WALK

This is a good place for a Function Walk of the graph of $y = 2^x$ (including negative integer values). The outdoor xy-coordinate system should be marked off, or ropes set up before class. For the axes, some teachers have used two pieces of rope wrapped with colored tape to mark the units. Alternatively, you could use chalk to mark and label the axes on the pavement. However you produce large-scale axes, the units should be about 30 inches apart to give students enough room to stand comfortably. The x-axis should include numbers from –5 to 5. The y-axis should include at least –5 to 16.

Before going outside you will need to give each group an integer and four different-colored index cards. Have students bring their calculators and pencils. Give each group an integer ($-4 \leq x \leq 4$), and have groups write their integer on their cards.

When you get outside, have those students with a red card (one from each group) stand on the x-axis facing toward Quadrants 1 and 2. The students who are observing should stand behind those with red cards and face the same direction because it corresponds to the standard orientation we use when graphing. Start with the function $y = 2^x$. Give the following directions: "Be sure you are standing with the mark that corresponds to your number between your feet. Start with a base of 2, and use your number as an exponent. Calculate the value using your calculator if necessary. Got it? When I say 'go,' take that many paces forward. Ready? Go!"

Mistakes will be made. Have students help each other out. In most cases the students will handle corrections themselves. Resist the urge to manage this too much.

After the students representing the integers are in place, have six more students stand on –2.5, –1.5, –0.5, 0.5, 1.5, and 2.5. Ask them to see if they can figure out, without their calculator, where they need to stand based on what they can observe about the graph so far. Then have them verify their guess with their calculators.

Have the rest of the students record the function (equation) and as much detail as they can about the resulting graph. Three members of each group will have information about each function, but no one will have all four. Then have those students with green cards graph $y = 3^x$, those with yellow cards graph $y = (3/2)^x$, and those with blue cards graph $y = (1/2)^x$. While each color group is "walking," the rest of the students should be recording the other three functions and their graphs. When the students get back inside with their groups, they should have enough information to solve the rest of the problems.

As part of gathering information about the functions, have the students check the domain and range during the Function Walk. First have students note the domain as they stand on the x-axis. After they have moved into position, have them make a 90° turn toward the y-axis and walk straight toward it. The space between 0 and 1 will become a bit crowded, but they now should be able to visualize the range.

FX-36. Your instructor will give you an integer and an index card. Write down the integer on one side of the index card. You will need this card, your calculator, and a pencil for this problem. To do this problem the class will be going outside. There you will see a set of *xy*-axes marked in chalk or represented with two ropes. All students holding cards of the same color will line up on the *x*-axis as in the illustration. The instructor will then give an equation. The first one is $y = 2^x$. The students standing on the *x*-axis then use their numbers in the place of *x* and calculate the results. When the instructor says, "Go!" each one walks forward the number of paces in his or her result. Then they stop. Their fellow group members will record the "graph" roughly on their index cards. Back in the classroom you will use the information to make tables and complete graphs in FX-37 and FX-38. The functions to be used in the "human graphing" exercise follow:

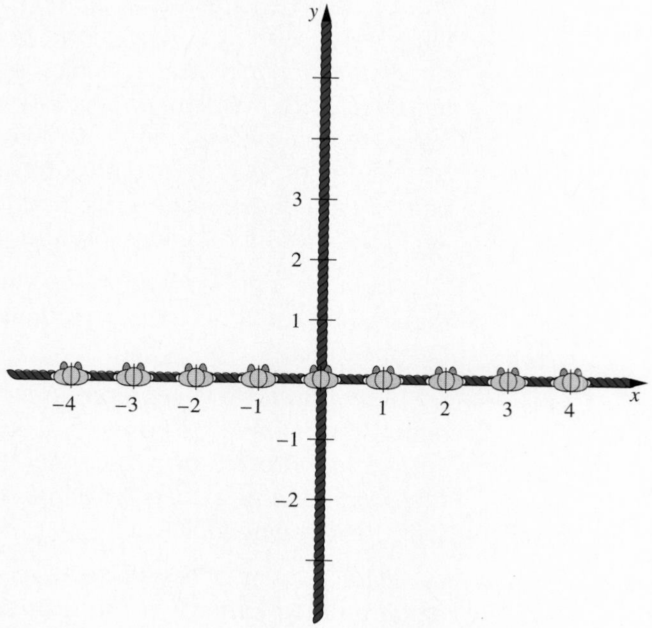

a. $y = 2^x$

b. $y = 3^x$

c. $y = \left(\frac{3}{2}\right)^x$

d. $y = \left(\frac{1}{2}\right)^x$

Your instructor will give you the rest of the directions.

FX-37. Make tables like the ones shown on the next page for each of the four functions from the Function Walk, and fill in as much as you can. Give *y* values in *fraction* form, and look for a pattern. Divide up the work among group members, but make sure everyone records all the results.

x	2^x	y
4	2^4	16
3	2^3	8
2	2^2	4
1	2^1	2
0	2^0	1
−1	2^{-1}	1/2
−2	2^{-2}	1/4
−3	2^{-3}	1/8
−4	2^{-4}	1/16

x	3^x	y
4	3^4	81
3	3^3	27
2	3^2	9
1	3^1	3
0	3^0	1
−1	3^{-1}	1/3
−2	3^{-2}	1/9
−3	3^{-3}	1/27
−4	3^{-4}	1/81

x	$(3/2)^x$	y
4	$(3/2)^4$	81/16
3	$(3/2)^3$	27/8
2	$(3/2)^2$	9/4
1	$(3/2)^1$	3/2
0	$(3/2)^0$	1
−1	$(3/2)^{-1}$	2/3
−2	$(3/2)^{-2}$	4/9
−3	$(3/2)^{-3}$	8/27
−4	$(3/2)^{-4}$	16/81

x	$(1/2)^x$	y
4	$(1/2)^4$	1/16
3	$(1/2)^3$	1/8
2	$(1/2)^2$	1/4
1	$(1/2)^1$	1/2
0	$(1/2)^0$	1
−1	$(1/2)^{-1}$	2
−2	$(1/2)^{-2}$	4
−3	$(1/2)^{-3}$	8
−4	$(1/2)^{-4}$	16

FX-38. Draw a graph for each of the four functions from the Function Walk by plotting points. You may want to check your graphs with a graphing calculator. Use the graphs and tables that you created for each function to write explanations for the following questions.

a. Give the domain and range of each function. *For all functions: D = all x; R: y > 0*

b. Use your graph to determine what 2^0, 3^0, $\left(\frac{3}{2}\right)^0$, $\left(\frac{1}{2}\right)^0$ represent. *the y-intercepts*

c. What are the x-intercepts for each function? *no x-intercepts*

d. When x is negative, which functions give y values less than one? *$y = 2^x$, $y = 3^x$, $y = 3/2^x$*

e. When x is negative, which functions give y values greater than one? *$y = 1/2^x$*

f. How does the value of the base affect the y outputs for negative x inputs?

g. Write an explanation in your Tool Kit of what a negative exponent does to a number.

h. Compare and contrast the graphs of the four functions. What is the same, and what is different?

The definition of an exponential function given in FX-39 is not complete, but that is OK for now. Later, when the questions arise naturally, students should complete the definition in their Tool Kits. Writing definitions is not a one-time process. They should get used to refining and sometimes revising their Tool Kit entries.

FX-39. Is your graph for $y = 2^x$ in FX-38 just a set of points, or did you draw a smooth continuous curve through all the points?

a. Draw a smooth curve connecting the points if you have not already done so.

> The function $y = 2^x$ is one simple example of an **exponential function**. An exponential equation has the general form $y = k(m^x)$, where k is the initial value and m is the multiplier. Be careful! The independent variable x has to be in the exponent. For example, $y = x^2$ is *not* an exponential equation, even though it has an exponent.

b. Add this graph to your Tool Kit, label it "exponential function," and on the graph write the general form for an exponential function. We will return to this definition to add some more details later.

c. Where have you seen equations of this type before?

FX-40. Use your graph of $y = 2^x$ to complete a table like the one shown here. You can use (1) the trace feature of the graphing calculator, (2) the method shown in parts a and b using the graphs you already made, or (3) the tables you made in FX-37 to find a good estimate and then check on a calculator.

x	2^x	y
4	2^4	16
3.1	$2^{3.1}$	9.1
3	2^3	8
2.7	$2^{2.7}$	6.8
2	2^2	4
1.4	$2^{1.4}$	2.8
1	2^1	2
0.6	$2^{0.6}$	1.52
0	2^0	1
−0.5	$2^{-0.5}$	0.7
−1	$(2)^{-1}$	0.5
−1.2	$2^{-1.2}$	6.8
−2	$(2)^{-2}$	0.25
−2.7	$2^{-2.7}$	6.8
−3	$(2)^{-3}$	0.125
−3.4	$2^{-3.4}$	0.09
−4	$(2)^{-4}$	0.06

Example of how to approximate $2^{0.5}$:

a. Locate the x value on the horizontal axis. Draw a verticle line at $x = 0.5$, intersecting the curve.

b. Next locate the corresponding y value on the vertical axis by drawing a horizontal line over to the y-axis. The corresponding y value is about 1.4.

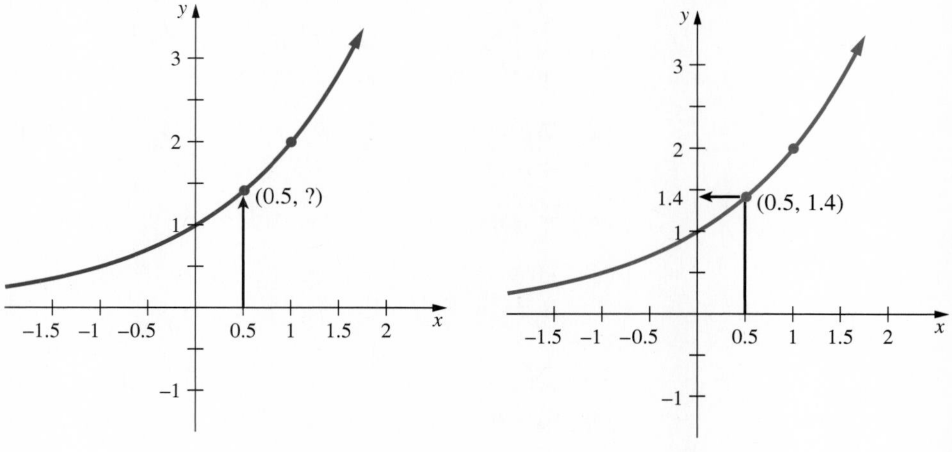

(a) **(b)**

EXTENSION AND PRACTICE

FX-41. Jim does not understand negative exponents. He thinks that $(4)^{-1}$ is a negative number. Kelly volunteers to help him. She tells him to graph $y = 4^x$. Does the graph include any negative results for y ? How can a graph help explain the meaning of both $(4)^{-1}$ and $(4)^{-2}$? Help Kelly out by finishing her explanation using the graph shown below. You must help Jim understand the correct meaning of negative exponents.

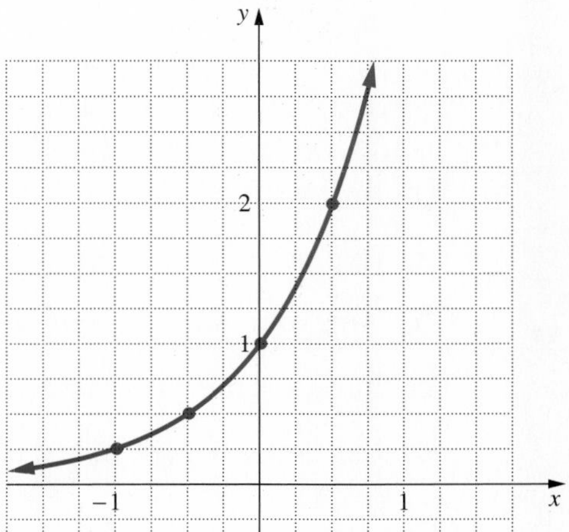

FX-42. On the graph for FX-41, find the result for $x = \frac{1}{2}$ and for $x = -\frac{1}{2}$. What did you notice about the results?
(1/2, 2), (−1/2, 1/2); they are positive.

FX-43. Manhattan Island (New York City) was purchased in 1626 for trinkets worth \$24. The real-estate value of Manhattan in 1994 was about \$34 billion. Suppose the \$24 had been invested instead at a rate of 6 percent compounded each year.

a. What is the multiplier? *1.06*

b. What is the initial value? *24*

c. What time value (number of years) is needed to find the value of the money in 1994? *368 years*

d. Write a function, $V(t) = ?$, to find the value V of the investment at any time t. $V(t) = 24(1.06^t)$

e. What would the investment be worth in 1994? $24(1.06^{368}) \approx 4.93 \times 10^{10}$

f. How does this value compare to the 1994 value of Manhattan? 3.4×10^{10}

FX-44. Write each of the following expressions in exponential form using the smallest integer base you can.

Example: $\frac{1}{16} = \frac{1}{2^4} = (2)^{-4}$

a. $\frac{1}{125}$ **b.** $\frac{1}{121}$ **c.** $\left(\frac{1}{9}\right)^x$ **d.** $\left(\frac{1}{32}\right)^{1-x}$

$(5)^{-3}$ $(11)^{-2}$ $(3)^{-2x}$ 2^{5x-5}

FX-45. Solve each equation for x:

a. $5^x = (5)^{-3}$ *−3* **b.** $6^x = 216$ *3*

c. $7^x = \frac{1}{49}$ *−2* **d.** $10^x = 0.001$ *−3*

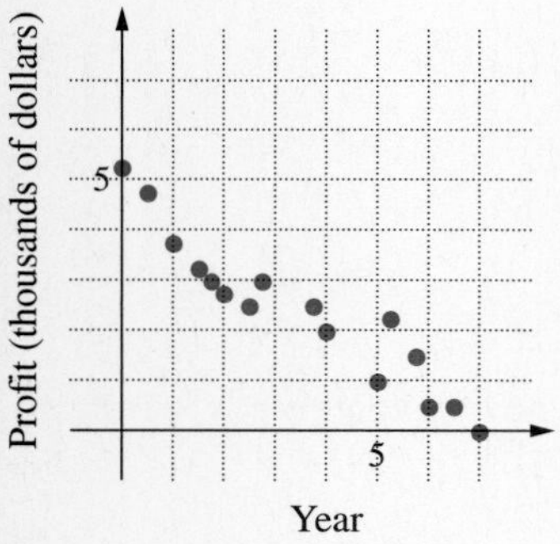

FX-46. For the past eight years, Laurel kept track of the profit figures for her new music studio, and plotted the data on a graph. She needs to find a line of best fit but has no idea how to do it. Explain to her by doing the following:

a. Explain what a line of best fit is.

b. Approximate a line of best fit, and give her an equation for the line.

c. Interpret the intercepts and slope of your line in terms of the data.

d. Explain what the slope of the line tells her about her business.

FX-47. For each of the following situations, identify the multiplier, the initial value, and the time value. Remember that the time must be in the same units as the multiplier.

Example: 3 percent raise per quarter for two years → multiplier = 1.03, time = 8 quarters.

a. A house purchased for \$126,000 has lost 4 percent of its value each year for the past 5 years. *0.96; 126,000; 5 years*

b. A 1970 Richie Rich comic book has appreciated at 10 percent per year, and originally sold for 35¢. *1.10; 0.35; 24 years in 1994*

c. A Honda Prelude depreciates at 15 percent per year. Six years ago it was purchased for \$11,000. *0.85; 11,000; 6 years*

FX-48. Solve the following system of equations.

$$y = -x - 2$$
$$5x - 3y = 22$$ *(2, –4)*

Treat the following sequence as we did in Chapter 2. The initial term is 7, and the first term after that is $6\frac{1}{3}$ is t(1).

FX-49. Consider the following sequence with initial value 7: 7, $6\frac{1}{3}$, $5\frac{2}{3}$, 5, ...

a. What kind of sequence is it? *arithmetic*

b. Write an equation that describes the sequence $t(n) = ?$ *7 – (2/3)n*

c. What is the fifteenth term after the initial term, $t(15)$? *–3*

d. Is –21 a term of the sequence? If so, which one? *42nd term*

FX-50. State the coordinates of the x- and y-intercepts for the graph of the following equation: $y = -x^2 + 3x - 5$ *no x-intercept; y: (0, –5)*

FX-51. Solve for x.

a. $x^3 = 27$ *3* **b.** $x^4 = 16$ *2, –2* **c.** $x^3 = -125$ *–5*

FX-52. An integer between 10 and 20 is selected at random. What is the probability that 3 is a factor of that integer? *3/9 = 1/3*

4.4 ROOTS AND FRACTIONAL EXPONENTS

Students should work on FX-53 individually and silently in their groups as you circulate around the room. They will struggle with the chart, and they'll probably keep you pretty busy. You'll also need to keep track of the numbers on the board to see whether you need to erase any of them.

Some instructors suggest that students continue with FX-54 working together in their groups; others recommend continuing with the Silent Board Game format.

FX-53. Silent Board Game Do this *individually* and *silently* (don't discuss it with your group). Figure out the rule for this function. The table below is also written on the board. As you figure out a number that fits, go up to the board and fill in the number (silently). If it's not the correct number, the instructor will erase it. Only one number from each person, please. Copy the table below onto a clean sheet of paper. You may use your calculator to help you guess and check. *You may need to slow down the process; otherwise the quicker students will finish the table before everyone gets a chance to think about it.*

x	25	$\frac{1}{4}$	25	64	16	$\frac{1}{25}$	49	81	0.25	0.36	−4	100	121	4	9
$r(x)$	5	$\frac{1}{2}$	*5*	*8*	*4*	$\frac{1}{5}$	*7*	*9*	0.5	*0.6*	*impossible*	*10*	*11*	2	*3*

Write the rule that the table illustrates. You can use words or symbols or both. *the square root of a number, or* $\sqrt{x}$

FX-54. Copy the following tables, and complete them.

a. Write the rule illustrated by each table. (You can use words or symbols or both.) *the cube root of a number, or* $\sqrt[3]{x}$

x	8	$\frac{1}{27}$	125	64	216	$\frac{1}{343}$	1	512	−8	−27	−64	1000
$r(x)$	2	$\frac{1}{3}$	*5*	*4*	6	$\frac{1}{7}$	*1*	*8*	*−2*	*−3*	*−4*	*10*

b. Write the rule illustrated by each table. (You can use words or symbols or both.) *the fourth root of a number, or* $\sqrt[4]{x}$

x	81	16	$\frac{1}{256}$	625	−16	$\frac{1}{2401}$	1	4096	1296	$\frac{1}{81}$	0.0016	0.0081
$r(x)$	3	*2*	$\frac{1}{4}$	*5*	*impossible*	$\frac{1}{7}$	*1*	*8*	6	$\frac{1}{3}$	*0.2*	0.3

FX-55. Compute the following without a calculator (that is; use guess and check):

a. $\sqrt[2]{49} = ?$ *7* **b.** $\sqrt[4]{16} = ?$ *2*

c. $\sqrt[4]{10{,}000} = ?$ *10* **d.** $\sqrt[5]{-32} = ?$ *−2*

FX-56. Write out in words what each of the following expressions says in symbols; then check them in your head or on your calculator. Part a has been done as an example.

a. Example: $13^2 = 169$ and $\sqrt[2]{169} = 13$

13 squared is 169, and the square root of 169 is 13.

b. $7^3 = 343$ and $\sqrt[3]{343} = 7$ *7 cubed is 343; and the cube root of 343 is 7*

c. $3^4 = 81$ and $\sqrt[4]{81} = 3$ *3 raised to the fourth power is 81; and the fourth root of 81 is 3*

FX-57. Use your graphing calculator to sketch the graphs of $y = \sqrt{x}$ and $y = x^{1/2}$.

a. Compare the graph of $y = \sqrt{x}$ to the graph of $y = x^{1/2}$. What conclusions can you draw?

b. The graph of $y = \sqrt[3]{x}$ is shown below. Graph $y = x^{1/3}$ on your graphing calculator. How does your graph compare to this one? What conclusions can you draw?

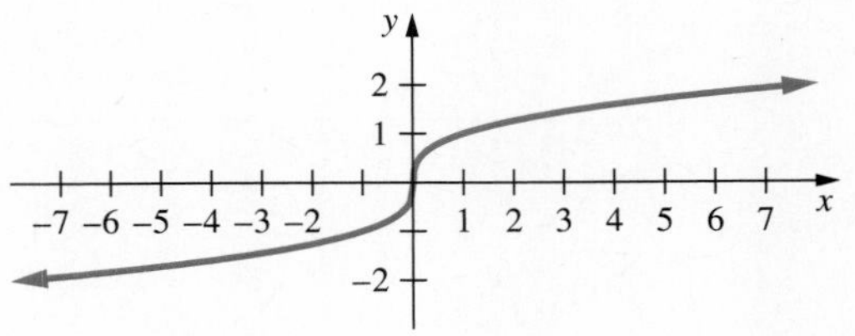

c. Using your conclusions, how could you rewrite $\sqrt[4]{x}$? $\sqrt[4]{x} = x^{1/4}$

d. Rewrite this relationship as a rule using $\sqrt[x]{b}$ and add it to your Tool Kit. $\sqrt[x]{b} = b^{1/x}$

FX-58. Use your graphing calculator to graph $y = \sqrt[3]{x^2}$, $y = \left(\sqrt[3]{x}\right)^2$, and $y = x^{2/3}$. Sketch the graphs.

a. Compare the graphs of $y = \sqrt[3]{x^2}$, $y = \left(\sqrt[3]{x}\right)^2$, and $y = x^{2/3}$. What conclusions can you draw?

b. Using your conclusions, how could you rewrite $\sqrt[n]{b^m}$ in exponential form? $b^{m/n}$

c. Rewrite this relationship as a rule, and add it to your Tool Kit.

FX-59. Rewrite each of the following expressions. Use fractional exponents instead of radicals and exponents.

Note: The radical notation ($\sqrt{\ }$) comes from an earlier time. Eventually it will probably be completely replaced by exponents. However, in the meantime, you will need to know how to get from one notation to the other because $\sqrt{\ }$ will still appear in textbooks and on standardized tests.

a. $\sqrt[3]{5^2}$ $\quad 5^{2/3}$ **b.** $\sqrt[4]{2^5}$ $\quad 2^{5/4}$ **c.** $\sqrt[2]{7^3}$ $\quad 7^{3/2}$

FX-60. Think of these as puzzles in which you are figuring out how to use the exponents. Solve each equation.

a. $\left(\sqrt[3]{125}\right)^2 = x$ $\quad x = 25$ **b.** $125^{2/3} = x$ $\quad x = 25$

c. $\left(\sqrt{x}\right)^3 = 125$ $\quad x = 25$ **d.** $x^{3/2} = 125$ $\quad x = 25$

e. What do you notice about the answers to parts a through d? How are the equations related to each other?

FX-61. Use one or more of the following methods to solve the problems in parts a through c:

1. Graphically, using a graphing calculator: show your graph and label your solution.

2. Guess and check using a scientific or graphing calculator: make a table showing your guesses and of course your solution.

3. Algebraically: show all your work.

a. $x^{1/4} = 2$ *16* **b.** $m^{1/3} = 7$ *343* **c** $r^{3/2} = 8$ *4*

EXTENSION AND PRACTICE

FX-62. Solve each of the following equations for x:

a. $2^{1.4} = 2^{2x}$ *$x = 0.7$*

b. $8^x = 4$ *$x = 2/3$*

c. $3^{5x} = 9^2$ *$x = 0.8$*

FX-63. Consider the following pattern:

$$\frac{1}{2^3} = \frac{1}{8},\ \frac{1}{2^2} = \frac{1}{4},\ \frac{1}{2^1} = \frac{1}{2},\ \frac{1}{2^0} = 1.$$

a. What are the values of

$$\frac{1}{(2)^{-1}},\ \frac{1}{(2)^{-2}},\ \frac{1}{(2)^{-3}},\ \frac{1}{(2)^{-4}}?$$

b. What is the value of $\frac{1}{(2)^{-n}}$?

c. Write a rule to describe this pattern, and add it to your Tool Kit.

FX-64. Which of the following equations are exponential functions? Explain.

a. $f(x) = 3x^2$

b. $y = 5(4)^x$ *yes*

c. $y = 5x^3$

d. $g(x) = 2.46^x$ *yes*

e. $y = (2^x)(3^x)$ *yes*

f. $y = \left(\frac{1}{2}\right)^x$ *yes*

FX-65. Three students are doing the same problem in which they have to find an equation for a pattern. They get three different answers: $y = (2)^{-x}$, $y = \frac{1}{2^x}$, and $y = 0.5^x$. Who is correct? Explain to these three students what has occurred. ***All equations are the same.***

FX-66. Find the multiplier and unit of time for each of the following:

a. a yearly increase of 1.23 percent in population. ***1.0123 for 1 year***

b. a monthly decrease of 3 percent on the value of a video. ***0.97 for 1 month***

c. the annual multiplier if there is a monthly decrease of 3 percent on the value of a video. ***$0.97^{12} \approx 0.694$, the multiplier for 1 year***

FX-67. Solve each of the following equations for x. You can guess and check, or you can be clever and remember such things as $8 = 2^3$.

a. $2^{x+3} = 64$ *x = 3* **b.** $8^x = 4^6$ *x = 4* **c.** $9^x = \frac{1}{27}$ *x = −3/2*

FX-68 When asked to solve $(x - 3)(x - 2) = 0$, Freddie gives the answer "$x = 2$." Samara corrects him, saying that the answer also should be $x = 3$. But Freddie says that when you solve an equation, you only have to find *one* value of x that works, and since 2 works, he's done. Do you agree with Freddie? Justify your answer.

FX-69. Solve each of the following equations for x:

a. $\frac{x+2}{x} = \frac{3}{7}$ *$x = -\frac{7}{2}$* **b.** $\frac{9}{x} = \frac{x}{4}$ *$x = \pm 6$*

FX-70. Solve each of the following equations for x:

a. $(3.25 \times 10^{27})^x = 1$ *x = 0* **b.** $\left(\sqrt[7]{239^3}\right)^x = 1$ *x = 0*

c. $\left(\frac{287{,}625}{1{,}119{,}628}\right)^x = 1$ *x = 0* **d.** $\left[\sqrt[3]{\pi}\left(\frac{4}{5}\right)^{-6}\right]^0 = x$ *x = 1*

FX-71. In *USA Yesterday* John saw a scatterplot of data (reproduced here) showing the cost to the government of collecting \$100 in taxes from 1989 to 1993.

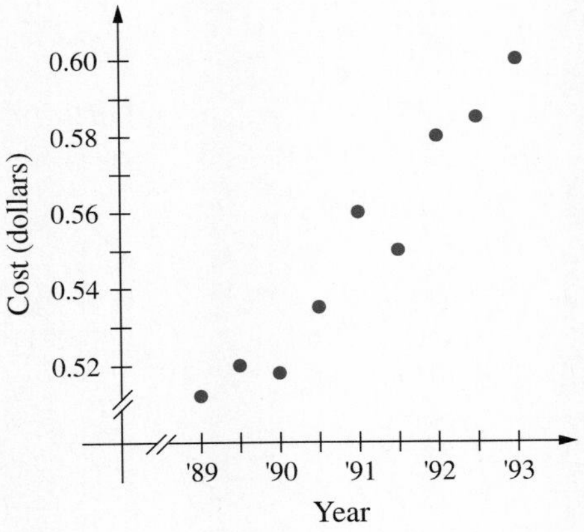

a. Estimate a line of best fit, and write its equation. ***Answers vary; y = 0.024x + 0.50 is reasonable.***

b. Use your answer to part a to estimate the cost of collecting \$100 in taxes in 1994. ***approximately \$0.64***

FX-72. On the spinner shown here, each "slice" is the same size. What is the probability that when you spin you will get:

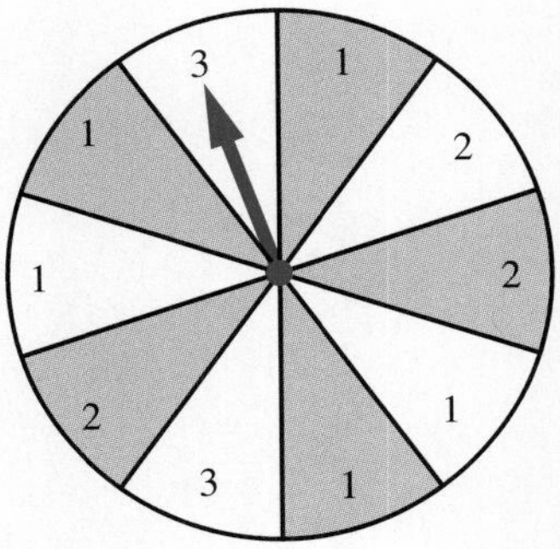

a. a 1? *5/10 = 1/2*

b. a 2? *3/10*

c. a 3? *2/10 = 1/5*

4.5 APPLICATIONS OF EXPONENTIAL FUNCTIONS

Sections 4.5 and 4.6 provide time for the students to absorb the ideas of the last several sections and to work on applications of exponential functions. There is a good variety of problems within these two sections. Our purpose is for students to see the usefulness and broad applicability of exponential functions.

FX-73. The following situations can be described by using exponential functions. They represent a small sampling of the situations where quantities grow or decay over some period of time. For Situations a through f do the following:

a. Find an appropriate time unit (days, weeks, years, and so on).

b. State the multiplier that would be used for that situation.

c. Identify the initial value.

d. Write an equation in the exponential form $V(t) = k(m)^t$ that represents the growth or decay. $V(t)$ represents the value at any given time t, and m is the multiplier, t is the time, and k is the initial value.

NOTE Check your Tool Kit to be sure you have good definitions for *multiplier, initial value,* and *time.*

Situation 1: A house purchased for \$120,000 has an annual appreciation of 6 percent per year. *years, 1.06, 120,000, $V(t) = 120{,}000(1.06)^t$*

Situation 2: The number of bacteria present in a colony is 180 at 12 noon, and the bacteria grows at a rate of 22 percent per hour. *hours, 1.22, 180, $V(t) = 180(1.22)^t$*

Situation 3: A 100-gram sample of a radioactive isotope decays at a rate of 6 percent every week. *weeks, 0.94, 100, $V(t) = 100(0.94)^t$*

Situation 4: The value of a car with an initial purchase price of \$12,250 depreciates by 11 percent per year. *years, 0.89, 12,250, $V(t) = 12{,}250(0.89)t$*

Students will need to adjust the multiplier in Situations 5 and 6. Let them discuss what it should be. It is important that the students do not round off the multiplier in Situation 6. It can make a great difference.

Situation 5: For an investment of $1000 the 6 percent annual interest is compounded monthly. *months, 1.005, 1000,* $V(t) = 1000(1.005)^t$

Situation 6: For an investment of $2500 the 5.5 percent annual interest is compounded daily. *days, 1.00015 . . . , 2500,* $V(t) = 2500(1.000015\ldots)^t$

FX-74. Choose one of the exponential equations you wrote in FX-73, and create a possible exam problem that could be solved using the equation.

A graphing calculator will certainly be useful for part e in FX-75.

FX-75. Suppose the annual fees for attending a campus of the University of California were $1200 in 1986 and the cost increased by 10 percent each year. Round answers to the nearest dollar.

a. Calculate the cost for the year 1997. *$3424*

b. What would you expect the cost to be four years from 1997? *$5013*

c. What was the cost in 1980? *$677*

d. Write an equation that describes how to find the cost. $y = 1200(1.10^x)$

e. Sketch a graph of this function.

f. By 1993 the fees totaled $3276. Locate this point in relation to your graphs. Is your model reasonable? Explain. *The fees are above the expected amount, which implies they are growing faster than 10% per year.*

FX-76. In 1993, the state legislature passed a law allowing the cost of fees for the state university system to increase by 5 percent per year without approval by the public or the legislature (larger increases have to be approved). Annual fees for the university were $1000 in 1993. The Smiths' baby girl was born in 1993, and they wanted to plan for her college education. They figured that, when she was 18, their daughter would be attending one of the state universities. Since 5 percent of $1000 is $50, they figured the increase in fees in 18 years would be 18 times $50, or $900; therefore, when Kelly started college, the annual fees would be $1900. Explain to the Smiths why they may need to save more than they are anticipating. How much could the actual cost be in 18 years? *The cost in 18 years could be $2400 per year, assuming no additional increases by the state legislature.*

FX-77. According to the U.S. Census Bureau, the population in the United States has been growing at an average rate of approximately 2 percent per year. The census is taken every 10 years, and the population in 1980 was estimated at 226 million people

a. How many people should the Census Bureau have expected to count in the 1990 census? *275.5 million*

b. How many people should the Census Bureau expect to count in the year 2000? *336 million*

c. If the rate of population growth in the United States continues to stay at about 2 percent, in about what year will the population in the United States reach and surpass one billion? *during the year 2055*

FX-78. If you purchase an item that costs $10.00 now and inflation continues at 4 percent per year (compounded yearly), when will the cost double? Show how you would compute this cost. Remember, you can use your calculator to guess and check, or you can use the graph, once you set up an equation. *$2 = 1.04^x$; about 18 years as solved by guess and check*

EXTENSION AND PRACTICE

FX-79. Solve each of the following equations for x. Notice the variety in these equations! Each one requires some different thinking. The purpose of these problems is to give you practice solving a variety of exponential equations and deciding which is which and what to do in each case.

a. $2^{x-1} = 64$ *$x = 7$* **b.** $4.7 = x^{1/3}$ *$x = 103.82$*

c. $8^{x+3} = 16^x$ *$x = 9$* **d.** $9^3 = 27^{2x-1}$ *$x = 1.5$*

e. $x^6 = 29$ *$x \approx \pm 1.75$* **f.** $25^x = 125$ *$x = 3/2$*

FX-80. Rewrite $(16)^{3/4}$ in at least four different-looking, but equivalent forms. *8, 2^3, $(16^3)^{1/4}$, $(16^{1/4})^3$, and several $\sqrt{\ }$ forms*

FX-81. Write and describe a situation that could fit the following function:

$$g(t) = 5.00(1.16^t)$$

FX-82. For each of the following problems, find the initial value.

a. A bond that appreciates at 4 percent per year will be worth \$146 five years from now. *\$120*

b. Dr. Hooper's car, which depreciates at 20 percent per year, will be worth \$500 seventeen years from now. *\$22,204*

FX-83. Factor each expression. Parts c and d require two steps. The example is a reminder of how to factor quadratic expressions.

Example: $2x^2 - 8x - 42 = 2(x^2 - 4x - 21) = 2(x - 7)(x + 3)$

a. $x^2 + 8x$ *$x(x + 8)$*

b. $6x^2 + 48x$ *$6x(x + 8)$*

c. $2x^2 + 14x - 16$ *$2(x + 8)(x - 1)$*

d. $2x^3 - 128x$ *$2x(x + 8)(x - 8)$*

FX-84. Consider the following sequence:

n	0	1	2	3	4
$t(n)$	10	5	2.5	1.25	0.625

a. Predict the next three terms. *0.3125, 0.15625, 0.078125*

b. Graph this sequence. What type of a sequence is it? *geometric*

c. Is your graph discrete or continuous? *Discrete; it is a sequence.*

FX-85. In order to check the solution of the following equation Jonathon substituted the x-value back into the original equation, but the result didn't make sense.

$$\frac{x-3}{4} = \frac{x+2}{5}$$

$$5x - 3 = 4x + 2$$

$$x = 5$$

a. Explain to him where he made his error, and show him how to solve for x correctly. *He did not use the distributive property correctly; $x = 23$.*

b. Jonathon then tells you his instructor showed the students how to arrive at the solution by using a graphical approach, but he can't recall how to do it. Write a brief explanation of how he could use two graphs to solve the problem. *Graph each side of the equation, and find the point of intersection of the two expressions. The x-coordinate of the point of intersection represents the solution.*

FX-86. When we say "$x = 3$" are we talking about a point or a line? Explain. ***It depends on the context; $x = 3$ could represent a point on a number line, a line in a plane, or a plane in space.***

*FX-87 is just for diagnostic purposes at this point. We will deal extensively with inequalities in Chapter 6. Now is **not** the time to stop and teach the graphing of inequalities. Do not assign FX-87 if you know your students don't know inequalities.*

FX-87. Graph each of the following inequalities. For example, in part a you can find the "edge" of the graph by graphing the line $y = 2x + 3$; then you'll need to decide what region to shade (above or below the edge). This problem is just to see what you know about inequalities. We will learn about them in Chapter 5.

a. $y < 2x + 3$ **b.** $y \geq -3x - 1$

FX-88. Solve the following equations for x.

a. $2x^2 + 6x - 7 = 0$ *$x = (-3 \pm \sqrt{23})/2$; or 0.898, −3.898*

b. $\frac{5}{x} - \frac{x}{3} = 8$ *$x \approx 0.610, -24.610$*

FX-89. On Lynn's last math exam, she had to graph the equations

$$y = \frac{5}{3}x - 2 \qquad \text{and} \qquad 6x + 3y = 16$$

to find their common solution. Lynn is not as fortunate as you. Since her instructor does not emphasize graphing, she doesn't know how to graph these equations. She must guess at the point of intersection from the three choices given:

$$\left(2, \frac{4}{3}\right) \quad \left(\frac{18}{11}, \frac{8}{11}\right) \quad \left(\frac{20}{11}, \frac{5}{3}\right)$$

a. What is the probability she chooses the correct answer? *1/3*

b. Graph the two equations for Lynn, and write the coordinates of the point of intersection. *(2, 4/3)*

c. Tell her how to find the point of intersection *without* graphing. Explain the advantages of this method.

4.6 FAST CARS — THE THEME PROBLEM

FX-90 is the same as FX-1. If you are keeping a portfolio, this problem would be a good assignment for it because FX-90 incorporates the majority of the important ideas from this chapter. (You may want to review the format for portfolio problems, which is discussed in Chapter 1.) Graphing calculators are useful to check the answers in Laps 3 and 5.

FX-90. Fast Cars As soon as you drive a new car off the dealer's lot, the car is worth less than what you paid for it. This situation is called **depreciation**, and it means you will sell the car for less than the price you paid for it. Most cars depreciate, some more than others (that is, at different rates). On the other hand, some classic older cars actually increase in value. This situation is called **appreciation**. Suppose you have a choice between buying a 1993 Mazda Miata for $17,000 that depreciates at 22 percent a year, or a used 1990 Honda CRX for $12,000 that depreciates at 18 percent a year. In how many years will their values be the same? Should you instead buy a 1967 Ford Mustang for $4000 that appreciates at 10 percent per year? Which car will have the greatest value in 4 years? in 5 years?

To help you solve this problem we have divided it into subproblems, this time identified as Laps since it's about cars.

Lap 1: What is the multiplier for the Miata? for the Honda? for the Mustang? *0.78, 0.82, 1.1*

Lap 2: Make a table like the one shown below, and calculate each car's value for each year.

Year	Mazda Miata	Honda CRX	Mustang
0	$17,000	$12,000	*4,000*
1	$13,260	$9,840	*4,400*
2	*10,343*	*8,069*	*4,840*
3	*8,067*	*6,616*	*5,324*
4	*6,293*	*5,425*	*5,856*
5	*4,908*	*4,449*	*6,442*
⋮			
10	*1,417*	*1,649*	*10,375*
⋮			
n	$17{,}000(0.78^n)$	$12{,}000(0.82^n)$	$4{,}000(1.10^n)$

Lap 3: Write two functions to represent the depreciation of the Miata and the CRX, and draw the graphs on the same set of axes. Are the graphs linear? How are they similar? How are they different?

Lap 4: When will the values of the two cars be the same?

Lap 5: Write an equation, and graph the results for the Mustang on the same axes with graphs for the Miata and the CRX.

For the Checkered Flag: Using the graph, which of the three cars—the Mazda, the Mustang, or the CRX—is worth the most after 4 years? after 5 years? after 10 years?

Victory Lap (Evaluation): After doing this problem what have you learned? Has this problem changed your view of buying cars? Pick one of the three cars, and explain why you would buy it.

Winner's Circle: Be sure you have included good definitions and examples of depreciation and appreciation in your Tool Kit.

As students are finishing FX-91, you may want to ask if the value of a car could be negative. This question could lead to a discussion of the limitations of mathematical models and the importance of considering the domain for which they make sense.

FX-91. The half-life concept applies to other situations besides radioactivity. It can apply to practically anything that is depreciating or decaying.

a. From the table of values in FX-90, estimate the half-life of the value of the Mazda and the Honda. *Mazda ≈ 3 years; Honda ≈ 3.5 years*

b. According to the mathematical model (though not necessarily corresponding to reality), when will each car have *no* value? *never*

Go back and check on how you solved parts b and e of FX-79. Be sure your group knows how to do these because you will need this idea today for FX-92 and FX-93.

FX-92. Find the annual rate of growth on an account that was worth $1000 in 1990 and $1400 in 1993. *$m = 1.118688942$, or approximately 11.9% growth*

FX-93. Find the monthly rate of decay on a radioactive sample that weighed 100 grams in May and 50 grams in November. *$m = 0.8908987$, so about 11% decay*

If students use the graphing calculator on FX-94, they may need to be reminded of how to use two graphs at once. For now it is valuable to have them estimate the intersection points from their graphs rather than use a calculator's intersect feature, so they can concentrate on the reasoning rather than the button sequence.

FX-94. Graph the following system of equations. In the first graph be sure to include a value between 0 and 1 and a value between –1 and 0.

$$y = x^3 \qquad \text{and} \qquad y = x$$

a. How many times do these functions cross each other? *3*

b. What are the coordinates of their intersections? *(1, 1), (–1, –1), and (0, 0)*

c. Solve the equation $x^3 = x$ for x. $x = 0, 1, \text{ or } -1$

EXTENSION AND PRACTICE

FX-95. Here are some more exponent puzzles. Solve each of the following equations for x.

a. $25^{x+1} = 125^x$ — $x = 2$

b. $8^x = 2^5 \cdot 4^4$ — $x = 13/3$

c. $27^{x/2} = 81$ — $x = 8/3$

FX-96. Factor each expression into three factors. Remember, this factoring takes two steps.

a. $2x^2 + 8x + 8$ — $2(x + 2)(x + 2)$

b. $6x^2 - 6x - 72$ — $6(x + 3)(x - 4)$

FX-97. To what power do you have to raise the given base to get the given result?

Example: 2, to get 128 *Solution*: 7, because $2^7 = 128$

a. 3, to get 27 *3*

b. 2, to get 32 *5*

c. 5, to get 625 *4*

d. 64, to get 8 *1/2*

e. 81, to get 3 *1/4*

f. 64, to get 2 *1/6*

g. x^3, to get x^1 *1/3*

h. x^3, to get x^{12} *4*

i. x, to get x^a *a*

FX-98. If $f(x) = 3(2)^x$, find:

a. $f(-1)$ *3/2*

b. $f(0)$ *3*

c. $f(1)$ *6*

d. What value of x gives $f(x) = 12$? *2*

e. Where does the graph of this equation cross the x-axis? the y-axis? *never, 3*

FX-99. Sketch a graph for each of the following situations:

a. the temperature of a hot cup of coffee left sitting in a room over a long period of time

b. the relationship between the amount of money you have in your savings account, and time

FX-100. Solve each of the following equations for x. Remember, you can solve the equation graphically, algebraically, or using guess and check.

a. $81 = 3^{2x}$ *$x = 2$* **b.** $x^5 = 243$ *$x = 3$* **c.** $(2x)^3 = -216$ *$x = -3$*

FX-101. Rewrite each of the following equations in y-form so that they are in the correct format for use on a graphing calculator.

a. $2x - 3y = 7$ *$y = (2x-7)/3$*

b. $2(x + y) = x - 4$ *$y = \left(-\frac{1}{2}\right)x - 2$*

FX-102. Solve each of the following equations for x:

a. $\frac{x+3}{x} = \frac{4}{5}$ *$x = -15$* **b.** $\frac{5}{x} = \frac{x}{10}$ *$x = \pm 5\sqrt{2}$*

FX-103. Solve the following systems of equations.

a. $2x + y = -7y$
$y = x + 10$ *(−8, 2)*

b. $3s = -5t$
$6s - 7t = 17$ *(5/3, −1)*

FX-104. Sketch the graphs of three different parabolas that have x-intercepts at (4, 0) and (8, 0). What can you tell about the coordinates of the vertex of each parabola? *The x-coordinate is 6.*

You may want to use FX-105 as an in-class group activity.

FX-105. Solve the following system of equations for D, E, and F. Think of it as a puzzle. What substitutions can you make?

$$F = -5ED^2$$
$$D = 3$$
$$6E = 2F - 32$$

$D = 3$, $F = 15$, $E = -1/3$

FX-106. The area of square A is 121 square units, the perimeter of square B is 80 units. Find the area of square C. *$31^2 = 961$*

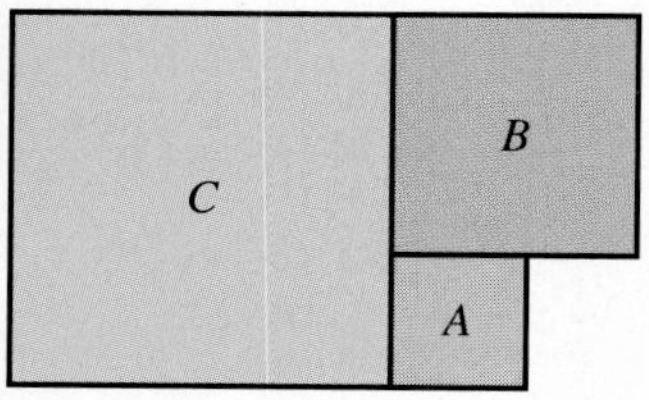

4.7 EXPONENTIAL INVESTIGATION AND SUMMARY

One purpose of FX-107 is to have students explore the amount of change in steepness of the curves when a is positive. Graphing calculators will get them started. Another purpose is to have them see that funny things happen when a is negative, and they may need to add a restriction to their Tool Kit definition for ***exponential function****. Remind them that this is a family of functions to investigate rather than just one. If you want the students to explore only the changes in steepness of the curves, have them focus on a > 0.*

Poster problems: *One suggestion to enhance the outcome of the investigations in FX-107 and 108 is to have each group prepare a poster to show their results. These posters can be done quickly with big colored pens on butcher paper so that the results can be put up and discussed, assessed, and revised on the spot. They also could be project posters used for assessment (after revision and discussion of the posters).*

FX-107. Investigate the family of functions $y = a^x$ for different values of a. Be sure that you tried enough different values of a.

In FX-108 we investigate the function $y = ax^2$ *to prepare for PG-2 in Chapter 5.*

FX-108. Investigate the family of functions $y = ax^2$. Be sure to use both negative and positive numbers for a. You'll need the results of this investigation for the beginning of Chapter 5.

FX-109. Summary In this chapter you have worked with exponential functions in many contexts. Your responses for the situation you create in part a should convince your group members and your instructor that you can apply what you have learned.

a. Make up a real situation that can be represented by an exponential function.

b. Give the equation that describes the situation, and tell how you arrived at the equation.

c. Draw the graph of the equation. Be sure to include all important information about the graph. How did you decide whether the points on the graph should be connected or not?

d. Make up a question that can be solved by using your equation or your graph. Clearly show how to answer the question.

FX-110. Self-Evaluation Write at least two paragraphs in which you provide an honest evaluation of "where you are" in the course so far. Consider the following:

a. Your progress through the material to date: Are you keeping up and understanding the material? What areas do you still feel weak in? What confuses you?

b. Your participation in your group's success: What roles have you assumed, and how has group problem solving affected your view of mathematics in general, this course in particular, or both?

EXTENSION AND PRACTICE

FX-111. ***Tool Kit Check*** Now would be a good time to review your Tool Kit and make sure you have included any new ideas about exponents that you learned in this chapter. When you have included new rules (such as $b^{1/x} = \sqrt[x]{b}$) that were based on patterns, write an explanation of why the rule should be true, and discuss whether it is true for all possible choices of the variables or whether there are some limitations. Also, include examples that will help you understand and remember what each topic means. Your Tool Kit should include at least the following items:

- zero power
- multiplier or base
- initial value
- time
- exponential function
- negative exponents
- fractional exponents
- roots—square roots, cube roots
- $b^{1/x} = \sqrt[x]{b}$
- $\sqrt[n]{b^m} = \left(\sqrt[n]{b}\right)^m = b^{m/n}$

FX-112 would make a good portfolio item.

FX-112. Many algebra students think that $(2)^{-2} = -4$. However, you know that

$$(2)^{-2} = \frac{1}{2^2} = \frac{1}{4}.$$

Explain, so that a beginning algebra student can understand, why.

$$(2)^{-2} = \frac{1}{4}$$

FX-113. Jonnique is still confused about the meaning of negative and fractional exponents, and now she has five problems to do that all look the same to her. In each case, explain in writing what the exponent is telling her to do; then show her the result.

a. $25^{1/2}$ *5* **b.** $(25)^{-2}$ $\frac{1}{625}$ **c.** $(25)^{-1/2}$ $\frac{1}{5}$

d. $(-25)^2$ *625* **e.** $(-25)^{1/2}$ *no real number*

FX-114. A sequence is given by $t(n) = 2(3)^n$.

a. What are $t(0)$, $t(1)$, $t(2)$, $t(3)$? *2, 6, 18, 54*

b. Graph this sequence. What are the domain and range?

c. On a separate set of axes, graph the function $f(x) = 2(3)^x$.

d. How are these graphs similar? How are they different?

FX-115. Solve the following equations for x. You might need to guess and check.

a. $1^x = 5$ *no solution* **b.** $2^x = 9$ *≈3.17*

FX-116. Solve each of the following for x and y where x and y are whole numbers. The fact that the prime factorizations of both 72 and 24 are all 2s and 3s should help.

a. $2^x \cdot 3^y = 72$ *$x = 3$, $y = 2$* **b.** $2^{x+y} \cdot 3^{x-y} = 24$ *$x = 2$, $y = 1$*

FX-117. Write each of the following expressions in an equivalent form that does not use negative or fractional exponents. Then give the value of the expression. You should be able to do all of these problems without your calculator!

a. $(5)^{-2}$ *$1/5^2 = 1/25$* **b.** $(4)^{-3}$ *$1/4^3 = 1/64$*

c. $9^{1/2}$ *$\sqrt{9} = 3$* **d.** $64^{2/3}$ *$\left(\sqrt[3]{64}\right)^2 = 16$*

FX-118. You have $5000 in an account that pays 12 percent annual interest. Compute the amount in the account at the end of one year if:

a. the interest is paid annually (once) *5600*

b. the interest is paid quarterly *5627.54*

c. the interest is paid monthly *5634.13*

FX-119. Find an equation of depreciation for a Pacinslosh Computer in the form $f(t) = km^t$ if the computer initially cost $3000 but three years later is worth only $500. *$f(t) = 3000(0.55)^t$*

FX-120. Rewrite each equation in y-form, and then find the point(s) of intersection of their graphs.

$$2x - 5y = 10 \quad \text{and} \quad 4(x - y) + 12 = 2x - 4$$

(–60, –26)

FX-121. Draw a quick sketch of the relationship between your test scores and the percentage of the assignments you complete.

FX-122. Find the coordinates of the x- and y-intercepts of the graph of $y = x^2 + 14x + 13$. *x: (–1, 0), (–13, 0); y: (0, 13)*

EXTRA PROBLEMS

FX-123. Two weeks ago a sample of bacteria weighed 4.2 grams. Last week it weighed 4.326 grams.

a. What is the multiplier? *1.03*

b. What is the rate of growth? *3%, for a multiplier of 1.03*

c. What is the weight of the sample *now*? *4.45578 g*

FX-124. Last month, Portia's car was worth $28,000. Next month it will be worth $25,270.

a. What is the rate of depreciation? *drop 5% per month; multiplier = 0.95*

b. What is her car worth now? *$26,600*

FX-125. Consider the following sequence:

n	0	1	2	3	4
$t(n)$	10	5	2	1	2

a. Graph this sequence. Is it arithmetic or geometric? *neither*

b. Is your graph discrete or continuous? Justify your choice. *Discrete—it is a sequence.*

c. Predict the next three terms. *5, 10, 17*

FX-126. Jonnique is writing a puzzle problem. She wants the values for x and y in the second equation to be the same as in the first. She originally wanted the values to be whole numbers so they could be guessed and checked, but the whole numbers she has tried don't work. Show a method, other than guess and check, for figuring out what numbers will work.

Jonnique's equations: $2^x \cdot 2^y = 64$ and $2^{3x} = 16^y$

(24/7, 18/7)

FX-127. Show two steps to calculate each of the following expressions *without* a calculator. Be careful: the order in which you use the information in the exponent can make the problem easy or ugly.

a. $8^{2/3}$ *4* **b.** $25^{3/2}$ *125* **c.** $81^{5/4}$ *243*

CHAPTER FIVE

THE GATEWAY ARCH

Parabolas and Other Parent Graphs

IN CHAPTER 5 YOU WILL HAVE THE OPPORTUNITY TO:

- become very familiar with an extended set of parent functions and their graphs (including the parabola, the square-root curve, a cubic curve, and a hyperbola);
- relate the numbers in an equation to the location and stretch of its parent graph;
- develop your ability to visualize graphs when you know their equations and to write equations when you know their graphs;
- develop your ability to ask questions to generate more information and to justify and explain your reasoning.

YOU WILL USE YOUR GRAPHING CALCULATOR DAILY AS A TOOL TO HELP YOU EXPERIMENT, ORGANIZE DATA, AND RECOGNIZE PATTERNS. AS IN A SCIENCE COURSE, YOU WILL WRITE UP YOUR OBSERVATIONS AND CONCLUSIONS IN LAB REPORTS.

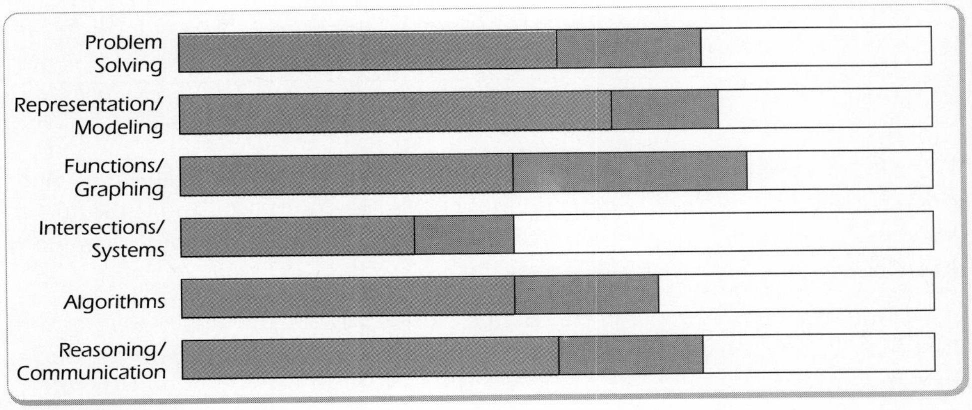

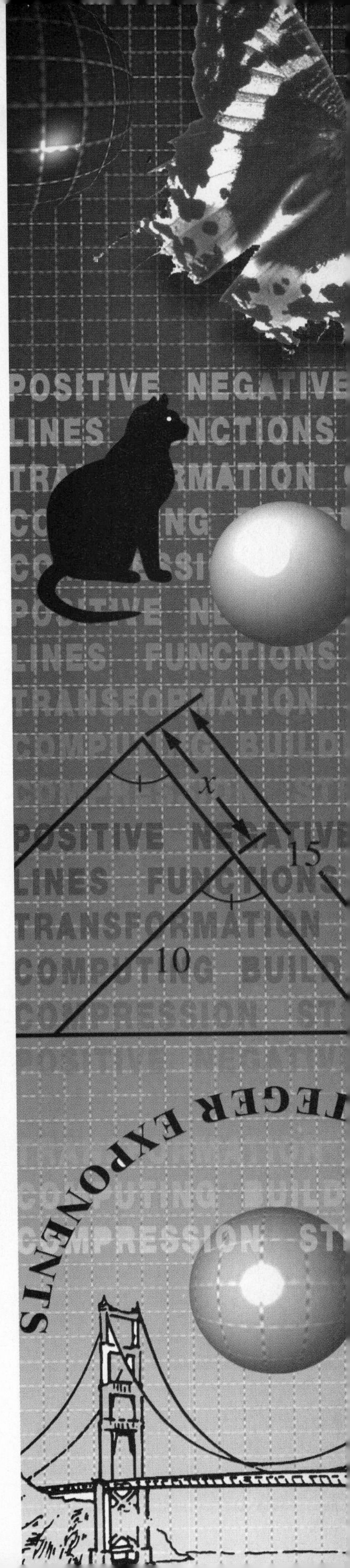

The first four sections of this chapter focus on parabolas. Sections 5.5 and 5.6 generalize what students have discovered about the location and orientation of parabolas to other functions, and Section 5.7 develops a method for determining the stretch factor. Students identify the simplest version of each function to create a Parent Graph Tool Kit. Section 5.8 adds one form of rational function to the set of parent graphs. This section can be optional. In Section 5.9 students use the ideas developed in the chapter to solve the theme problem, which is followed by other summarizing activities.

5.1 THE SHRINKING TARGETS LAB

We will return to PG-1 later in the chapter. For now, we expect students only to figure out how to flip the graph of $y = x^2$ upside down, not how to shift it or stretch it appropriately. The Parabola Investigation in PG-15 will deal with these issues.

Note: the arch is a catenary, better modeled with the hyperbolic cosine function, but for our purposes it provides an interesting example that is roughly a parabola.

The Gateway Arch in PG-1 is the theme problem for this chapter. For now, just read it and answer parts a and b. We will return to solve part c later in the chapter.

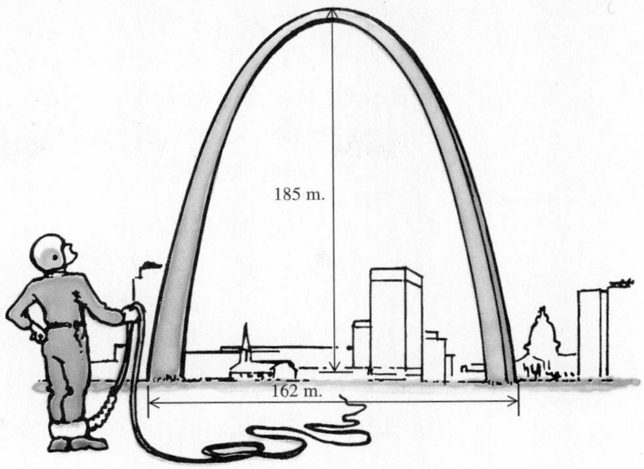

PG-1. The Gateway Arch Would you ever go bungee jumping? Beginners jump off bridges over water, or off cranes over a big air cushion. But true daredevils (some call them crazy) dream of jumping off the Gateway Arch in St. Louis, Missouri. The Arch has no soft landing area in case the bungee cord breaks, and no one has ever done this, so we are not recommending it. But analyzing the mechanics of such a feat makes for a challenging problem.

The most daring jumper might jump off the very top of the Arch, but this is 185 meters high. Jumpers who are more sensible might prefer to start at a lower point on the Arch. They would want a rule telling them how high the Arch is at *any* point, so they'll know how long a bungee cord to use.

a. What kind of curve could be used to approximate the Arch? That is, what kind of equation has a graph with a shape like this? *parabola, quadratic*

b. The Arch is approximately the shape of an upside-down parabola. (*Note*: A better model is a catenary or hyperbolic cosine, but we can approximate it with a parabola.) What is the simplest equation for a parabola that you know of? $y = x^2$

c. Think back to when you investigated $y = ax^2$ in FX-108 in Chapter 4. How can you change the simplest equation for a parabola from part b to get an equation to represent the Arch? You may not yet have the information you need to answer. We do not expect you to solve it completely right now, but discuss your ideas with your group and be ready to share them with the class in a few minutes. We will return to this problem at the end of the chapter.

For the investigation in PG-2, you will again need scales, cardboard, and compasses. Old file folders work as well as the cardboard. Folders are easily weighed on a triple-beam balance scale. Photocopying concentric circles onto tag board can save some time in class. Make it clear to the students that they will be using their data on other days so they must keep their results organized and handy. Students can use the same lab report format found in the Resource Pages for the Sharpening Pencils Lab in Chapter 2 (EF-1 to EF-7).

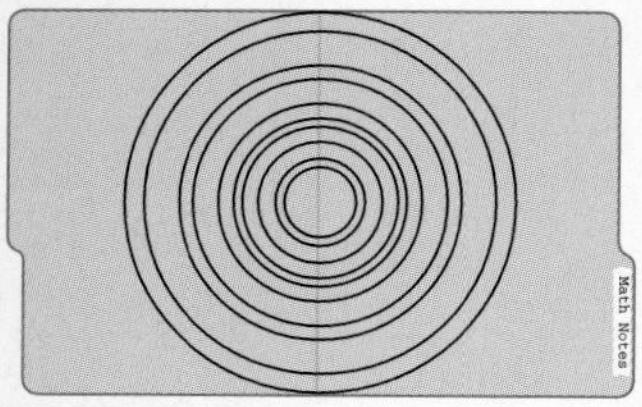

PG-2. The Shrinking Targets Lab (PG-2 to PG-4) On a sheet of cardboard or an old file folder, find and mark the *exact* center. With a compass, make the largest circle you can on the cardboard, and cut it out. Draw about nine more concentric circles on this circular disk (*not* equally spaced), but *don't* cut them out yet!

a. Set up a table with headings, "Length of Radius (cm)" and "Mass of Disk (g)." Record the length of the radius of each circle in the first column. (You are going to cut out each circle starting from the outside, and weigh the disk of cardboard each time, recording the information in a table. When your table is complete, you will graph the results.)

b. Before you graph the results, what do you think the graph will look like?

c. Carefully graph your results and sketch in the smooth curve that seems to fit the points best (a **"best-fit" curve**).

d. What kind of curve does your graph seem to be? What kind of equation will it have? Save this graph. We will return to it later to determine its equation more accurately. *half of a parabola.*

e. Predict the mass of a circle with a radius twice as large as your largest circle. Explain how you figured this out. *It should be four times the mass of their largest circle.*

PG-3. For the data shown in your table in part a of PG-2, what is the domain of x-values? (That is, what are acceptable lengths for the radius?) What is the range for y-values? *x-values: 0 to the largest radius possible on cardboard; y-values: 0 to mass of largest circle*

PG-4. Does your graph in part c of PG-2 have x- or y-intercepts? If so, what are they and what do they represent? If not, explain completely why not. *Students should mention that when the length of the radius is zero, the mass is zero.*

EXTENSION AND PRACTICE

PG-5. Your results from this problem will be useful in the Parabola Investigation, which you will do in PG-15 and PG-16.

a. Draw the graph of $y = (x - 2)^2$. If you are doing the graph by hand, be sure to use at least the values from -1 through 5 for x.

b. How is this graph different from the graph of $y = x^2$? What difference in the equation accounts for the difference in the graphs? *shifted to the right 2 units.*

c. Based on your observations in part b, write an equation for a graph that "sits on" the x-axis at the point (5, 0).

PG-6. Consider the sequence with the initial value 256, followed by the terms 64, 16, . . .

a. Write the next three terms of this sequence. Then find a rule for the sequence. *4, 1, 0.25, $t(n) = 256(0.25^n)$*

b. If you were to continue writing terms of this sequence, what would happen to the terms? *The terms would get smaller and smaller.*

c. Sketch a graph of the sequence. What happens to the points as you go farther to the right? *They get closer to zero.*

PG-7. Write the equation for each graph.

a.

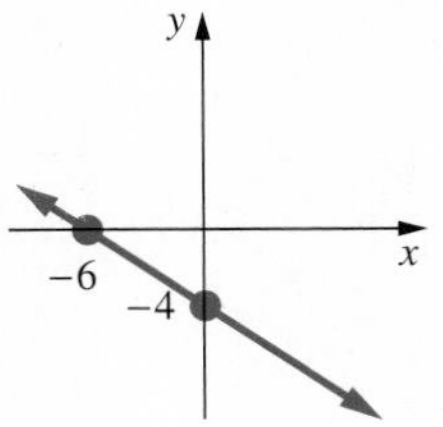

$y = (-2/3)x - 4$

b.

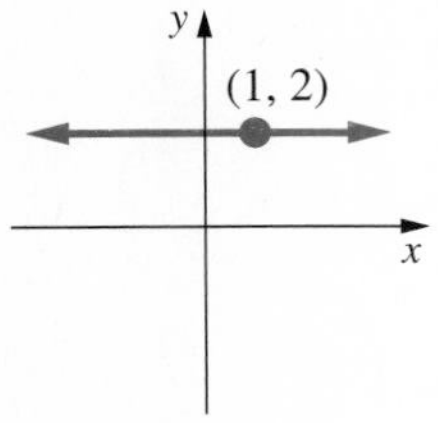

$y = 2$

c.

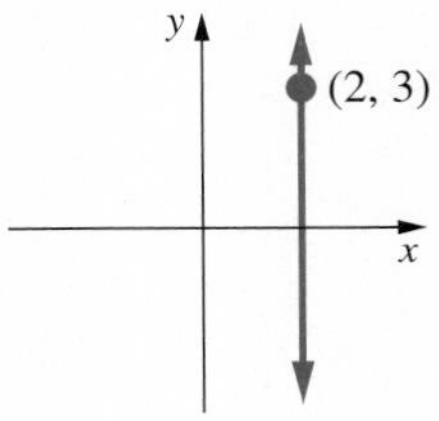

$x = 2$

d.

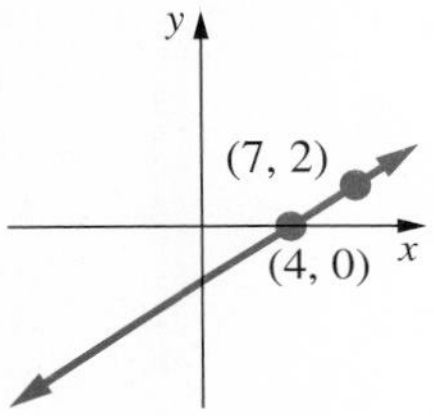

$y = (2/3)x - 8/3$

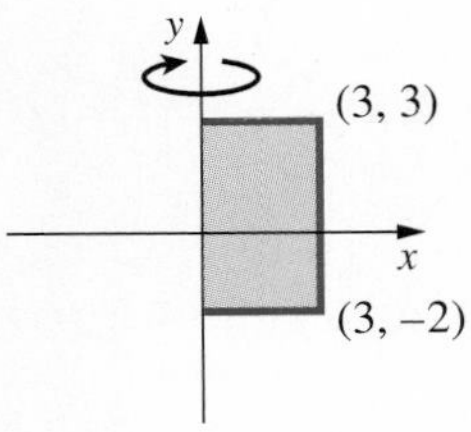

PG-8. Imagine spinning the rectangle in the diagram about the y-axis.

a. Draw the shape you'd get. *a cylinder*

b. Find the volume using the formula $V = \pi r^2 h$. *$45\pi \approx 141.37$*

PG-9. Remember function machines? Each of the following pictures shows how the same machine changes the given x-value into a corresponding $f(x)$ value. Find the rule for this machine. *$f(x) = x^2 + 1$*

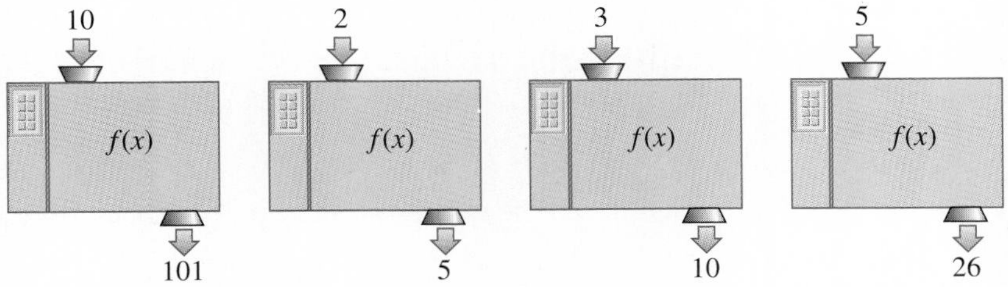

PG-10. Factor each of the following expressions:

a. $4x^2 - 9y^2$
$(2x - 3y)(2x + 3y)$

b. $8x^3 - 2x^7$
$2x^3(2 + x^2)(2 - x^2)$

c. $x^4 - 81y^4$
$(x^2 + 9y^2)(x - 3y)(x + 3y)$

d. $8x^3 + 2x^7$
$2x^3(4 + x^4)$

PG-11 previews the concept of line symmetry, which students will use in Section 5.3.

PG-11. What is a **line of symmetry**? The figure shows an example of one line of symmetry.

a. Draw something that has a line of symmetry.

b. Draw something that has *two* lines of symmetry.

c. What basic geometric shape has an *infinite number* of lines of symmetry? *a circle*

PG-12. Consider the two sequences $t(n) = 3n$ and $s(n) = 2^n$.

a. Fill in the table for each sequence.

n	$t(n)$
0	*0*
1	*3*
2	*6*
3	*9*

n	$s(n)$
0	*1*
1	*2*
2	*4*
3	*8*

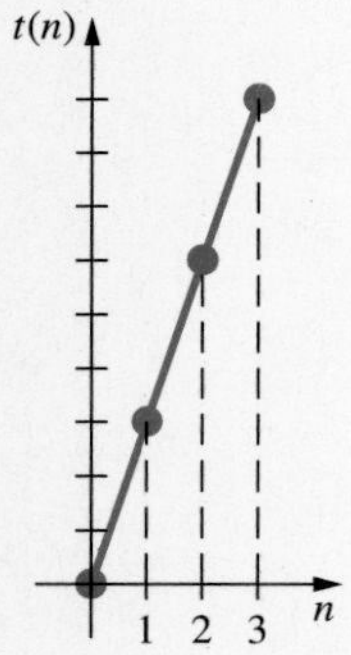

b. Roger graphed the sequence $t(n)$, shown here. He noticed he could form a sequence of right triangles that increase in size and all touch the point (0, 0). Roger thinks the triangles are similar. Is he right? Justify your answer. *Yes. The points lie on a line (since the sequence is arithmetic), so they have a constant slope. Thus the ratio of the sides of the triangles is the slope.*

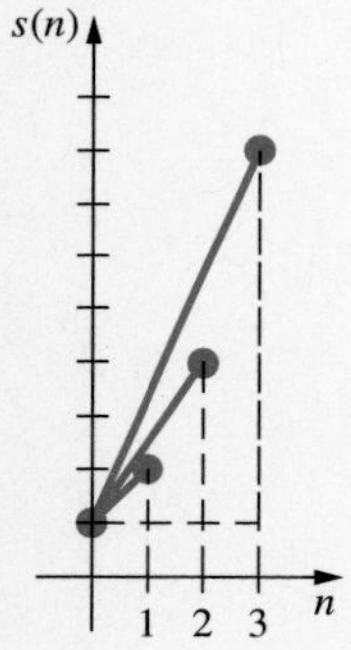

c. Nicola graphed $s(n)$ and drew in triangles too. Are her triangles similar? Explain. *No. The points aren't collinear, so the slope and therefore the ratios of the sides vary.*

PG-13. Find the point where the graph of $y = 3x - 1$ intersects the graph of $2y + 5x = 53$. *(5, 14)*

PG-14. Right now it probably costs an average of $150 each month for your food.

a. Five years from now, how much will you be spending on food each month if you're eating about the same amount and inflation runs at about 4 percent per year? *about $182.50*

b. Write an equation that represents your monthly food bill x years from now if both the rate of inflation and your eating habits stay the same. $y = 150(1.04)^x$

c. Use the equation from part b to calculate your monthly food bill in the year 2020. *The answer depends on the year in which this problem is done.*

5.2 THE PARABOLA INVESTIGATION

The Parabola Investigation in PG-15 and PG-16 is central to the rest of the course. You should encourage students to struggle and discover on their own. Although it may seem to take forever, letting them discover for themselves now will save lots of reteaching and frustration later. Encourage experimentation and testing of conjectures. Let students share their discoveries across groups if necessary. As you circulate you may need to remind students to take good notes because the relationships they see now will lead to some important Tool Kit entries later.

*You may want to bring a rubber band to class to help students visualize what **stretching** and **compressing** mean.*

PG-15. The Parabola Investigation (PG-15 and PG-16) For this investigation we have not provided a worksheet because we want you to gain some practice in efficiently keeping and effectively organizing your notes. This will help you recognize patterns and describe them. As you attempt to answer each question, be sure to keep track of all the equations you try and their results. An equation that doesn't work in one case may be the key to a question that comes later.

a. Graph the parabola $y = x^2$. Make an accurate sketch of the graph. Be sure to label any important points on your graph. In addition to x- and y-intercepts be sure to label the lowest point, which is called the **vertex**.

b. Find a way to change the equation to make the same parabola *open downward*. The new parabola should be congruent to (the same shape and size as) $y = x^2$, with the same vertex, except it should open downward so that its vertex will be its highest point.

Record the equations you tried along with their results. Write down the results even when they were wrong—they may come in handy later.

c. Find a way to change the equation to make the $y = x^2$ parabola *stretch vertically* (it will appear narrower). The new parabola should have the same vertex and orientation (that is, opening up) as $y = x^2$.

Record the equations you tried, along with their results and your observations.

d. Find a way to change the equation to make the $y = x^2$ parabola *compress vertically* (it will appear wider).

Record the equations you tried, their results, and your observations.

e. Find a way to change the equation to make the $y = x^2$ parabola *move 5 units down*. That is, your new parabola should look exactly like $y = x^2$ but with the vertex at $(0, -5)$.

Record the equations you tried, along with their results. Include a comment on moving the graph up as well as down.

f. Find a way to change the equation to make the $y = x^2$ parabola *move 3 units to the right*. That is, your new parabola should look exactly like $y = x^2$, except that the vertex should be at the point $(3, 0)$. If you need an idea to get started, look back to PG-5.

Record the equations you tried, along with their results. Comment on how to move the parabola to the left as well as how to move it to the right.

g. Finally, find a way to change the equation to make the $y = x^2$ parabola *vertically compressed (wider), open down, move 6 units up, and move 2 units to the left*. Where is the vertex of your new parabola?

Record the equations you tried, their results, and your comments on how each part of the equation affects its graph.

PG-16. For each equation below, predict (i) the vertex and (ii) the orientation (open up or down?), and (iii) tell whether it is the same as a vertical stretch or a compression of $y = x^2$. (iv) Sketch a quick graph based on your predictions.

a. $y = (x + 9)^2$ **b.** $y = x^2 + 7$ **c.** $y = 3x^2$

d. $y = \frac{1}{3}(x - 1)^2$ **e.** $y = -(x - 7)^2 + 6$ **f.** $y = \frac{5}{2}(x - 2)^2 + 1$

g. $y = 2(x + 3)^2 - 8$ **h.** $4y = -4x^2$ **i.** $y = 4x - 4$ *a line!*

j. Check your predictions for the equations in parts a through i on your graphing calculator. If you made any mistakes, correct them and briefly describe why you made the mistake (what incorrect idea you had). Then make a neat and accurate graph for each function.

PG-17. Take out your Tool Kit. Write some notes for yourself on what changes in the basic equation $y = x^2$ will move, flip, stretch, or compress the basic parabola.

Students' Tool Kit entries at this point will not be complete or thorough. Don't push for thoroughness yet. That should come in a few days. For now, take note of which relationships they ***do*** *see.*

EXTENSION AND PRACTICE

PG-18. Your friend is taking an algebra class at a different school, where she is not allowed to use a graphing calculator. Explain to her how she can get a good sketch of the graph of the following function without a calculator *and* without having to substitute a lot of different numbers for x and do a lot of calculations.

$$y = 2(x + 3)^2 - 13$$

a. Be sure to explain how to locate the vertex, whether the parabola should open up or down, and how its shape is related to the shape of $y = x^2$.

b. Your friend also needs to know the x- and y-intercepts. Show her how to get those without having to draw an accurate graph or use a graphing calculator. (–5, 0), (–1, 0), (0, 5)

PG-19. Solve for z:

a. $4^{2z/3} = 8^{z+2}$ *–18/5* **b.** $5^{(z+1)/3} = 25^{z/12}$ *–2*

PG-20. Lettie just got her driver's license, and her friends have nicknamed her "Leadfoot" because she always does at least 80 mph (miles per hour) on the freeway.

a. At this speed, how long will it take her to travel 50 miles? ***0.625*** *hr or 37.5 min*

b. How long would it take her if she traveled the 50 miles at 65 mph? ***0.77*** *hr or 46.2 min*

c. Speeding tickets carry fines of about \$200. If Lettie gets a ticket on this 50-mile trip, what is her cost per minute of time saved? *\$23.53 per min*

PG-21. Daniela, Kieu, and Duyen decide to go the movies one hot summer afternoon. The theater is having a summer special: "Three Go Free" (if they each buy a large popcorn and a large soft drink). They take the deal and end up spending \$19.50. The next week, they go back again, only this time, they each pay \$2.50 to get in, they each get a large soft drink, but they share one large bucket of popcorn. This return trip also costs them a total of \$19.50.

a. Find the price of a large soft drink and the price of a large bucket of popcorn. *popcorn, \$3.75; soft drink, \$2.75*

b. Did you write two equations, or did you use another method? If you used another method, write two equations now and solve them. If you already used a system of equations, skip this part.

PG-22. Multiply these expressions:

a. $2x^3(3x + 4x^2y)$ $\quad 6x^4 + 8x^5y$

b. $(x^3y^2)^4(x^2y)$ $\quad x^{14}y^9$

There are several subproblems to do in part c before multiplying. Think: "factor, factor, factor."

c. $\dfrac{x^2 - 9}{2x} \cdot \dfrac{x^2 + x}{x^2 - 2x - 3}$ $\quad \dfrac{x+3}{2}$

PG-23. Solve for x: $ax + by^3 = c + 7$ $\quad x = \dfrac{-by^3 + c + 7}{a}$

PG-24. Consider the function $g(x) = 2(x + 3)^2$:

a. Find $g(-5)$. *8*

b. Find $g(\pi)$. *about 75.44*

c. If $g(x) = 32$, figure out what number x can be. *−7 or 1*

d. If $g(x) = 0$, figure out what number x can be. *−3*

Could PG-25 be the real parent graph? This is one of those "backwards" graphs that is anti-intuitive because the way we set it up is not chronological.

PG-25. When a pregnant woman goes into labor, her contractions start out short and mild. These contractions grow progressively in intensity and duration. When they start, the contractions are far apart (30 minutes to an hour) and at the time of delivery, they are happening one right after the other. Sketch a graph showing the relationship between the length of time between contractions and the intensity of the contractions.

PG-26. Solve for the indicated value:

a. $x =$ ___ $\sqrt{61}$

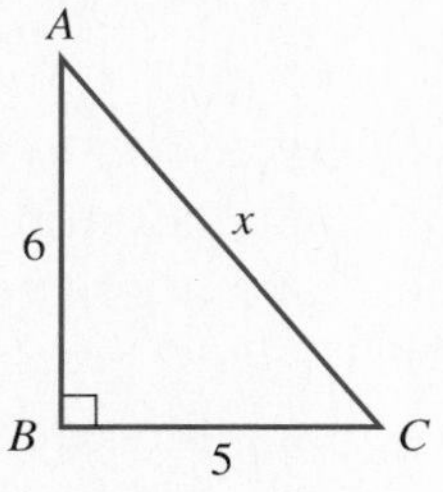

b. $BC =$ ___ $3\sqrt{3} \approx 5.20$

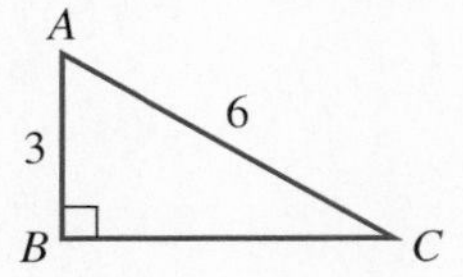

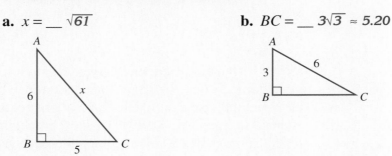

Problem PG-27 will give you the opportunity to assess students' understanding of randomness.

PG-27. You are making a dartboard. It is a square with sides 30 centimeters long. Place a 5-cm circle as a target somewhere within the square to maximize the probability that it will be hit when a dart hits the board *at random*. Where do you plan to place the circle? Explain your thinking about the placement of the circle. ***It does not matter.***

5.3 AVERAGING THE INTERCEPTS: TWO FORMS FOR THE EQUATION OF A PARABOLA

Start this section by having students discuss their Tool Kit entries on equations of parabolas with their groups and having them compare their graphs for PG-15 and PG-16 if they haven't already had a chance to do that. Then ask for reports from the groups on their conjectures. At this point, just facilitate the discussion of what they have found out. Don't feel compelled to fill in all the pieces yet.

The focus of the in-class assignment is an informal development of completing the square. Finding the vertex is an important skill that the students should be able to accomplish in a variety of ways. The method introduced in this chapter is averaging the intercepts. This method will allow students to complete the square for any quadratic function with real x-intercepts. It will also work when there is no real x-intercept, but we'll get to that later after complex numbers are introduced. If your course syllabus requires that students complete the square by traditional algebraic methods, there is a development of that approach in Appendix B.

PG-28. Graphing Form and Standard Form As you have worked with quadratic functions, equations, and expressions, you have regularly seen two forms. One is known as the graphing (or vertex) form, the other as standard form.

A quadratic equation in **graphing or vertex form** looks like this:

$$y = a(x - h)^2 + k$$

For example, the equation

$$y = 3(x - 1)^2 - 5$$

is in graphing form where $a = 3$, $h = 1$, and $k = -5$.

NOTE: The equations you found in the Parabola Investigation (PG-15) should be in graphing (or vertex) form.

The following quadratic equation represents the same curve as $y = 3(x - 1)^2 - 5$, but it is written in standard form.

$$y = 3x^2 - 6x - 2$$

A quadratic equation in **standard form** is written as

$$y = ax^2 + bx + \mathrm{c}$$

For example, the equation

$$y = 3x^2 - 6x - 2$$

is in standard form, where $a = 3$, $b = -6$, and $c = -2$.

a. Use your graphing calculator to verify that the two equations $y = 3(x-1)^2 - 5$ and $y = 3x^2 - 6x - 2$ are equivalent.

b. Show algebraically that these two equations are equivalent by starting with the graphing form and demonstrating step by step how to get the standard form.

PG-29. As you saw in PG-28, changing from graphing form to standard form is a straightforward algebraic process. To go the other way—from standard form to graphing form—is algebraically trickier, so we'll use the graph to help us out.

Consider the function $y = x^2 - 8x + 7$.

a. What are the coordinates of the x-intercepts? *(1, 0), (7, 0)*

b. Find the average value of the x-intercepts. $x_{ave} = 4$

c. What is the y value that corresponds to this average x value? $f(x_{ave}) = -9$

d. Make a sketch of this function, label the intercepts and the vertex, and draw in a line of symmetry. *vertex (4, −9)*

e. Rewrite the equation in graphing form. What relationships do you see between this equation and any points on your graph? $y = (x - 4)^2 - 9$

PG-30. For each of the following two functions, do parts a through e:

$$f(x) = x^2 + 3x - 10 \qquad g(x) = x^2 - 4x - 2$$

a. What are the coordinates of the x-intercepts? *(−5, 0), (2, 0);* $(2 \pm \sqrt{6}, 0)$

b. Find the average value of the x-intercepts. $x_{ave} = -3/2;\ 2$

c. What is the y-value that corresponds to each average x-value? $f(-3/2) = -49/4;\ g(2) = -6$

d. Make a sketch of each function, label the intercepts and the vertex, and then draw in a line of symmetry. *vertices: (−1.5, −12.25), (2, −6)*

e. Rewrite each equation in the form you found in the Parabola Investigation (PG-15 and PG-16), in other words in graphing form.
$f(x) = (x + 3/2)^2 - 49/4;\ g(x) = (x - 2)^2 - 6$

f. Discuss with your group any relationships you see between the average of the intercepts, the equations, and your graphs.

PG-31. Consider the function $h(x) = -4x^2 + 4x + 8$.

a. What are the coordinates of the x-intercepts of the graph of $h(x)$? *(2, 0), (−1, 0)*

b. Find the average value of the x-intercepts. $x_{ave} = 1/2$

c. What is the y-value that corresponds to the average x-value? $h(1/2) = 9$

d. Make a sketch of each function, label the intercepts and the vertex, and then draw in a line of symmetry. *vertex: (1/2, 9)*

e. Rewrite the equation in graphing (or vertex) form. $h(x) = -4(x - 1/2)^2 + 9$

PG-32. Based on your work with PG-29 through PG-31, describe a possible method for finding the vertex without drawing the graph. With your group, make up another example, and show how to use your method.

PG-33 could be a discussion opportunity. Also, this might be a good time to discuss the differences between (1) an algebraic solution, (2) a geometric solution, and (3) a solution arrived at by using tables and guess and check. Are all three methods equally valid for any problem, or are some methods better for some problems? What do we mean by "better"?

PG-33. Explain what the differences are between an accurate sketch and a careful graph.

EXTENSION AND PRACTICE

PG-34. Solve:

a. $y^2 - 6y = 0$ $y = 0, 6$

b. $n^2 + 5n + 7 = 7$ $n = 0, -5$

c. $2t^2 - 14t + 3 = 3$ $t = 0, 7$

d. $\frac{1}{3}x^2 + 3x - 4 = -4$ $x = 0, -9$

e. What do all of the equations in parts a through d have in common? *All have zero as a solution.*

PG-35. Find the vertex of each parabola by averaging the x-intercepts; then rewrite the equation in graphing form. If the equation is already in graphing form, write it in standard form.

a. $y = (x - 3)(x - 11)$
$(7, -16)$; $y = (x - 7)^2 - 16$

b. $y = (x + 2)(x - 6)$
$(2, -16)$; $y = (x - 2)^2 - 16$

c. $y = x^2 - 10x + 16$
$(5, -9)$; $y = (x - 5)^2 - 9$

d. $y = (x - 2)^2 - 1$
$(2, -1)$; $y = x^2 - 4x + 3$

> The vertex of a parabola locates its position in relation to the axes. The vertex serves as a **locator point** for a parabola. Other types of graphs we will investigate in this course also have locator points. These points have different names but the same purpose for each different type of graph.

PG-36. Add *locator point* to your Tool Kit. Use a parabola and its vertex as one example. It would be a good idea to leave room for another example in case you find one later.

Parts c and d of PG-37 require students to square both sides of the equation. The student answer section suggests squaring in case students do not think of it as a possibility. These problems are just a preview and a reminder about squaring and parentheses. We will return to problems like these later when we also deal with the fact that the graphing form students get by squaring may not be equivalent to the original equation.

PG-37. Rewrite each equation so that you could enter it into the graphing calculator.

a. $x - 3(y + 2) = 6$ $y = \frac{1}{3}x - 4$ **b.** $\frac{6x - 1}{y} - 3 = 2$ $y = \frac{6}{5}x - \frac{1}{5}$

c. $\sqrt{y - 4} = x$ $y = x^2 + 4$ **d.** $\sqrt{y + 4} = x + 3$ $y = x^2 + 6x + 5$

e. Use the calculator forms of the equations in parts a and b to find the x- and y-intercepts for their graphs. *(a) x: (12, 0), y: (0, –4); (b) x: (1/6, 0), y: (0, –1/5)*

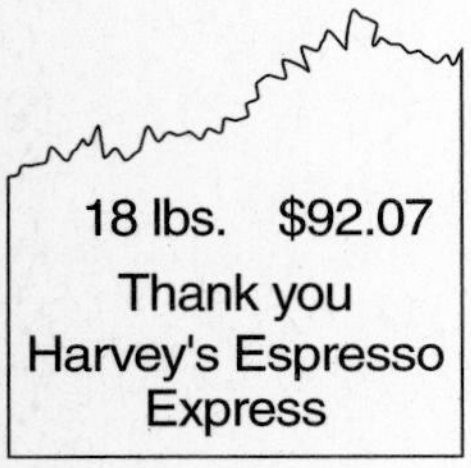

PG-38. Harvey's Espresso Express, a drive-through coffee stop, is famous for its great house coffee, a blend of Colombian and Mocha Java. Their arch rival, Jojo's Java, sent a spy to steal their ratio for blending beans. The spy returned with a torn part of an old receipt that showed only the total number of pounds and the total cost, 18 pounds at $92.07. At first Jojo was angry, thinking the spy hadn't stolen enough information. Then he realized he knew the price per pound of each kind of coffee ($4.89 for Colombian and $5.43 for Mocha Java), and that was enough. Show how he could use equations to figure out how many pounds of each type of beans Harvey's used. *10.5 lb; 7.5 lb*

PG-39. Can the quadratic formula be used to solve $4x^3 + 23x^2 - 2x = 0$? Show how or explain why you can't. *Students may say no, since the expression is a cubic. Support this observation, but then encourage them to find a way to solve it. You might have to suggest factoring and a common factor. The quadratic formula is necessary after all!* $(-23 \pm \sqrt{561})/8$ *and 0*

PG-40. Scientists can estimate the increase in carbon dioxide in the atmosphere by measuring increases in carbon emissions. In 1991 the annual carbon emission was about 7 gigatons (1 gigaton is a billion metric tons). Over the last several years annual carbon emission has been increasing by 1 percent per year.

a. At this rate how much carbon will be emitted in the year 2000? *7.656 gigatons*

b. Write a function $C(x)$ to represent the amount of carbon emitted in any year after the year 2000. $C(x) = 7(1.01)^{x+9}$ *or* $7.656(1.01)^x$

PG-41. Lilia wants to have a circular pool put in her backyard. She wants the area around the pool to be cement.

a. If her yard is a rectangle 50-foot by 30-foot, what is the largest-radius pool that can fit in her yard? *15 ft*

b. If the cement is to be 8 inches thick and it costs \$1.00 per cubic foot, what is the cost of putting in the cement? (Remember: volume = base area · depth.) *\$528.76*

5.4 PARABOLAS AS MATHEMATICAL MODELS

PG-42 and PG-43 are real situations that can be modeled by parabolas. In each of these problems, groups must decide for themselves where to put the x- and y-axes in order to generate an equation.

At this point, the students probably do not have the skill to find an exact solution. We expect that they will guess and check with their calculators and try to come close to a reasonable equation and find an approximate stretch factor.

PG-42. In our everyday lives, we're surrounded by parabolas. Parabolas are good models for all sorts of things in the world. Indeed, many animals and insects jump in parabolic paths.

This diagram shows a jackrabbit jumping over a 3-foot-high fence. In order to clear the fence, the rabbit must start its jump at a point four feet from the fence.

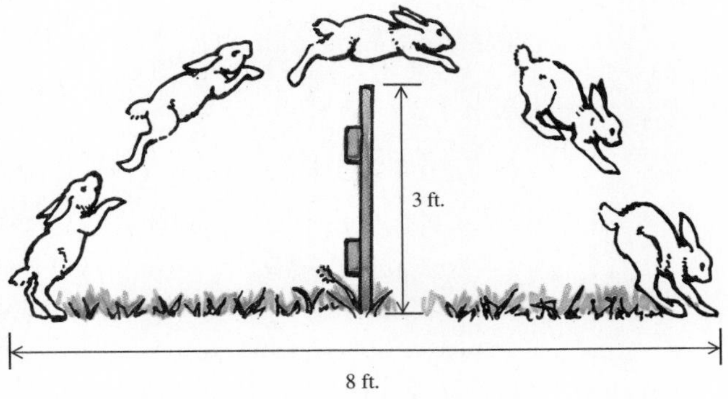

a. Given that you will be writing an equation to represent the rabbit's jump, where is the best place to put the x- and y-axes? As a group, decide where to put the x- and y-axes on the diagram. *Two possibilities for the origin are at the base of the fence or at the point where the rabbit starts its jump.*

b. Find the equation of a parabola that models the jackrabbit's jump. *Approximations of $y = -\frac{3}{16}x^2 + 3$ or $y = -\frac{3}{16}(x - 4)^2 + 3$ are possibilities. Students will probably use a decimal approximation for 3/16 and may use a number slightly larger than 3 to make sure the rabbit clears the fence.*

c. What do the dependent and independent variables represent in this situation? ***Answers depend on placement of axes, but in general the independent variable is position and the dependent variable is height.***

d. What are the range and domain of your equation? What parts of the domain and range are appropriate for the actual situation? ***Answers depend on placement of axes.***

PG-43. A fireboat in the harbor assists in putting out a fire in a warehouse along the pier. Use the same process as in PG-42 to find the equation of the parabola that models the path of the water from the fireboat to the fire, if the distance from the barrel of the water cannon to the roof of the warehouse is 120 feet and the water shoots up to a maximum height of 50 feet above the barrel of the water cannon.

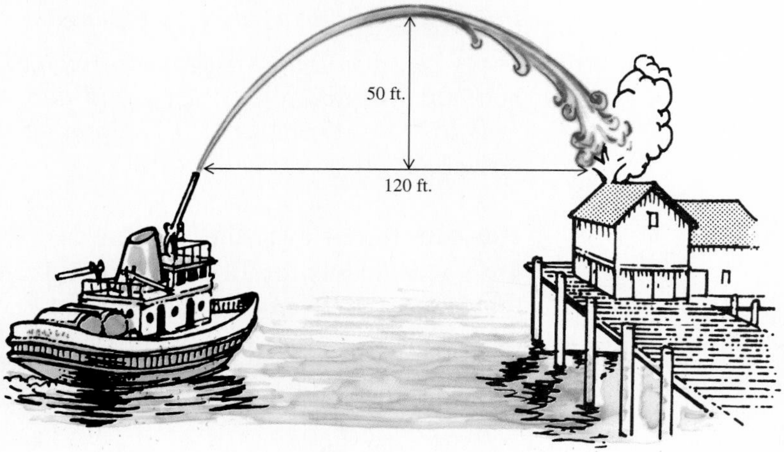

PG-44. Consider the equations $y = 3x - 5$, $y = 3x^2 - 5$, and $y = \frac{1}{3}x^2 - 5$.

a. Draw accurate graphs of these equations on the same set of axes.

b. In the equation $y = 3x - 5$, what does the 3 tell you about the graph? ***slope***

c. Is the 3 in $y = 3x^2 - 5$ the slope? Explain. ***No, because only lines have (constant) slopes. This 3 is the stretch factor.***

PG-45. Do the sides of a parabola ever curve back in, as the figure at right does? Give a reason for your answer. ***No—but don't expect an answer based on a function definition (although the input/output idea may come up).***

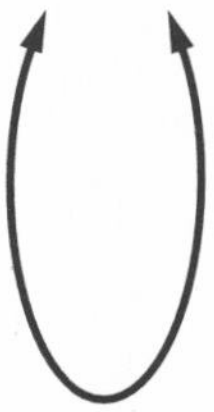

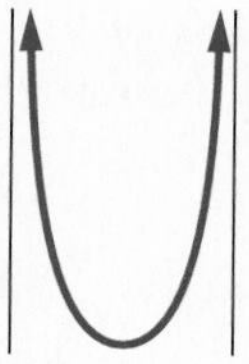

PG-46. Do the sides of the parabola approach straight vertical lines, as shown in the picture at left? (In other words, do parabolas have **asymptotes**?) Give a reason for your answer. ***No, because the domain is unlimited (any number can be squared).***

PG-47. Tabari's group wrote the following equation for the jumping jackrabbit in PG-42.

$$y = -0.2(x - 4)^2 + 3$$

a. Graph this equation on your graphing calculator, and then make an accurate sketch.

b. What part(s) of the graph represent the jackrabbit's jump? What parts don't?

EXTENSION AND PRACTICE

PG-48. Find the x- and y-intercepts of the graphs of the following functions:

a. $y = 2x^2 + 3x - 5$
x: (1, 0), (–5/2, 0), y: (0, –5)

b. $y = \sqrt{2x - 4}$
x: (2, 0), y: none

PG-49. If $g(x) = x^2 - 5$, find the following:

a. $g(0.5)$
$g(1/2) = -4.75$

b. $g(h + 1)$
$g(h + 1) = h^2 + 2h - 4$

PG-50. If $g(x) = x^2 - 5$, find the value(s) of x that make the following statements true:

a. $g(x) = 20$ *$x = \pm 5$*

b. $g(x) = 6$ *$x = \pm\sqrt{11}$*

PG-51. While watering your outdoor plants, you notice that the water coming out of your garden hose follows a parabolic path. Write the equation of a parabola that describes the path of the water from the hose to the plant. *$y = -\frac{8}{25}(x - 5)^2 + 8$ standing at (0,0)*

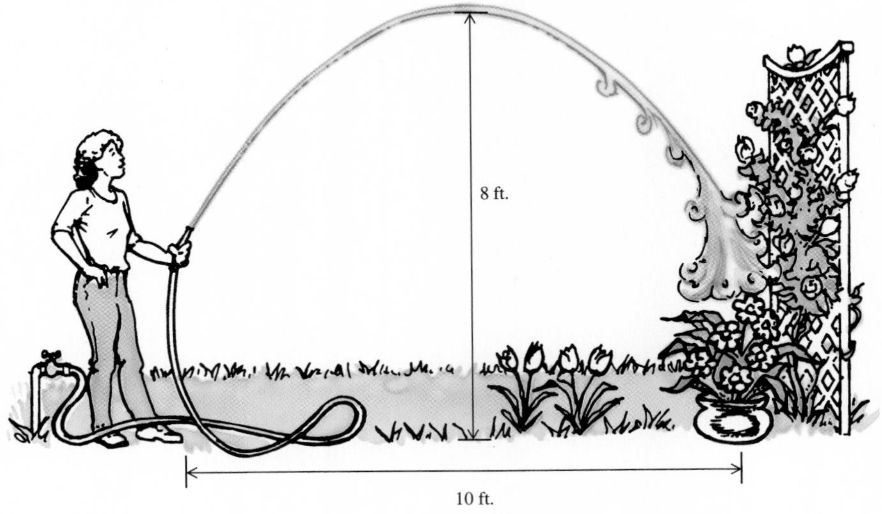

PG-52. Andrew likes to make up weird sequences and name them after himself. He made up these two sequences: $A(n) = 2(n - 1) + 3$ and $a(n) = 2(n - 1)^2 + 3$.

a. Are his sequences arithmetic, geometric, or something else? ***A(n) is arithmetic, but a(n) is quadratic.***

b. What would their graphs look like? ***A(n) is a line; a(n) is a parabola.***

In PG-53, the student who is not sure what to write should think of drawing a graph before trying to write the equation. This is a way of assessing whether students are starting to think of these things themselves. Be sure to check with them on this. Some students may need an interpretation for $-\infty < x < \infty$.

PG-53. Write an equation of a parabola that fits this description:

domain: $-\infty < x < \infty$ range: $y \geq 2$

There are many possible answers. All you have to be sure of is that the parabola does not contain a point where the y-value is less than 2.

PG-54. Solve each equation for x (that is, put in x-form):

a. $y = 2(x - 17)^2$

$x = \pm\sqrt{\frac{y}{2}} + 17$

b. $y + 7 = \sqrt[3]{x + 5}$

$x = (y + 7)^3 - 5$

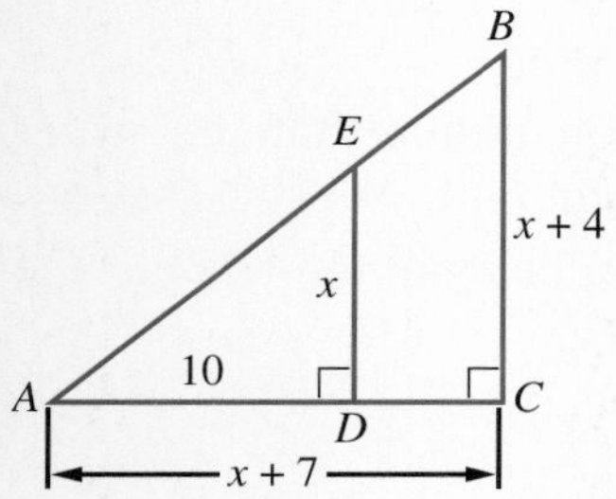

PG-55. $\Delta ABC \sim \Delta AED$ in the diagram shown.

a. Solve for x. ***8***

b. Find the perimeter of ΔADE. ***≈30.8***

In PG-56, note that the graph in part a isn't very intuitive—a distance vs. loudness graph starts high and decreases. Many students will graph a bell shape. If so, this is a good time for some whole-class graphical interpretation. The bell shape could be argued as appropriate if the student sees his or her position as the origin and the negative side of the x-axis as representing direction. Students should talk about what they are visualizing. Seeing themselves with the distance first decreasing then increasing is different from the way we generally graph distance on the x-axis, small to large.

PG-56. You are standing on the corner, waiting to cross the street, when you hear a booming car stereo approaching.

a. Sketch a graph that shows the relationship between how far away from you the car is and the loudness of the music.

b. Which is the dependent variable, and which is the independent variable? ***Loudness depends on distance.***

5.5 MOVING OTHER FUNCTIONS

This section may not be a part of your required curriculum, but it is a valuable exercise for generalizing what students have learned about parabolas and using it in a new context. If you do decide to skip this section, you'll need to watch out for follow-up problems throughout the problem sets.

The goal of PG-57 is to build the idea that the methods used to move parabolas around work for **any** *function. Thus the idea of a parent graph is useful—we can start with one simple version of a familiar function and generate a whole family of functions just by applying the transformations we already know. This concept leads to a quick-sketching technique for many functions, and to the reverse process—starting from a graph and recreating the equation.*

The goal of PG-57 is for students to see that graphs with similar shapes have related equations. Each student will need a copy of the Resource Page. Originally we had students use their calculators and sketch each graph. This is probably a better method, but it takes a lot of time. The Resource Page allows students to predict and match equations and graphs, then check their predictions with their graphing calculators. Circulate to correct any inaccurate work. You probably will need to remind some groups about the need for parentheses.

PG-57. Each person needs a copy of the Resource Page. Working as a group, compare each of the following equations with the graphs on the Resource Page, and try to match each equation with its graph. Then check, your predictions with your graphing calculators. On the graphs, label the coordinates of intercepts or any other important points that are relatively easy to find. You may split up the work in any way you think is fair, but each person should keep his or her own record of each graph with its important points labeled and its equation written next to it. (For parts g, h, and i, be sure to insert parentheses when entering these equations into the calculator.

a. $y = x^3$ **b.** $y = 2^x$ **c.** $y = \sqrt{x}$

d. $y = -4 + \sqrt{x}$ **e.** $y = (x - 1)^3$ **f.** $y = 2^x - 4$

g. $y = 2^{x-4} - 3$ **h.** $y = \sqrt{x+3} + 1$ **i.** $y = 2\sqrt{x-3}$

j. $y = (x - 2)^3 + 1$ **k.** $y = 2^{x+3}$ **l.** $y = \frac{1}{2}(x+2)^3$

PG-58. On the Resource Page the graphs are clearly grouped by their shape.

a. What are the similarities and differences among the *equations* of the graphs that have the same shape?

b. Which graph, in each group, has the simplest equation? Keep these graphs with your answers to this problem. You will need them later in this chapter.

PG-59. Look at your graph of $y = x^3$. This curve is called a **cubic**, and it doesn't have a vertex. What could you use for a locator point?

PG-60. This problem explores domains and ranges.

a. Most of the graphs in PG-57 have unlimited domains; in other words, the independent variable x can represent any real number. For which of the graphs are there numbers that can't be used to replace x? Why is this?

b. The ranges are limited for two of the three types of graphs. That is, the ranges do not include all real numbers. What is it about their equations that causes this?

PG-61. In PG-15 you explored how to change the equation of $y = x^2$ to shift the graph up or down and left or right. Explain the relationship between that lab and today's graphs.

Now is a good time to stop and synthesize. We want students to say, "What's the big deal? The graphs of almost all these equations are shifts of the graphs of the first three. You don't have to graph each separately—you can just graph the first three equations and then trace them in their new locations."

EXTENSION AND PRACTICE

PG-62. Sketch a graph of each of the following equations. Include the y-intercept and locator point for each, and approximate the x-intercepts where appropriate.

a. $y = x^3 + 5$ **b.** $f(x) = (x - 10)^3$ **c.** $g(x) = \sqrt{x} + 7$

d. $y = \sqrt{x + 8}$ **e.** $y = 2^x + 7$ **f.** $h(x) = (x - 2)^3 - 6$

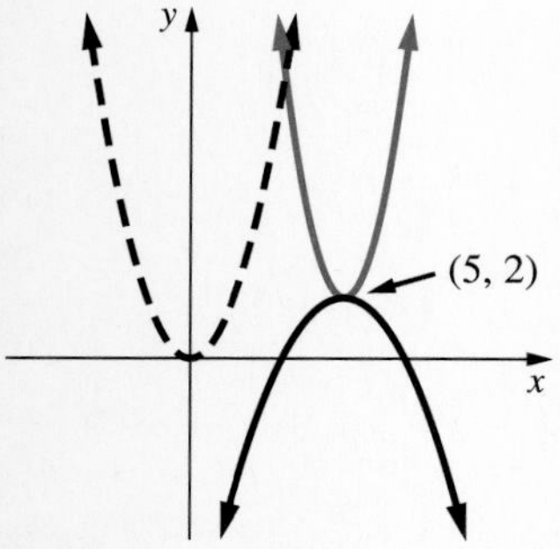

PG-63. The graph of $y = x^2$ is shown in dashed lines. Estimate the equations of the two other parabolas. Be sure to include in your equation a number that approximates the stretch or compression of the graph of $y = x^2$. *$y \approx 2(x - 5)^2 + 2$ and $y \approx -\frac{1}{2}(x - 5)^2 + 2$*

PG-64. Use the locator points shown on each graph to write a possible equation for the graph.

a.

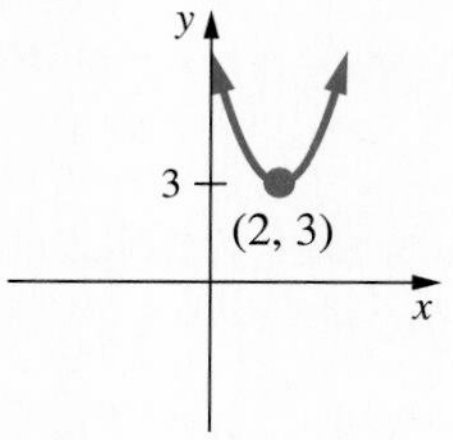

$y = (x - 2)^2 + 3$

b.

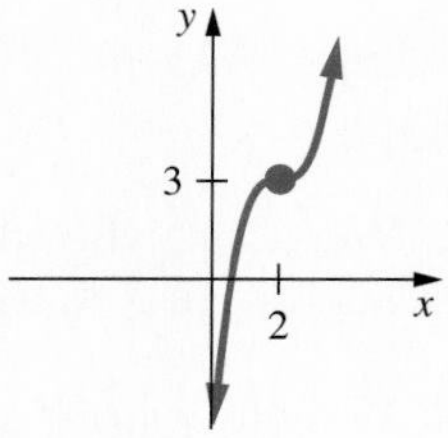

$y = (x - 2)^3 + 3$

c.

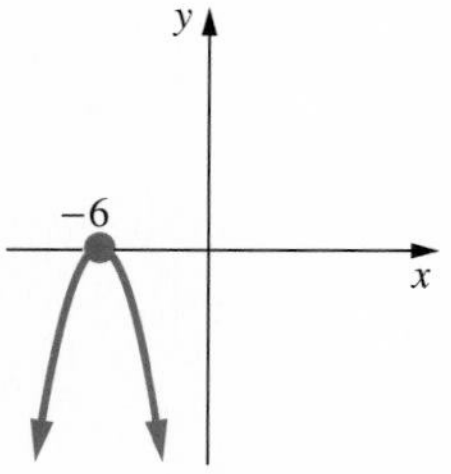

$y = -2(x + 6)^2$

PG-65. Find the domain and range for each graph in PG-64.

PG-66. Find the x- and y-intercepts and the vertex (locator point) of $y = x^2 + 2x - 80$. Then sketch its graph. *x: (–10, 0), (8, 0), y: (0, –80); vertex: (–1, –81)*

PG-67. If $h(x) = (x + 2)^{-1}$, find the value of each expression:

a. $h(3)$ *$h(3) = 1/5$*

b. $h(-3)$ *$h(-3) = -1$*

c. $h(a - 2)$ *$h(a - 2) = 1/a$*

PG-68. Multiply to remove parentheses in each expression.

a. $(x - 1)(x + 1)$ *$x^2 - 1$*

b. $2x(x + 1)(x + 1)$ *$2x^3 + 4x^2 + 2x$*

c. $(x - 1)(x + 1)(x - 2)$ *$x^3 - 2x^2 - x + 2$*

PG-69. Find the x- and y-intercepts of the graph of $y = (x - 1)(x + 1)(x - 2)$. *x: (±1, 0), (2, 0); y: (0, 2)*

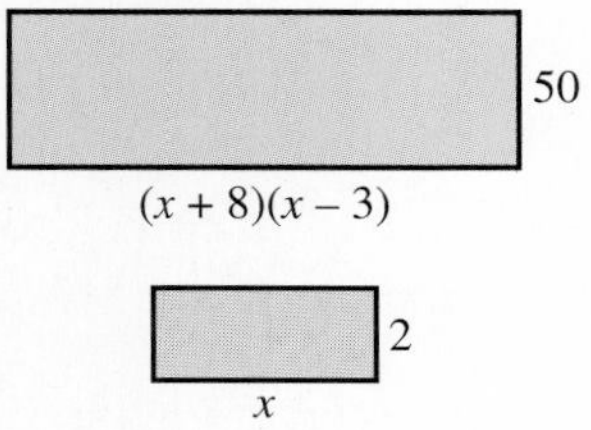

PG-70. The two rectangles shown here are similar. Find the length of each. *$x \approx 21.14$, so the other length is 528.60. There is another solution to the resulting equation, but it is negative.*

PG-71. Kristin's grandparents started a savings account for her when she was born. They invested \$100 in an account that pays 8 percent interest compounded annually.

a. Write an equation to express the amount of money in the account on Kristin's xth birthday. *$y = 100(1.08^x)$*

b. How much is in the account on her 16th birthday? *\$342.59*

c. What are the domain and range for the equation that you wrote in part a? *D: $x \geq 0$; R: $y \geq 100$*

5.6 PARENT GRAPHS AND GENERAL EQUATIONS

*In this section, students work with the idea of a parent graph and transformation ideas from the previous section. Since **parent graph** is not a well-defined term, there is no right answer to the question of which exponential equation to use. Students should decide based on what works best for their understanding and use of the idea. Students will need several Parent Graph Tool Kit Pages for PG-72.*

PG-72. What is the simplest equation of a parabola you can think of? Most people would probably say that $y = x^2$ is the simplest equation of a parabola. We use the term **parent graph** to describe the simplest version of a whole family of graphs. So the graph of $y = x^2$ is the parent graph for all parabolas. We can get the equation of any other parabola just by moving, flipping, or stretching the graph of $y = x^2$.

a. What do you think is the equation for the parent graph for all lines? for exponentials? for square roots? for cubics? Discuss each with your group and be sure everyone agrees.

Some families of functions are so important that they will come up over and over again throughout this course. So it will come in handy to have a **Parent Graph Tool Kit**. A special Tool Kit Page on which to keep a catalog of parent graphs and their equations is in the Resource Pages for you to either photocopy or use as an outline.

b. Put the information about each of the functions you identified as parent graphs in part a on a Parent Graph Tool Kit Page. For each one, include its name, a careful graph that is clearly labeled, a domain and range, and all the other things involved in investigating a function.

PG-73. Some groups of students were debating whether $y = 2^x$ should be the parent of all exponentials. One student suggested, "Since the parent is supposed to be the *simplest* equation you can get, and since $y = 1^x$ and $y = 0^x$ are both simpler than $y = 2^x$, one of those two should be the parent of all exponentials." What do you think? ***Both are horizontal lines, so they are clearly not very representative of exponential functions.***

> The term **parent graph** is an informal description for a representative graph that has the simplest equation. Since it is not a strictly defined term, there is some leeway in deciding what graph should represent all exponentials. For now $y = 2^x$ is a good representative.
>
> We may find later that some cubics may not be well represented by $y = x^3$, but for the cubics we are seeing now and those we will see in the near future, it will do.

PG-74. For each of the following functions:

1. State the equation of the parent graph.
2. Create a reasonable sketch of the graph, including any intercepts. You should try sketching this without a graphing calculator, but feel free to use one to check your ideas.
3. Give the domain and range.
4. Find the coordinates of any locator point.

a. $y + 1 = x^2$ **b.** $y + 4 = x^3$ **c.** $y + 6 = (x + 2)^3$

d. $y + 2 = 2^x$ **e.** $y = \sqrt{x - 3}$ **f.** $y = \sqrt{x} + 2$

g. $y - 3 = (x - 4)^3$ **h.** $y + 2 = (x - 3)^2 + 1$ **i.** $y - 5 = 2^{x+4}$

Note: Some of these functions may be easier to figure out if you put them all in y-form.

PG-75. General Forms for Equations of Parabolas One way of writing a general equation for a parabola is

$$y = a(x - h)^2 + k.$$

This equation tells us how to start with the parent graph of $y = x^2$ and shift, stretch (compress), or flip it to get any other parabola. This equation is in graphing form.

Another general equation for a parabola is

$$y = ax^2 + bx + c.$$

This equation is in standard form.

a. The equation $y = a(x - h)^2 + k$ tells us how to start with the parent graph $y = x^2$ and shift, stretch (or compress), or flip it to get any other parabola. Explain what each letter (a, h, and k) represents in relation to the graph of $y = x^2$. Add this information to your Parent Graph Tool Kit.

b. The standard form of the equation does not tell us much about its graph. What are the only things we can easily determine about the graph by just looking at the standard form? *stretch factor and direction if we know a, and y-intercept*

PG-76. Use the idea for the general equation of a parabola in PG-75 to write a general equation in graphing form for a *cubic* using the parent $y = x^3$. Explain how each letter in your general equation affects the stretch or location. Make sure everyone in your group agrees, and be ready to present your ideas to the class.

If you notice that not all groups have an appropriate general equation and explanation, pull the class together and have several groups present their ideas. Otherwise wait until groups finish the next problem.

PG-77. Sketch a graph of each of the following cubics without using a graphing calculator. Then use your calculator to check your thinking.

a. $f(x) = 3(x - 2)^3 + 2$ **b.** $y + 3 = \frac{1}{3}(x + 1)^3$

PG-78. Take out your Parent Graph Tool Kit. As a group, write **general equations** for *each* parent graph. Be ready to explain how your general equations work; that is, tell what effect each part has on the graph: stretch and direction, horizontal location (left/right shift), and vertical location (up/down shift). Include two specific equations and their graphs to illustrate how the numbers that replace a, h, and k in the general equation affect the location and stretch of the graphs.

EXTENSION AND PRACTICE

We expect the students to get most of the shifts in PG-79 and to make reasonable guesses on the stretches.

PG-79. Write a possible equation for each of the following graphs. Assume that each graph paper unit is one unit.

a.

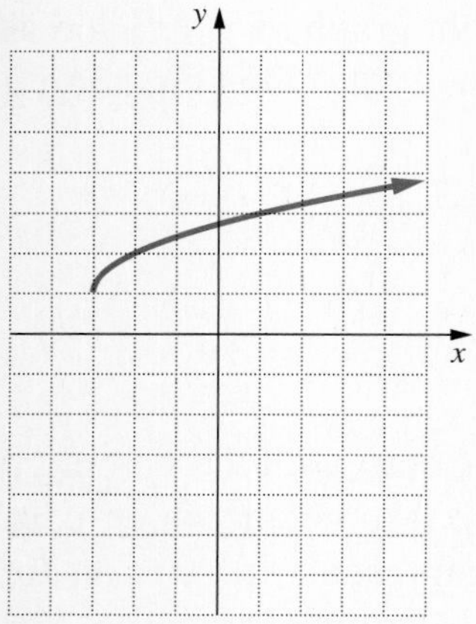

$y = \sqrt{x + 3} + 1$

b.

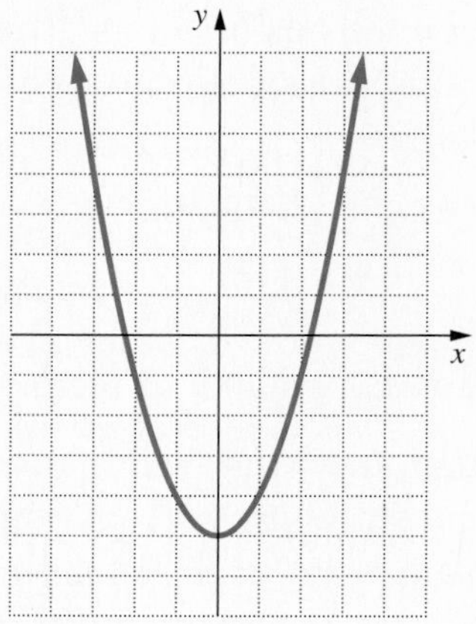

$y = x^2 - 5$

c.

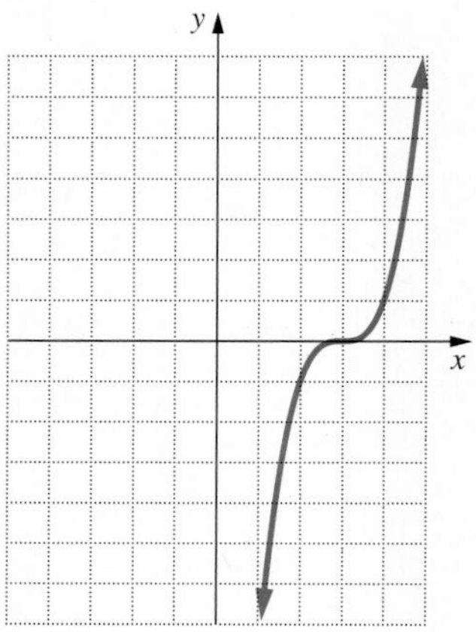

$y = (x - 3)^3$

d.

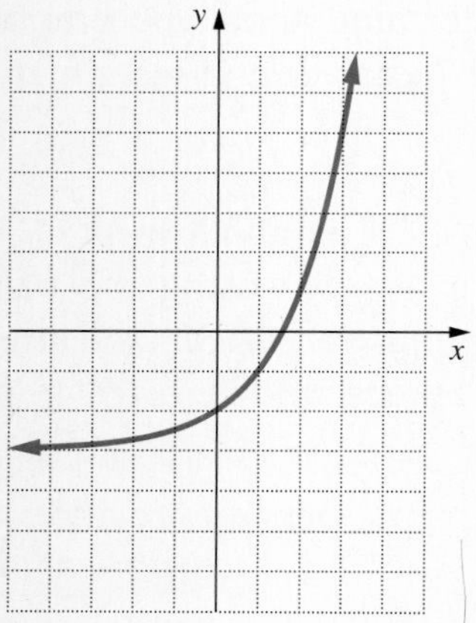

$y = 2^x - 3$

e.

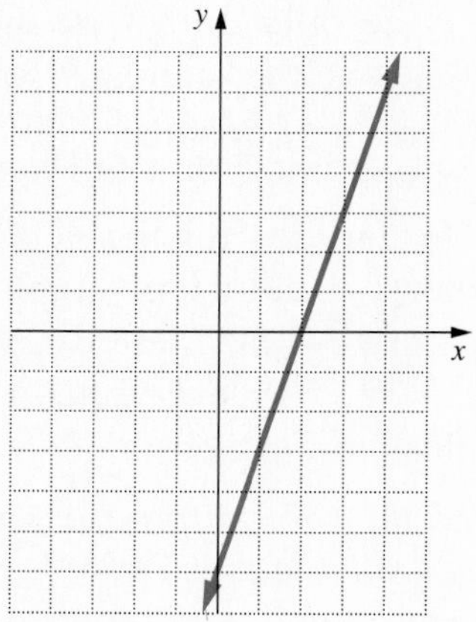

$y = 3x - 6$

f.

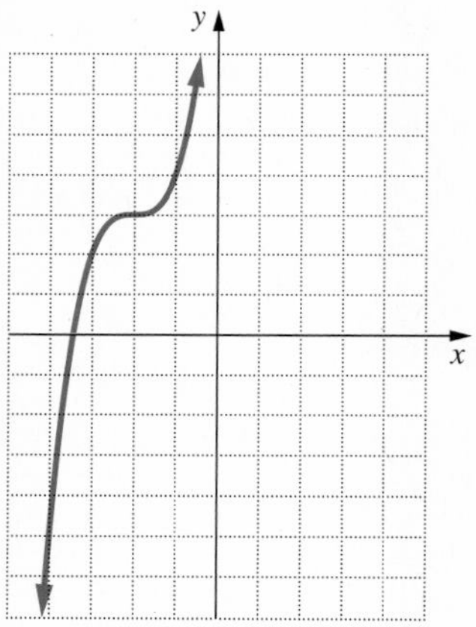

$y = (x + 2)^3 + 3$

g.

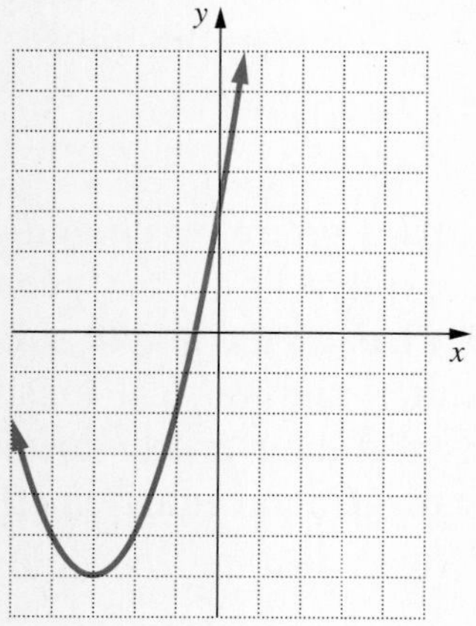

$y = (x + 3)^2 - 6$

h.

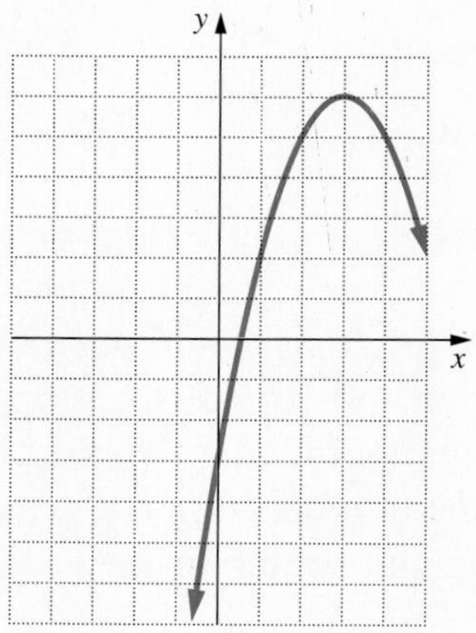

$y = -(x - 3)^2 + 6$

i.

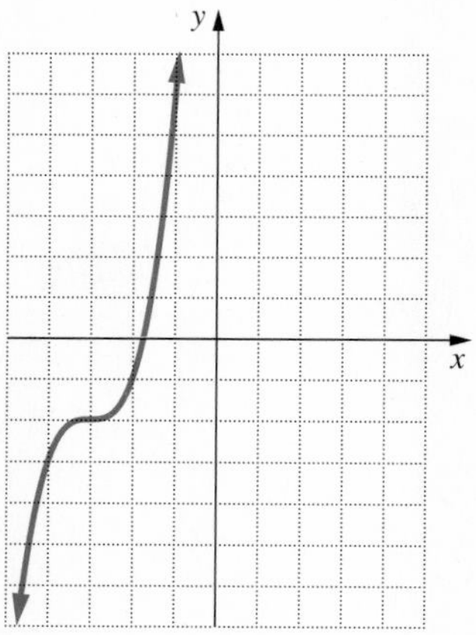

$y = (x + 3)^3 - 2$

PG-80. Write the equations of three different parabolas with the vertex at (4, 0). *examples: $y = (x - 4)^2$; $y = 5(x - 4)^2$; $y = -3(x - 4)^2$*

PG-81. Explain the difference between the graphs of $y = \sqrt{x}$ and $y = 4\sqrt{x+5} + 7$. *The second graph shifts the first 5 units to the left and 7 units up, and also stretches it by a factor of 4.*

PG-82. How is $y = 2^x$ different from $y = -(2^x)$? Sketch the graph of $y = -(2^x)$. *The second graph is a reflection of the first over the x-axis.*

PG-83. Sketch the graph of $y = 2(x - 1)^2 + 4$.

a. Now rewrite the equation $y = 2(x - 1)^2 + 4$ without parentheses. That is, perform whatever algebra is necessary to get rid of the parentheses. Remember the order of operations! *$y = 2x^2 - 4x + 6$*

b. If you were to draw the graph of the new form of the equation, how would the graph differ from the graph of the original? Are there any differences between the graphs of the two equations? Explain your reasoning. *none*

c. What is the parent graph of $y = 2(x - 1)^2 + 4$?

d. What is the parent graph of $y = 2x^2 - 4x + 6$?

PG-84. For $(x + y)^2$, Caren wrote the expression $x^2 + y^2$. Her teacher marked it wrong but didn't tell her how to answer the problem correctly. Write an explanation of what is wrong, and show Caren how to square the sum of x and y.

PG-85. The gross national product (GNP) was 1.665×10^{12} dollars in 1960, and then the GNP increased at the rate of 3.17 percent per year until 1989.

a. What was the GNP in 1989? *4.116×10^{12}*

b. Write an equation to represent the GNP t years after 1960, assuming that the rate of growth remains constant. *$y = 1.665(10^{12})(1.0317)^t$*

c. Do you think the rate of growth remains constant? Explain.

PG-86. Is the graph of $y = x^2$ the parent graph of $y = 9 + 6x - x^2$? Explain how you decided.

PG-87. Find the coordinates of the x- and y-intercepts for each of the following functions:

a. $g(x) = (x + 3)^3$
x: (–3, 0); y: (0, 27)

b. $y - 1 = 3^x$
x: none; y: (0, 2)

PG-88. You will need your graph and data from the Shrinking Targets Lab (PG-2) at your next class. Find it now (or recreate it!), and review it before class.

5.7 DETERMINING THE STRETCH FACTOR

This section focuses on finding rather than guessing and checking a specific value for the factor **a** *in order to write an equation as a model. It would be a good idea to have data from the Shrinking Targets Lab (PG-2 and PG-3) and a sketch of the graph available in case some students do not have their results from that Lab handy. We also introduce the nonfunction $x = y^2$ in order to clarify the definition of a function.*

PG-89. The Shrinking Targets Lab Revisited Look back at your Shrinking Targets Lab graph from PG-2. When you decided what kind of equation could represent the data, did you say $y = ax^2$? You might have said $y = x^2$, which would have been sufficient at that time, but now we can be more accurate and get a better version of the equation to represent the curve you drew.

a. Find a convenient point on the curve you drew and substitute the x- and y-coordinates for x and y in the equation $y = ax^2$. Solve for a. Then use that result to write a more accurate equation for your graph.

b. Graph your equation from part a on your graphing calculator. Compare this graph with your graph of the actual data. Explain the similarities and differences in terms of the difference in the domain for the real situation and the domain for the purely mathematical representation of the equation.

PG-90. The Fireboat Again The equation below is one possibility for the parabola formed by the water cannon on the fireboat in PG-43:

$$f(x) = a(x - 60)^2 + 50$$

a. Will the value of a be positive or negative? Explain.

b. Calculate the value for a by substituting the coordinates of some known point other than the vertex. Then write the complete equation.
$f(x) = -0.0139(x - 60)^2 + 50$

c. Why wouldn't it work to substitute the coordinates of the vertex in part b? If you're not sure, try it.

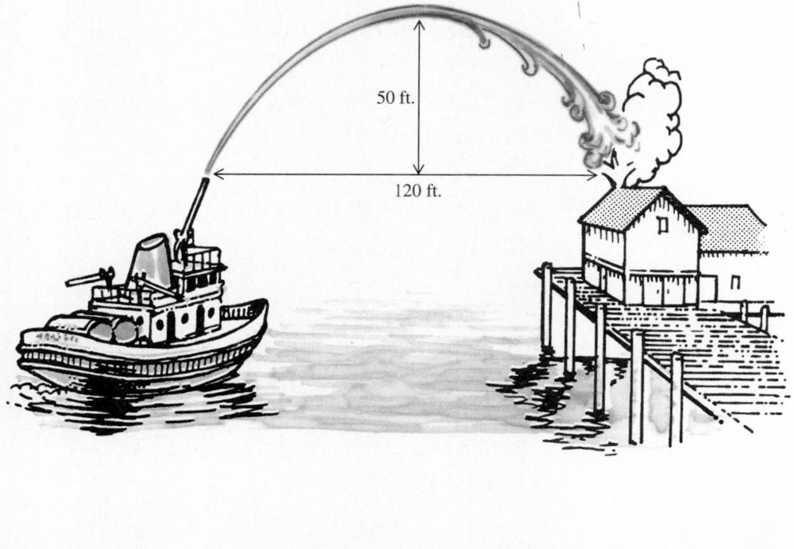

PG-91. Curtis, who lives on a farm and likes cows, has invented a weird type of machine, which he calls Curtis's Cow Machine. Each picture represents the same machine.

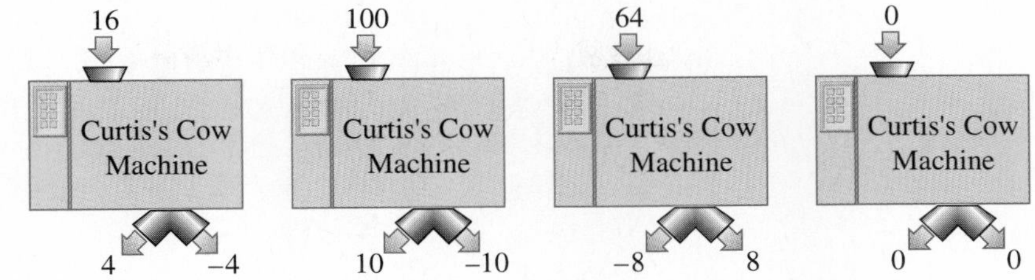

a. What makes this machine different from all the function machines you have seen in this text? Explain. *This machine has two outputs for each input, but a function machine has only one output per input.*

b. Make a table and draw a graph that represents his machine. Be sure to include x-values like 1, 4, and 9, and remember that the numbers at the top of the machines are the inputs and the numbers that come out of the cow machine are the outputs.

c. Write a rule for Curtis's Cow Machine. *$x = y^2$ or $y = \pm\sqrt{x}$*

PG-92. If you haven't already done so, try to find a way to graph $x = y^2$ on your graphing calculator. Since your graphing calculator only graphs functions in y-form, you will have to be clever. Talk to your group and come up with a way of getting $x = y^2$ to graph on your calculator. If your group is completely stumped, read parts a, b, and c.

a. How is $x = y^2$ different from all the others you've graphed? *It's not a function, but students won't say it in this way. They'll probably say something like, "it's the y that's squared," or "you can't graph it on the graphing calculator," or "you need to substitute for y, not x, to graph it."*

b. Have you seen a graph that looks like half of $x = y^2$? What is its equation? *$y = \sqrt{x}$*

c. What equation will give the other half? Use your graphing calculator to check. *$y = -\sqrt{x}$*

PG-93. Remember Carmichael B. Pierce's original function machine that gave out rope lengths for Digger's leash? Take a look at it again. Whenever C. B. Pierce plugged a number into his machine, he got *only one* number out. Whenever a relationship has this property—for each input there is one and only one output—the relationship is called a **function**.

a. Look at your graph of $y = x^2$. Is it a function? Explain how you know.

b. Is $x = y^2$ a function? Why?

c. Add the definition of a function to your Tool Kit. Include an example of one graph that is a function and one that is not, and label which is which.

PG-94. Of the parent graphs listed in your Parent Graph Tool Kit, which are graphs of functions? Explain.

EXTENSION AND PRACTICE

PG-95. The cables that hold up a suspension bridge (like the Golden Gate Bridge) also have the shape of a parabola. If the tower rises 100 feet above the roadbed, find an equation to model the section of cable between the towers of the bridge in the diagram. *$y = 0.01(x - 100)^2$ or $y = 0.01x^2 - 2x + 100$*

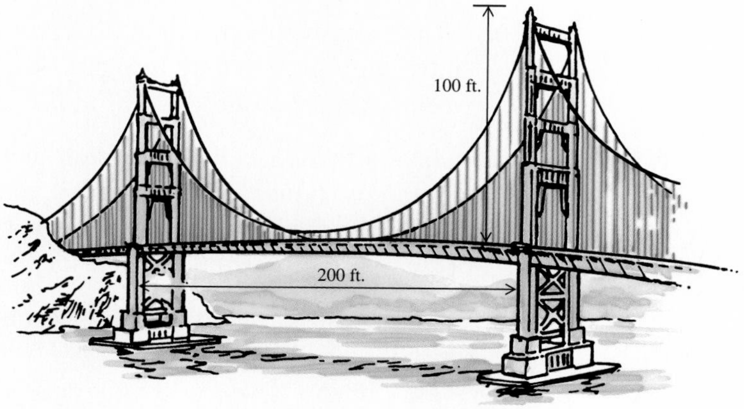

PG-96. The amount of profit (in millions) made by Scandal Math, a company that writes math problems based on tabloid articles, can be found by the equation $P(n) = -n^2 + 10n$, where n is the number of textbooks sold (also in millions). Find the maximum profit and the number of textbooks Scandal Math must sell in order to attain this maximum profit. *Maximum profit is \$25 million when $n = 5$ million.*

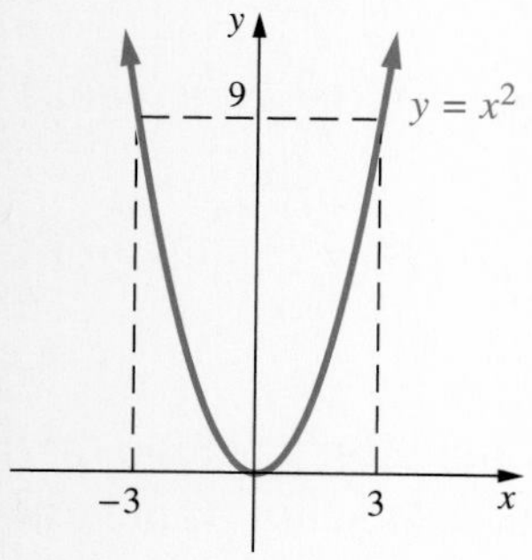

PG-97. Anne says that $y = x^2$ is not a function, because when she substitutes 3 for x, she gets 9, and when she substitutes −3 she also gets 9. Two different inputs give her the same output. Brooke says that Anne has the function idea all mixed up. Help! Explain who's right. Is $y = x^2$ a function or not? Why? *Brooke is right.*

PG-98. By mistake Jim graphed $y = x^3 - 4x$ instead of $y = x^3 - 4x + 6$. What should he do to his graph to get the correct one? *Move it up 6 units or redraw the axes 6 units lower.*

PG-99. If $x^2 + kx + 18$ is factorable, what are the possible values for k? *$\pm 11, \pm 9, \pm 19$*

PG-100. Consider this system of equations:

$$3y - 4x = -1$$
$$9y + 2x = 4$$

a. Solve this system. *(x = 1/2, y = 1/3)*

b. Find where the two lines intersect. *(1/2, 1/3)*

c. Explain fully the relationship between what you did in part a and what you did for part b.

PG-101. Write the description for a situation that could fit the following function:

$$f(t) = 2000(0.91)^t$$

PG-102. Write an equivalent expression without parentheses for each of the following expressions.

a. $(2x^2y)^3 = ?$
$8x^6y^3$

b. $(5x + 0.5)^2 = ?$
$25x^2 + 5x + 0.25$

c. $5(2s - 7)(2s + 7) = ?$
$20s^2 - 245$

d. $(5d^{-5})^3(-3d^3)^5 = ?$
$-30{,}375$

PG-103. Find the x- and y-intercepts and the locator points of the following functions.

a. $y - 7 = 2x^2 + 4x - 5$
x: (−1, 0); y: (0, 2);
V: (−1, 0)

b. $x^2 = 2x + x(2x - 4) + y$
x: (0, 0), (2, 0); y: (0, 0);
V: (1, 1)

PG-104. Find the equation of a parabola that has a vertex (2, 3) and contains (0, 0). $y = -\frac{3}{4}(x - 2)^2 + 3$

PG-105. Use the fact that the triangles are similar to solve for x:

a.

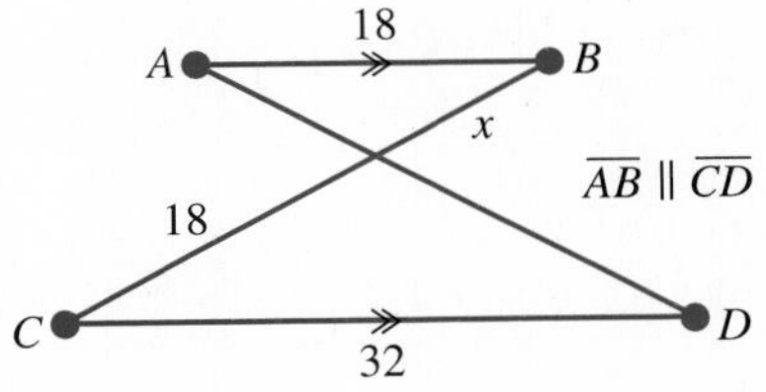

y = 10.125

b.

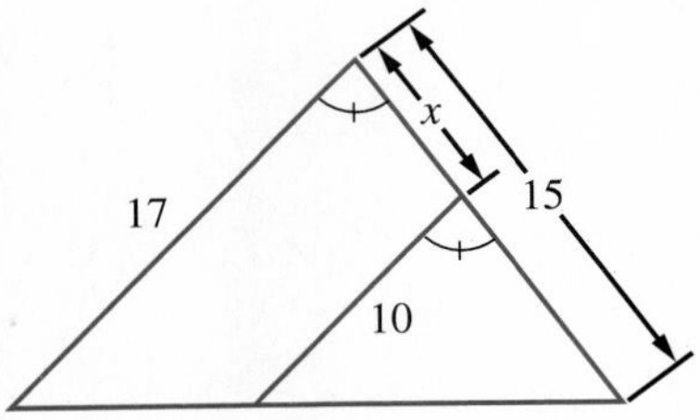

x ≈ 6.18

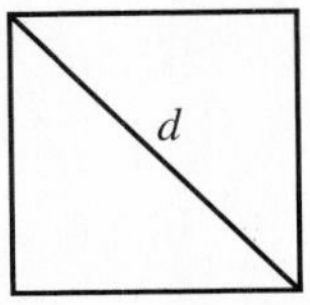

PG-106. The area of a square is 225 square centimeters.

a. Make a diagram and list any subproblems that you would need to solve to find the length of the diagonal d.

b. What is the length of the diagonal? $15\sqrt{2} \approx 21.21$ cm

5.8 AN INTRODUCTION TO RATIONAL FUNCTIONS

If rational expressions are a part of your course syllabus, you may need to do this section in which students investigate variations on the function $y = 1/x$. Otherwise this section can be skipped completely.

In this section we investigate functions with asymptotes. Students will need to get started on these problems right away and use groups to their advantage when doing the calculations (dividing the work). Unfortunately some graphing calculators connect the last pixel point where x is less than 3 with the first one where x is greater than 3 (particularly the TI-82 and TI-83, unless they are on "zoom, decimal"). Be sure to check with the groups to clear up this problem if it arises. When you look at the groups' problems, you'll know instantly whether they are beginning to understand the concept of asymptote.

PG-107. When you use a graphing calculator for this problem, you need to be careful when typing in the equation of the function. The calculator follows the same order of operations you learned in earlier math courses. For the function

$$h(x) = \frac{1}{x-3},$$

do you want to take the reciprocal of just the x or of the $x - 3$? What can you do to make the calculator do what you want?

a. Use a full sheet of graph paper, and set up a pair of axes with the origin near the center, vertically, but only about 2 inches in from the left side of the page. On both axes make each graph paper unit represent 0.25 units. Now make an accurate graph. Use a calculator to make the substituting and calculating easier, but make an extensive table; don't just copy the picture of the graph shown on your calculator. Some calculators will give an extra vertical line in the middle of the graph. That line should not be part of the graph. The graph should have two unconnected parts.

b. What seems to be happening to the graph near $x = 3$? Did you use values like 3.5, 3.25, 2.5, and 2.75? If you didn't, do it now, and explain what is happening.

x	3.5	3.25	2.5	2.75
y				

c. Find each of the following values and be sure to plot each on your graph (if possible).

i. $h(2.9)$ *−10* **ii.** $h(2.99)$ *−100* **iii.** $h(2.999)$ *−1000*

iv. $h(3.1)$ *10* **v.** $h(3.01)$ *100* **vi.** $h(3.001)$ *1000*

d. What happens, on the graph and algebraically, when you try $h(3)$? Give a good explanation for this phenomenon. *Graph "blows up." We cannot divide by zero.*

e. What are the domain and range for the function h? Explain. *D: all numbers except 3; R: all numbers except 0.*

f. Where does the graph of h cross the x-axis? Compare this to your picture and justify your answer. *Never.*

*Here you need to have a discussion about what really is happening at the point $x = 3$. Give other simple examples if you think students need them. You can use the idea that there is an imaginary slit in the graph paper along the line $x = 3$ that the graph can get very close to, but cannot cross, as if the paper has been cut along the line $x = 3$. When students are ready, introduce the term **asymptote**. More importantly, they need to believe that 3 is **not** in the domain and 0 is **not** in the range. If they don't believe it, tell them you'll give them 100 extra credit points to find the y value that corresponds with $x = 3$ and another 100 points for the x-value that gives $y = 0$.*

PG-108. As a group, come up with another equation that will have a different graph, but the same phenomenon as

$$h(x) = \frac{1}{x-3}.$$

Test it out. Give it to another group to graph, and find the appropriate viewing screen to be able to see the complete graph.

PG-109. Sketch a graph of each of the following equations, and identify each graph with its equation. Be sure to label the coordinates of intercepts or any other important points on each graph that are relatively easy to find. For each equation, notice which value can't be used to replace x. Remember that the fraction bar is a grouping symbol on paper, so be sure to use parentheses when entering the equations for parts c and d in the calculator.

a. $y = \frac{1}{x}$

b. $y = \frac{1}{x} - 4$

c. $y = \frac{1}{x-2} + 1$

d. $y = \frac{2}{x+4}$

*PG-110 introduces the term **asymptote** but does not provide a definition. That should come later as students become more familiar with the idea.*

PG-110. For each graph in PG-109 draw one vertical line and one horizontal line that do not touch the graph, although the graph will come very close. These lines are called **asymptotes**.

a. Write down the equations of the asymptotes for each of the graphs, and give the coordinates of the point where the asymptotes intersect.

b. These functions are called **rational functions**. Rational functions can include addition, subtraction, multiplication, and division. What is the parent graph for this group of rational functions?

c. What point would you use for a locator point for these graphs?

d. How are the coordinates of each locator point related to the equation for each graph?

e. Record the important information about asymptotes and rational functions in your Tool Kit.

PG-111. The functions in PG-109 represent just one group of rational functions. There are many other groups. Use your graphing calculator and sketch a graph of each of the following rational functions to see a few examples of some other things that can happen when division is involved.

a. $y = \dfrac{1}{x^2 - 4}$

b. $y = \dfrac{x}{x^2 - 4}$

c. $y = \dfrac{x+1}{(x-2)(x+3)}$

d. $y = \dfrac{5}{(x^2+1)} - 1$

NOTE: Be sure to use parentheses when entering these equations in the calculator, and be sure to use a calculator window that does not draw vertical lines where asymptotes are supposed to be. Asymptotes are very useful in giving you information about the graph, but they are not part of the graph itself.

---

e. Why is there no vertical asymptote for the graph in part d but two vertical asymptotes for each of the other graphs?

PG-112. Make up another rational function. Quickly sketch what you think it will look like, and then check your prediction with your calculator. Each group member should make up a different one and check to see whether the others can predict the result.

EXTENSION AND PRACTICE

We expect the students to get most of the shifts on PG-113 and to assume no stretch or compression of the parent graph of 1/x.

PG-113. Write a possible equation of each of these graphs. Assume that one mark on each axis is 1 unit.

a.

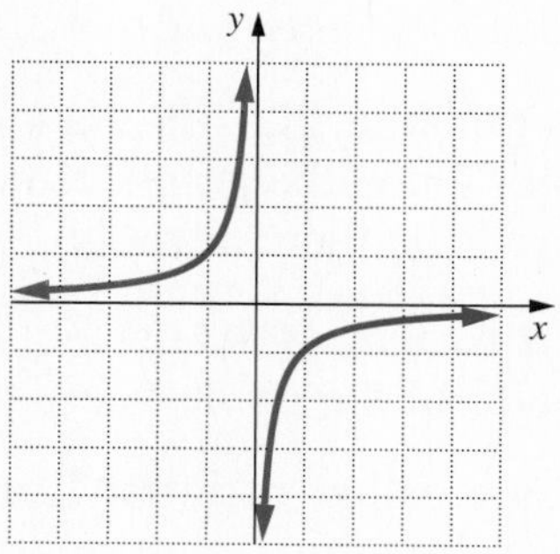

b.

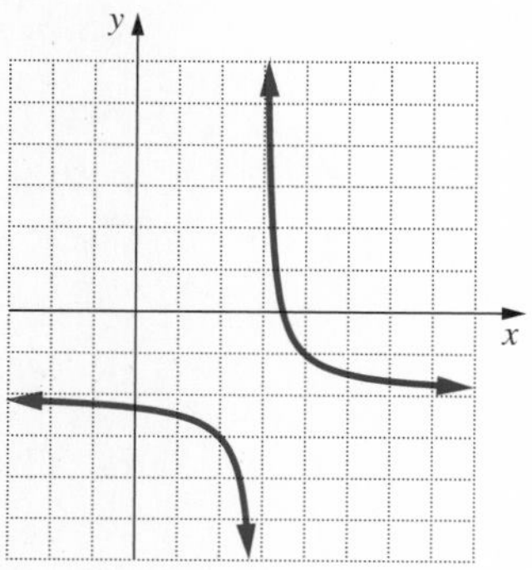

PG-114. Explain the difference between the graphs of

$$y = \frac{1}{x} \quad \text{and} \quad y = 4\left(\frac{1}{x+5}\right) + 7$$

The second graph shifts the first 5 units left and 7 units up, and also stretches it by a factor of 4.

PG-115. Tasha is experimenting with her graphing calculator. She claims that the graph of $y = \frac{1}{x}$ is the parent graph of $y = \frac{4}{x}$. Do you agree? If not, tell why not. If you do agree, what change (shift, stretch, or compress) could you make to the graph of $y = \frac{1}{x}$ to get the graph of $y = \frac{4}{x}$? *yes, stretch factor of 4*

The following problem might be too time consuming without a graphing calculator.

PG-116. Carefully graph $y = x^3 - 4x^2 - 3x + 18$ for $-4 < x < 4$.

a. How many x-intercepts does it have and what are their coordinates? *only two—at (–2, 0) and a double root at (3, 0)*

b. What *small* change could you make in the equation so that the graph would have:

i. one x-intercept? *add a positive number*

ii. three x-intercepts? *add a negative number*

iii. four x-intercepts? *impossible*

PG-117. Based on your experience stretching the graph of $y = x^2$, what would a stretched version of the graph of $y = x^3$ look like?

a. Draw an accurate graph of $y = x^3$. (x-values from -2 to 2 will be sufficient!)

b. Now imagine *stretching* this graph just as you stretched a parabola. Sketch in what you think the stretched graph would look like.

c. Imagine *compressing* $y = x^3$. Sketch in what you think the compressed graph would look like.

d. Describe how would you change the equation $y = x^3$ so that the graph is stretched or compressed.

PG-118. Describe the graph of $y = -x^3$ in relation to its parent.

PG-119. Consider a line with a slope of 3 and a y-intercept at (0, 2).

a. Sketch the graph of this line.

b. Write the equation of the line. $y = 3x + 2$

c. Find the first four terms of the sequence $t(n) = 3n + 2$. Plot the terms on a new set of axes next to your graph from part a. *2, 5, 8, 11*

d. Explain the similarities and differences between the graphs and equations in part a through c. Are both continuous? *One is continuous and one is discrete.*

PG-120. Imagine a sphere being inflated so that it gets larger and larger, but remains a perfect sphere. Consider the function defined this way: the inputs are the radius of the sphere, and the outputs are the volume. You may remember from a previous math course that the formula for the volume of a sphere is

$$V(r) = \frac{4}{3}\pi(r)^3.$$

a. What is the parent graph of this function? Explain how you decided. *$y = x^3$*

b. What is the stretch factor? *$4\pi/3$*

c. If the radius is measured in centimeters, what are the units for the volume? *cm^3*

d. Make a sketch of this function.

e. What are the domain and range? *D: $r \geq 0$; R: $V(r) \geq 0$*

PG-121. Draw the graph of $y = 2x^2 + 3x + 1$.

a. Find the x- and y-intercepts. *x: (−1/2, 0) (−1, 0); y: (0, 1)*

b. Where is the line of symmetry of this parabola? Write its equation. *$x = -3/4$*

c. Find the coordinates of the vertex. *(−3/4, −1/8) or (−0.75, −0.125)*

d. Write the equation in graphing form. $y = 2\left(x + \frac{3}{4}\right)^2 - \frac{1}{8}$

PG-122. Change the equation in the previous problem so that the parabola has only one x-intercept. *Move it up 0.125 units to get: $y = 2x^2 + 3x + 1.125$.*

PG-123. Shuneel claims he has an easier way of changing the equation so that the parabola has only one x-intercept. He simply changed $y = 2x^2 + 3x + 1$ to $y = 0x^2 + 3x + 1$. What do you think of his method? *It's not a parabola anymore.*

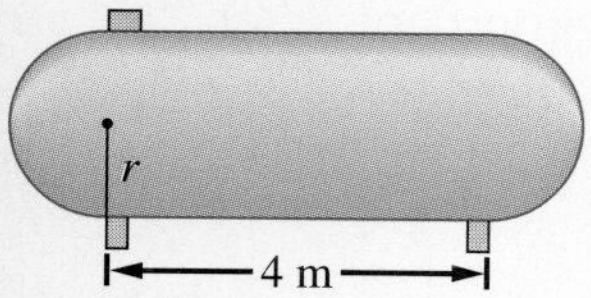

PG-124. People who live in isolated or rural areas often have their own tanks for gas to run appliances like stoves, washers, and water heaters. Some of these tanks are made in the shape of a cylinder with two hemispheres on the ends (a *hemisphere* is half a sphere, or half a ball).

The Insane Propane Gas Tank Company (motto: "*We're Crazy about Gas*") wants to make tanks with this shape, but in different models of different sizes. The cylindrical part of all the tanks will be 4 meters long. But the radius r will vary among different models. Remember, the volume of a sphere is

$$V = \frac{4}{3}\pi r^3$$

and the volume of a cylinder is $V = \pi r^2 h$.

a. One of their tanks has a radius of 1 meter. What is its volume? *$4\pi + (4/3)\pi \approx 16.755$ m^3*

b. When the radius doubles (to 2 m), will the volume double? Explain. Then figure out the volume of the larger, 2-m tank. *No, the volume of the cylindrical part will be four times as large, and the volume of the hemispheres eight times as great. volume = $(80\pi)/3 \approx 83.776$ m^3.*

c. Write an equation that will let Insane Propane Gas Tank Company determine the volume of a tank with any size radius. *$V = (4/3)\pi r^3 + 4\pi r^2$*

PG-125. Rewrite the equation for the gas tank in part c of PG-124 using x and y instead of V and r, and if you have a graphing calculator, sketch its graph. What part of your graph represents possible volumes for the gas tanks? What does the rest of the graph represent?

5.9 THE GATEWAY ARCH: SUMMARY

The Gateway Arch returns as a possible portfolio problem, or you might want to consider it part of the chapter summary.

PG-126. Summary (PG-126 to PG-129) Remember why we wanted to move parabolas around in the first place? We wanted the specific equation for a parabola that could approximate the Gateway Arch. Now we can do it. Go back and read PG-1 again, and create an equation. Write out your work in good detail, explaining your ideas where necessary. Your solution should be clear enough to convince your group members and your instructor that you understand the mathematics.

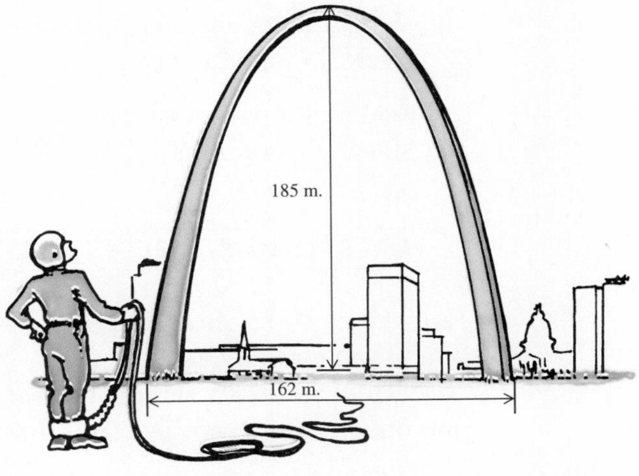

PG-127. A student from another algebra class says that he can use a method called "completing the square" to write the equation $y = 2x^2 - 12x + 11$ in graphing form. Explain to this student, step by step as you work through the problem, how you can find the x-intercepts of the graph and then use them to write the graphing form of the equation.

x-intercepts: $\left(\dfrac{12 \pm \sqrt{56}}{4}, 0\right)$, $y = 2(x-3)^2 - 7$

PG-128. Write an explanation describing how a graph is related to its parent graph so that someone new to this class could understand how to tell whether the stretch factor of its equation was:

a. positive or negative

b. less than 1 or greater than 1

PG-129. Each of the general equations in your Parent Graph Tool Kit has a locator point.

a. For each parent graph, describe what part of the graph the locator point locates.

b. For each parent graph, write an equation for a new graph using the locator point (3, –4), and sketch each graph.

PG-130. *Self-Evaluation* Look back over your work on the labs and investigations that you did in this chapter. Write several paragraphs summarizing what you've discovered. What new mathematics did you learn? What were the key ideas? What conclusions, conjectures, and generalizations can you make? What unanswered questions do you still have?

PG-131. *Tool Kit Check.* Be sure that your Tool Kit has at least the following items in it. For each of these, include an example and any other information that will be helpful for you.

- flipping graphs
- shifting graphs
- stretching or compressing graphs
- the graphing and standard forms of quadratic equations
- finding the vertex of a parabola starting from standard form
- parent graphs
- locator point
- function
- rational function
- asymptote

EXTENSION AND PRACTICE

PG-132. *Portfolio Growth over Time (Problem No. 1)* This is the second time around for this growth-over-time problem. Be sure you include everything you did the first time and more. Remember, this is a problem to do on your own and save for future reference.

On a separate sheet of paper (you will be handing this problem in separately or putting it into your portfolio), explain everything you know about the following equations:

$$y = x^2 - 4 \quad \text{and} \quad y = \sqrt{x+4}$$

PG-133. For each of the following equations, explain what d does to the shape, size and/or location of the equation's graph:

a. $y = d\sqrt{x}$ **b.** $y = 3x^2 - d$

c. $y = (x - d)^2 + 7$ **d.** $y = \frac{1}{x} + d$

PG-134. Sketch a graph of $g(x) = x^2 - 2x$.

a. Is it a function? How do you know? *yes*

b. What are the coordinates of its vertex? *(1, –1)*

c. What are its domain and range? *D: all real numbers; R: $y \geq -1$*

PG-135. Consider the functions $f(x) = x^3 + 1$ and $g(x) = (x + 1)^2$.

a. Sketch the graphs of the two functions.

b. Find $f(3)$. *28* **c.** Solve for x if $f(x) = 9$. *2*

d. Find $g(0)$. *1* **e.** Solve for x if $g(x) = 0$. *–1*

f. Solve for x if $f(x) = -26$. *–3* **g.** Solve for x if $g(x) = -16$. *no solution*

PG-136. Find the x- and y-intercepts and the locator point of each function in PG-135. Then find the equations of any lines of symmetry.

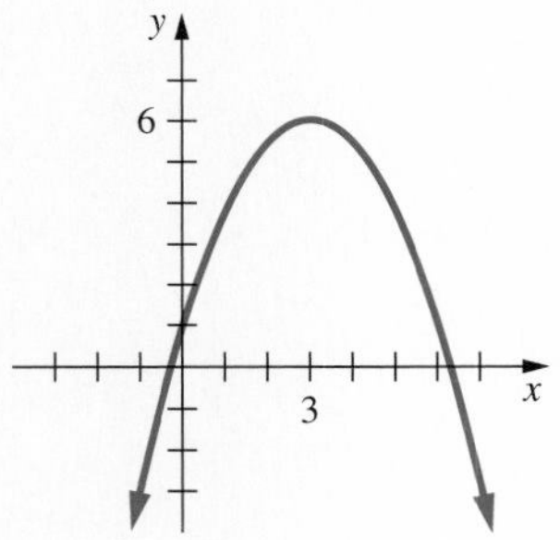

PG-137. Write an equation for this graph, assuming that it has y-intercept (0, 1). *$y = -\frac{5}{9}(x - 3)^2 + 6$*

PG-138. An archer shoots an arrow at a target that is 40 meters away. If the arrow strikes the target at the same height above the ground as it was released (about 1.3 meters above ground, or at "eye level"), choose a reasonable maximum height for the flight of the arrow, and find a possible equation to model the path based on the maximum that you chose. There is not enough information to get the precise equation of the parabola, so you will need to draw a reasonable trajectory and estimate. *Results will vary depending on what people choose for a reasonable maximum height.*

PG-139. Sketch a graph of each function:

a. $g(x) = (x + 3)^3$ **b.** $y - 1 = 3^x$ **c.** $y = -\sqrt{7 - x} + 3$

PG-140. Consider the function $p(x) = x^2 + 5x - 6$:

a. Find the point where the graph of $p(x)$ intersects the y-axis *(0, –6)*

b. Find the points where the graph of $p(x)$ intersects the x-axis. *(–6, 0) and (1, 0)*

c. Now suppose $q(x) = x^2 + 5x$. Find the intercepts of $q(x)$, and compare the graphs of $p(x)$ and $q(x)$. *x: (0, 0), (–5, 0); y: (0, 0). The graph of p(x) is 6 units lower than that of q(x).*

CHAPTER SIX

THE TOY FACTORY

Linear Systems

IN THIS CHAPTER YOU WILL HAVE THE OPPORTUNITY TO:

- see how graphs of systems of inequalities form the basis for a method to solve more complicated application problems;
- extend your knowledge of solving systems of equations to problems involving three variables and three equations;
- develop renewed appreciation for the value of group work in solving problems.

THE CHAPTER PROVIDES SEVERAL PROBLEMS THAT REQUIRE PULLING TOGETHER YOUR KNOWLEDGE ABOUT SOLVING AND GRAPHING IN ORDER TO ADDRESS THE KINDS OF QUESTIONS POSED BY BUSINESS PROFESSIONALS AND BIOLOGISTS. THE ACTUAL QUESTIONS ARE DESIGNED TO BE SOLVABLE WITHOUT THE USE OF ELABORATE TECHNOLOGY, SO THEY ARE NECESSARILY OVER SIMPLIFIED BOTH IN SIZE OF NUMBERS AND IN COMPLICATION. BUT THE TOOLS AND THE THINKING YOU USE TO SOLVE THEM ARE REPRESENTATIVE OF THE TOOLS AND THINKING NEEDED IN THE REAL SITUATIONS.

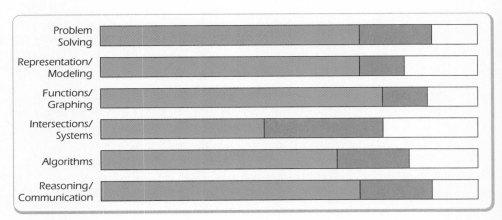

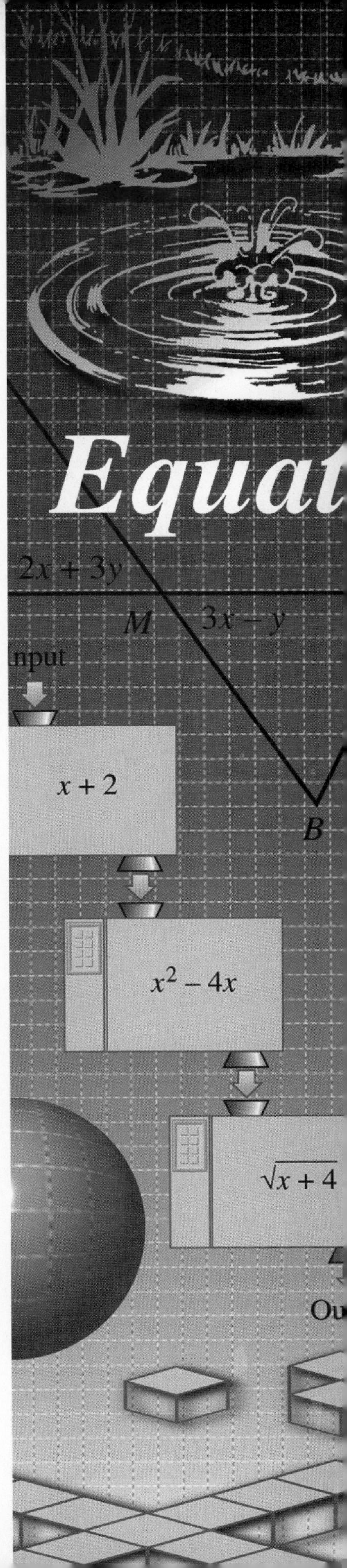

MATERIALS	CHAPTER CONTENTS		PROBLEM SETS
	INEQUALITIES AND THEIR APPLICATIONS		
LS-1 Resource Page	**6.1**	Linear Inequalities	LS-1 – LS-21
	6.2	Systems of Linear Inequalities	LS-22 – LS-35
Transparencies (one per group)	**6.3**	Applications of Inequalities	LS-36 – LS-48
	6.4	Sandy Dandy Dune Buggies, Inc.	LS-49 – LS-60
	SYSTEMS OF EQUATIONS IN THREE VARIABLES		
Graphing Calculators	**6.5**	Three Equations, Three Variables	LS-61 – LS-76
	6.6	Building Equations: Solving for the Coefficients	LS-77 – LS-92
	6.7	Applications of Parabolas, Summary	LS-93 – LS-114

In this chapter, we want students to build on their understanding that the graph of a function represents all those points that satisfy the equation. To do this, we explore graphing inequalities and solving systems of equations, which will lead to the development of general systems in Chapter 8. Throughout this chapter, be sure to have graphing calculators readily available.

This chapter can be split depending on what topics your department emphasizes and requires. Sections 6.1 through 6.4 cover inequalities and their applications. (Section 6.4 is an assessment day covering inequalities.) Sections 6.5 through 6.8 continue with solutions of systems, but now extend to systems of three equations and three variables. Section 6.7 also ***includes*** *an assessment covering solving systems and applications. As mentioned earlier, changing the way we assess students is essential to truly changing the way we teach and the way students learn mathematics.*

This chapter provides an opportunity to try an alternative form of assessment. If you have been using traditional chapter exams to evaluate your students, give LS-93–LS-96 a try. This assessment should be done ***instead*** *of an exam. If you are looking for understanding, you will be able to tell which students have it from their work on these items.*

In Section 6.5, where we look at solving three equations with three unknowns, we mention to the students that these systems can be solved with a graphing calculator, and we refer interested students to Appendix D. In this appendix, we explain ***generally*** *how to solve Ax = B, where A is a coefficient matrix and x and B are column matrices. Appendix D is not calculator-specific, so you will need to figure out how to do this on your calculator if you decide to take this approach with your class. With just a few examples students can understand how to solve these systems, and some can even understand why it works. For those students who find that small errors keep them from answering the bigger questions correctly, allowing them to use the graphing calculator to solve these systems lets them get past the number crunching and move on to the more interesting questions.*

6.1 LINEAR INEQUALITIES

We start with a graphical approach to inequalities. The first problem uses a number line to reinforce the concepts ***less than*** *and* ***greater than*** *and to show graphically why and when we reverse the inequality.*

LS-1. On your paper, draw a number line as shown below or use a copy of the Resource Page containing eight handy number lines.

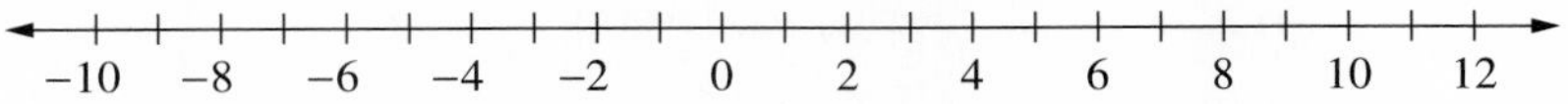

On the number line it is easy to see that −2 is less than 6 because it is to the left of 6. Algebraically, we represent the fact that −2 is less than 6 as $-2 < 6$.

a. Place a point at −2 and another at 6 on the number line. If we add 3 to each of these numbers, we can represent the results on the number line as follows:

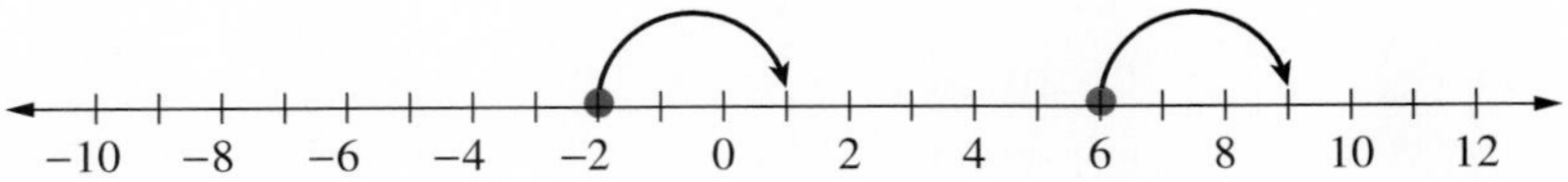

$-2 < 6$
$-2 + 3 < 6 + 3$
$1 < 9$

The new points, 1 and 9, maintain the same order: $1 < 9$. We say that the inequality is **preserved**. Will the inequality be preserved if we multiply rather than add a number to each side? *Answers will vary.*

b. Using the points −2 and 6, multiply each number by $\frac{1}{2}$ and indicate the result with arrow "jumps" on your number line like those shown in part a. Use a new number line for this. Is the inequality still preserved? *Yes.*

c. Using the points −2 and 6, multiply each by −1. What happened? Represent this on the number line and with an inequality. *$-2 < 6$, $2 > -6$; the inequality is not preserved.*

d. Start with a new number line. Mark −5 and −1 on the number line. Multiply both numbers by 2, and indicate the result with jumps on the number line. Is the inequality preserved? *Yes*

e. Start again with −5 and −1, but this time multiply by −2. Show this on the number line. What happened? Is the inequality preserved? *$-5 < -1$, $10 > 2$; no*

f. This time locate −5 and 1 on a new number line and write an inequality. Now multiply by −2. Show the jumps on the line, and write the inequality that represents the results. *$-5 < 1$, $10 > -2$*

g. Locate points at −2 and −6 on a number line. Divide both by −2, and indicate this result on the number line. What happened? *$-2 > -6$, $1 < 3$*

LS-2. In LS-1 we saw that multiplying both sides of an inequality by a negative number changes the original relationship. How does it change it? *reverses it*

a. What statement would you make to show the relationship between 2 and −10? $2 > -10$

b. Multiply both numbers by a negative number and write the inequality for the result. Show two more examples by choosing any two numbers, writing an inequality, multiplying by a negative number, and writing the inequality for the result.

LS-3. Your analysis in LS-1 and LS-2 will help you accurately transform inequalities into *y*-form (that is, solving for *y*).

a. Substitute $y = 1$ into the inequality $-2y < 6$. Is the resulting inequality true or false? *True.*

b. Write the original inequality in *y*-form. Substitute $y = 1$ into your transformed inequality. Is it still true? If not, what must you do to the inequality to make it true again? $y > -3$*; still true only when inequality reversed*

c. Repeat this test on the following inequalities: choose a *y* value that makes the original inequality true, write the inequality in *y*-form, then test the value again to be sure that you have adjusted the inequality correctly so it represents an equivalent statement.

i. $-3y + 1 > 4$ $\quad y < -1$

ii. $2y + 7 > 5$ $\quad y > -1$

iii. $6 - 4y < 0$ $\quad y > 3/2$

d. Write a rule about multiplying and dividing both sides of inequalities by negative numbers. Be sure to include this rule and several examples in your Tool Kit. *Reverses the order of the inequality.*

e. If you were to add a negative number to both sides of an inequality, would you have to reverse the relationship? Give at least two examples, and explain why or why not.

LS-4. When solving an inequality, when and why do you have to reverse the inequality symbol? When is it *not* necessary to reverse it? *multiplying or dividing by a negative number*

LS-5. Remember that when drawing the graph of $y > 3x - 1$, we begin by graphing the line $y = 3x - 1$ with a dashed line. It is customary, in other words at some point people decided, to use a dashed line instead of a solid line for problems like this one. Why would it be important to use something different from a regular solid line? *We are not including the line* $y = 3x - 1$*, since points on this line are not in the solution set.*

a. Graph the line $y = 3x - 1$ (draw it dashed), and plot the points from the following table in addition to your line. Label each point with its coordinates.

Test Point	Location (Above, Below, or On)	$y > 3x - 1$ (True or False)
$A(0, 0)$	*Above*	*True*
$B(3, 4)$	*Below*	*False*
$C(-1, -4)$	*On*	*False*
$D(-2, 3)$	*Above*	*True*
$E(\frac{-2}{3}, -4)$	*Below*	*False*
$F(5, 14)$	*On*	*False*
$G(-3, 9)$	*Above*	*True*

b. Complete the table. Which of these points make $y > 3x - 1$ true? ***A, D and G***

c. Which region of your graph should be shaded to represent all the points for which $y > 3x - 1$ is true? Describe the region *and* shade it. ***any points above*** $y = 3x - 1$

d. From the table you can see that even though test points C and F are on the line, they do not make $y > 3x - 1$ true. That is why you use a dashed line for the graph of $y > 3x - 1$. This dashed line indicates that points on the line $y = 3x - 1$ are *not* part of the solution for $y > 3x - 1$. Be sure to include this information about why the lines for some inequalities are dashed in your Tool Kit .

LS-6. Graph the line $y = \frac{1}{2}x + 1$.

a. How many points do you need to check to decide which side of the line to shade to represent $y \le \frac{1}{2}x + 1$? Explain, and shade the appropriate side. ***We need to check only one point***

b. Shade the appropriate area above or below the line, and explain why you chose to shade this side. ***Shade the side that will make the inequality true.***

c. Should the line you graphed originally for this problem, $y = \frac{1}{2}x + 1$, be solid or dashed? Explain. ***solid***

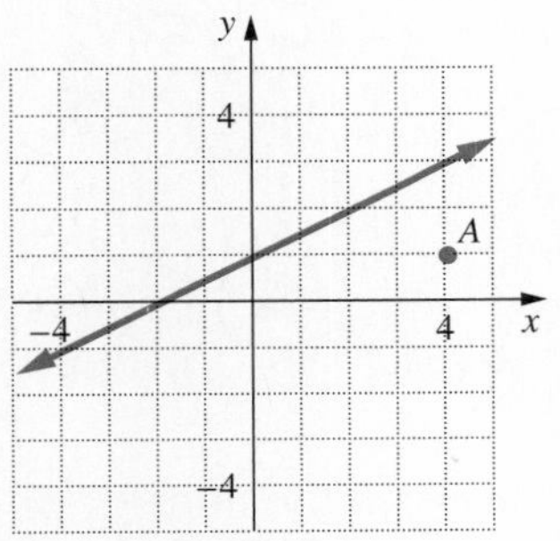

LS-7. To determine which side of the line $y = \frac{1}{2}x + 1$ must be shaded, we can use a test point. Here's how: select any point *not* on the line. In the graph shown here, we have chosen point A at (4, 1) to be our test point. Substitute the x- and y-coordinates into the inequality to see if the result is true or false.

$$1 \le \frac{1}{2}(4) + 1$$

$$1 \le 2 + 1$$

$$1 \le 3$$

True!

Since the point A makes the inequality true, *every* point below the line will make the inequality true. Why? Explain this to your group. We shade the side of the line that makes the inequality true, so which side do we shade? Why? *Shade below the line because we want the points that are less than $\frac{1}{2}x + 1$.*

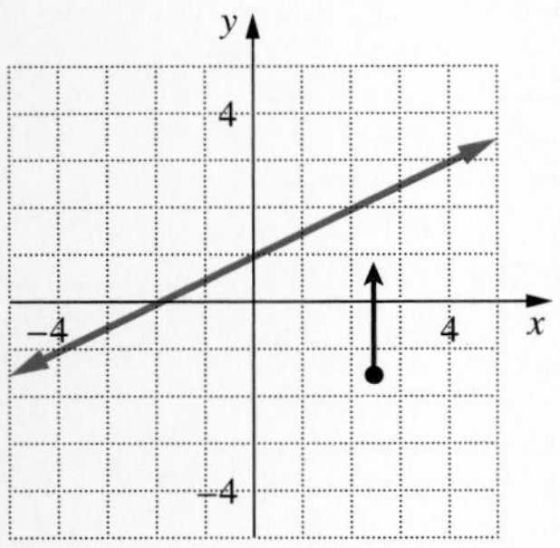

LS-8. Hobbs says he has a faster way to decide which side of the line $y = \frac{1}{2}x + 1$ to shade. "Instead of trying out the numbers that represent a point on one side of the line, I just shade above the line when it says '$y >$' and below the line when it says '$y <$'."

Calvin is doubtful. "But how can you tell what is above or below a steep line?" he asks.

"It's easy" says Hobbs. "Look at the graph I drew. Just put your pencil point somewhere on the graph paper, but not on the line. Visualize a vertical line through your point. To get to the graph, do you have to slide up the vertical line or down? If you answered 'down,' then you are *above* the line and the y value of your point is higher ($y >$) than the point on the line. If you said 'up' you're *below* the line and the y value is less ($y <$)."

"Oh yeah," says Calvin. "I think I get it."

a. Try Hobbs's method for $y \geq \frac{1}{3}x + 2$.

b. Now try it for $y < 3x - 7$.

c. People who "think visually" tend to like this method. Others prefer checking the numbers in the inequality. What is your preference?

LS-9. Graph the solution to each of the following inequalities on a different set of axes. Label each graph with the inequality as given *and* with its y form. Choose a test point, and show that it gives the same result in both forms of your inequality.

a. $3x - 3 < y$ *$y > 3x - 3$* **b.** $3x - 2y \leq 6$ *$y \geq \frac{3x}{2} - 3$*

LS-10. Simone has been absent and does not know the difference between the graph of $y \leq 2x - 2$ and the graph of $y < 2x - 2$. Explain thoroughly so that she completely understands what points are excluded from the second graph. *All points where $y = 2x - 2$*

LS-11. Since she had been doing so well in her algebra class and her roommate is very talkative, Lorrel decided to reward herself with her own phone line. Lorrel's boyfriend lives in the San Francisco Bay area, and she needs to choose a long-distance plan. The phone company offers two plans: Normal Rate, which costs \$12 for each hour of use, or High-Use Rate, which has a monthly fee of \$12 plus \$10 for each hour of use.

a. Write an equation, one for each plan, which will represent Lorrel's long-distance cost based on the number of hours she calls her boyfriend. *Normal: $C = 12h$; High: $C = 10h + 12$*

b. Graph each of the equations you wrote for part a on the same set of axes. Under what conditions is it better for Lorrel to use the Normal Rate? When is it better to use the High-Use Rate? *Use High-Use for more than 6 hr, Normal for less than 6 hr*

c. When could either company's plan be used for the same price? What is the significance of this value? How does it relate to the graph? *Same for 6 hr; it is the solution to the system.*

EXTENSION AND PRACTICE

LS-12. Angela and Zack are going to a Rolling Stones concert. They agreed that Angela would pay $105 for the tickets. Zack would pay for the limo rental. A Cadillac rents for $50 plus $6 per hour. A Lincoln rents for $25 plus $12 per hour. Write an equation to represent each company's cost using hours as the independent variable. Graph both equations on the same set of axes. Be sure to label the axes.

a. For what number of hours can Zack rent either car for about the same cost? (Find your answer to the nearest 1/2 hour.) How much is the cost? *≈ 4 hr; cost ≈ $73–74*

b. If Zack rents the car for seven hours, which car should he use? How much is the cost? *Cadillac = $92*

c. Zack wants to spend the same amount as Angela. Which car should he rent to get the most time? How long will they have the car? *Cadillac = 9 hr and 10 min*

LS-13.

a. Sketch and describe their graphs of the equations $y = 3$ and $x = -2$:

b. Where do these graphs cross? *(–2, 3)*

LS-14.

a. Graph the following on the same set of axes:

i. $y = 3$ **ii.** $y = 2$ **iii.** $y = 1$

b. What would be the equation of the x-axis? *$y = 0$*

c. What would be the equation of the y-axis? *$x = 0$*

LS-15. Solve the following system of equations.

$$2x + 6y = 10$$
$$x = 8 - 3y$$

a. Describe what happened. *no solution.*

b. Draw the graph of the system.

c. How does the graph of the system explain what happened with the equations? Be clear. *Parallel lines do not intersect.*

d. What might be a similar situation in three dimensions? *parallel planes*

LS-16. Jameela needs to find the point of intersection for the lines $y = 18x - 30$ and $y = -22x + 50$. She takes out a piece of graph paper and then realizes she can solve this problem without graphing. Explain how Jameela is going to accomplish this, and find the intersection point. *She will solve the system algebraically; (2, 6)*

LS-17. Find the x- and y-intercepts:

a. $2x - 3y = 9$
x: (9/2, 0); y: (0, –3)

b. $3y = 2x + 12$
x: (–6, 0); y: (0, 4)

LS-18. Solve for w:

a. $w^2 + 4w = 0$
0, –4

b. $5w^2 - 2w = 0$
0, 2/5

c. $w^2 = 6w$
0, 6

LS-19. In 1976, the world's largest ice cream sundae was made in New York City's Central Park. It used 1500 gallons of ice cream and weighed 7250 pounds, of which 25 pounds were cherries. One average-sized cherry weighs approximately 6.5 grams, and 1 kilogram is approximately equal to 2.2 pounds.

a. Approximately how many cherries were used for the sundae? *≈1750*

b. If the sundae was divided up evenly so that each person received one cherry, what would be the weight of an individual serving of the sundae? *4.14 lb*

LS-20. Sketch the graphs of $y = 3x$ and $y = 3x - 1$ on the same set of axes. Where do the curves cross? Justify your answer. *They don't cross.*

LS-21. When a pebble is thrown into a pond, it creates ripples that move away in concentric circles from the point where the pebble entered the water. Sketch a graph showing the relationship between the area of the circle formed and the distance as the ripple moves away from the pebble.

a. What are the independent and dependent variables for this relationship?

b. What are the domain and range for this relationship? *D: 0 to the distance to the nearest edge of the pond; R: 0 to πr^2, where r is the shortest distance to the edge of the pond.*

6.2 SYSTEMS OF LINEAR INEQUALITIES

In this section we want students to practice graphing systems of inequalities and to start thinking about what the shaded regions represent. It is mostly a day to consolidate ideas. As you circulate among the groups, it may be helpful to remind them that one quick way to graph a line is to use the intercepts.

Solution:

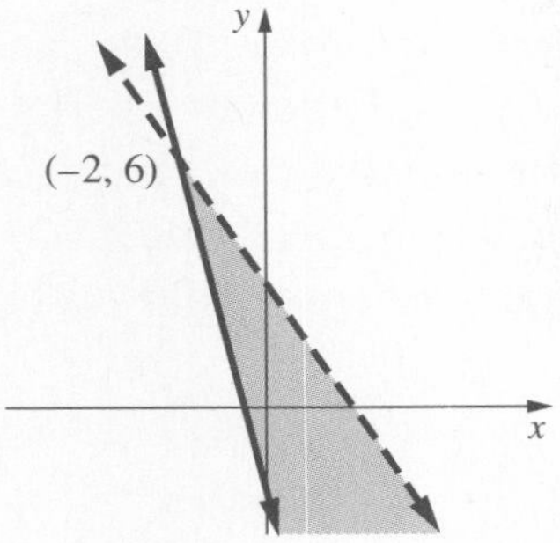

LS-22. Consider the following inequalities:

a. Graph them on the same set of axes. Lightly shade the solution of each.

$$y \geq -4x - 2$$
$$3x + 2y < 6$$

b. Test the following points in *both* inequalities, and label them on your graph. Indicate which ones are solutions to *both* inequalities.

$$A(0, 4) \quad B(0, 0) \quad C(-1, -1) \quad D(4, 3)$$

B only

c. What does the area where the graphs overlap represent? Choose a point in this region and justify your answer. *It shows where both inequalities are true.*

LS-23. Graph the following four inequalities on the same set of axes. Before you begin, it would be a good idea for the group to discuss the most efficient way to graph a line. Be very careful to compare your answers as you work through the parts!

1. $2y \geq x - 3$ **2.** $x - 2y \geq -7$

3. $y \leq -2x + 6$ **4.** $-9 \leq 2x + y$

a. What type of polygon is formed by the solution of this set of inequalities? Write a convincing argument to justify your answer. *rectangle; constructed from perpendiculars*

b. Find the vertices of the polygon. If your graph is very accurately drawn, you will be able to determine the points from the graph. If it is not, you will need to solve the systems (pairs) of equations that represent the edges of your graphs. *(1, 4), (−3, −3), (−5, 1), (3, 0)*

LS-24. Find the area of the polygon that you graphed in LS-23. *30 sq. units*

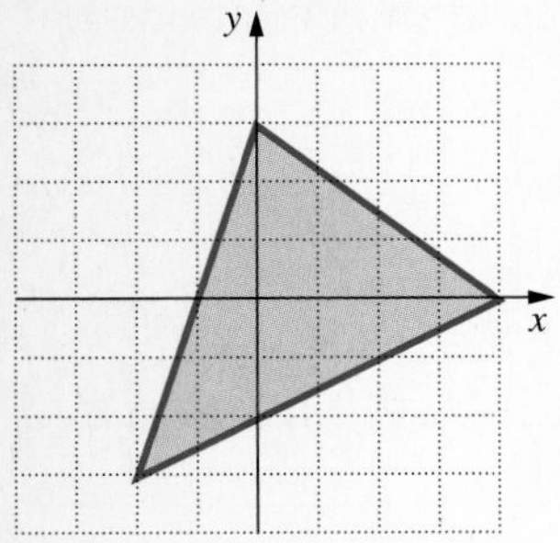

LS-25. Write the three inequalities that will form the triangular region shown here. Is it possible for more than one set of inequalities to represent this region? Explain. *$y \le 3x + 3$, $y \ge 0.5x - 2$, $y \le -0.75x + 3$; No. The graph of an inequality is the shading above or below a line, and only one line can go through two points.*

EXTENSION AND PRACTICE

LS-26. Use these three equations to solve for x, y, and z:

$$3x + 8 = 2$$

$$7x + 3y = 1$$

$$\frac{1}{2}x + y - 8z = 8$$

$x = -2$, $y = 5$, $z = -1/2$

LS-27. The Alvarez family plans to buy a new air-conditioner. They can buy the Super Cool X140 for $800, or they can buy the Efficient Energy 2000 for $1200. Both models will cool their home equally well, but the Efficient Energy model is less expensive to operate. The Super Cool X140 will cost $60 a month to operate, while the Efficient Energy 2000 costs only $40 a month to operate.

a. Write an equation to represent the cost of buying and operating the Super Cool X140. *$C = 800 + 60m$*

b. Write an equation to represent the cost of buying and operating the Efficient Energy 2000. *$C = 1200 + 40m$*

c. How many months would the Alvarezes have to use the Efficient Energy model to compensate for the additional cost of this model? *20 months*

d. Figuring they will use it only four months each year, how many years will the Alvarezes have to wait to start saving money overall? *5 years*

LS-28 is difficult. You should work through it in advance to decide whether or not your students should spend the time it will take to organize and understand it.

LS-28. Macario's salary is increasing by 5 percent each year. His rent is increasing by 8 percent each year. Currently, 20 percent of Macario's salary goes to pay his rent. Assuming that Macario does not move or change jobs, what percentage of his income will go to pay rent in 10 years? Think about this problem, and if you see a way to solve it, do it. If not, try parts a, b, c, and d. *26.5%*

a. If x represents Macario's current salary, write an expression to represent his salary 10 years from now. $x(1.05)^{10}$

b. Use x in an expression to represent Macario's rent now. $0.2x$

c. Use what you wrote in part b to write an expression for the rent ten years from now. $0.2x(1.08)^{10}$

d. Now use the expressions that you wrote in parts a and c to write a ratio that will help you answer the question.

$$0.2\left(\frac{1.08}{1.05}\right)^{10} = \frac{p}{100}$$

LS-29. Consider the pattern shown in the figure. Each cube is 1 centimeter on a side.

a. Based on the pattern, sketch or describe the missing figure III.

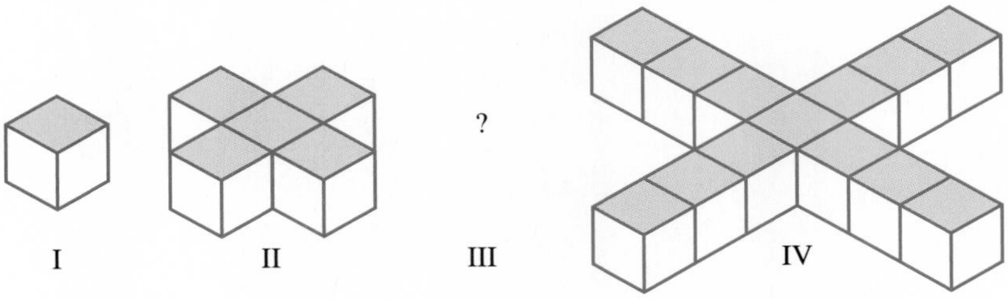

b. Find the volume of each of the four figures.

c. Assume that the pattern continues with figures V, VI, VII, and so forth. Write an expression to represent figure number N. What kind of sequence is this? *$4N - 3$; arithmetic*

LS-30. A 3-centimeter cube is first painted red and then cut into 27 unit cubes (each side is 1 cm long). All of the cut cubes are placed in a bag, and one is drawn out at random. What is the probability that the cube that is drawn has paint on the following number of faces?

a. 0 *1/27* **b.** exactly 2 *2/9* **c.** exactly 2 *4/9*

d. exactly 3 *8/27* **e.** 4 or more *0*

LS-31. Sketch, compare, and contrast the graphs of $y > x^2$ and $y > (x - 4)^2 + 2$. *The second one is shifted to the right 4 units and up 2 units.*

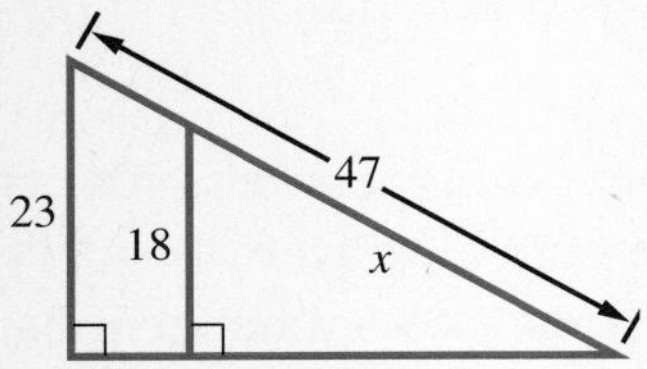

LS-32. Using the relationships shown in the figure, solve for x. It might be helpful to redraw the figure as two separate triangles. *$x \approx 36.78$*

LS-33. Ramin is trying to evaluate an expression and he cannot get the negative sign to work on his calculator! Explain to Ramin how he can simplify

$$\frac{7\left(4^{-2001}\right)}{2\left(4^{-1997}\right)}$$

without using a calculator. *Rewrite as* $\left(\frac{7}{2}\right)\cdot(4)^{-4}$ *or* $\frac{7}{\left(2\cdot 4^{4}\right)}$

LS-34. The cost of food has been increasing about 4 percent per year for many years. To find the cost of an item 15 years ago, Juanita said, "Take the current price and divide by 1.04^{15}." Her friend Alisa said, "That might work, but it's easier to take the current price and multiply by 0.96^{15}!" Explain who is correct and why. *Juanita is correct; if the original price was p and the new price is N, the relationship is described by the equation, $N = p(1.04)^{15}$, which is solved by dividing by 1.04^{15}. Alisa is using the relationship $p = N(0.96)^{15}$, which gives a different result.*

LS-35. Examine the data points plotted in this graph.

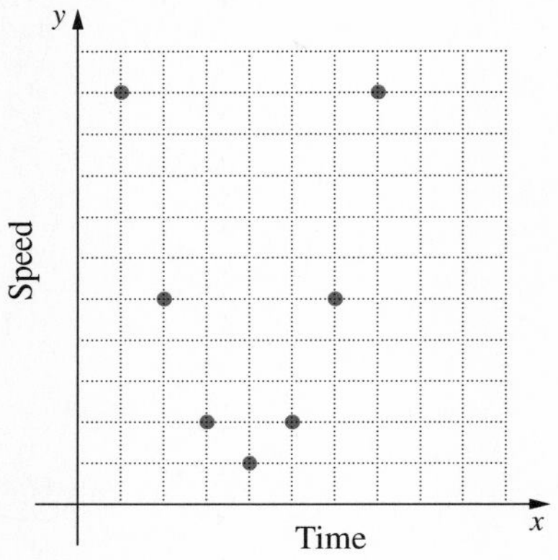

a. If the independent variable is time and the dependent variable is the speed of a skater, what does this graph tell you? *The skater's speed slows and then speeds up again.*

b. Find an equation that, when graphed, will pass through thesepoints. *$y = (x - 4)^2 + 1$*

c. Describe another situation that might produce a graph like this. *The skater could have skated over a hill.*

6.3 APPLICATIONS OF INEQUALITIES

The applications in this section are excellent problems for tying ideas together. We have found that students value the opportunity to pull a lot of ideas together and to see how algebra might be used in context.

*These are **big** problems, so allow sufficient time. Make sure students are working in groups to make this work simpler. These are not homework problems, and we don't expect students to have reached the point of being able to do them on their own. But if they are patient and are willing to go through each part carefully and work with their group, they will be able to do them together. Remember, we are not looking for mastery on these applications; we* *are providing an opportunity to practice representing, graphing, and equation-solving skills in a new context. Help each group as they need it.*

Each group of students will need an overhead transparency or a piece of (hamburger) patty paper on which to draw profit lines in the next problems. Students should work as partners on this, periodically checking with their whole group.

LS-36. The Toy Factory Otto Toyom builds toy cars and toy trucks. Each car needs four wheels, two seats, and one gas tank. Each truck needs six wheels, one seat, and three gas tanks. His storeroom has 36 wheels, 14 seats, and 15 gas tanks. He needs to decide how many cars and trucks to build so he can maximize the amount of money he makes when he sells them. He makes \$1.00 on each car and \$1.00 on each truck he sells. (This is a very, very small business, but the ideas are similar for much larger enterprises.) We will divide this problem into subproblems:

a. Otto's first task is to figure out what his options are. For example, he could decide to make no cars and no trucks and just keep his supplies. On the other hand, because he likes to make trucks better, he may be thinking about making five trucks and one car. Would this be possible? Why? What are all the possible numbers of cars and trucks he *can* build, given his limited supplies? This will be quite a long list. An easy way to keep your list organized and find some patterns is to plot the points that represent the pairs of numbers in your list directly on graph paper. Use the x-axis for number of cars and the y-axis for number of trucks. Make a fairly large, neat first-quadrant graph, and plot the points that represent the possible numbers of cars and trucks. (Why only the first quadrant?) We will need to use this graph later in this problem and in the next.

b. In part a you figured out all the possible combinations of numbers of cars and trucks Otto could make. Which of these give him the greatest profit? Explain how you know your answer is right. You have to convince Otto, who likes trucks better. ***6 cars and 2 trucks for a profit of \$8***

c. New scenario: Truck drivers have just become popular because of a new TV series called *Big Red Ed*. Toy trucks are a hot item. Otto can now make \$2.00 per truck, though he still gets \$1.00 per car. He has hired you as a consultant to advise him, and your salary is a percentage of the total profits. What is his best choice for the number of cars and the number of trucks to make now? How can you be sure? Explain. ***3 cars and 4 trucks for a profit of \$11***

Some groups may need help with the first inequality in part a of LS-37.

LS-37. In LS-36 you probably had to carry out a lot of calculations in order to convince Otto that your recommendation was correct. In this problem we'll take another look at Otto's business using some algebra and graphing tools we already know.

a. The first task is to write three inequalities to represent the relationship between the number of cars x, the number of trucks y, and the number of:

i. wheels	**ii.** seats	**iii.** gas tanks
$4x + 6y \leq 36$	*$2x + y \leq 14$*	*$x + 3y \leq 15$*

b. There are two more inequalities that should be included, $x \geq 0$ and $y \geq 0$. Explain why we can assume that $x \geq 0$ and $y \geq 0$.

c. Carefully graph this system of five inequalities on the same set of axes you used for LS-36. Shade the region of intersection lightly.

d. What are the vertices of the pentagon that outlines your region? Explain and show exactly how and why you could use the five equations below to find those five points. In many problems these points are difficult to determine from the graph alone. *(0, 0), (0, 5), (3, 4), (6, 2), (7, 0)*

$x = 0 \qquad 4x + 6y = 36 \qquad x + 3y = 15$

$y = 0 \qquad 2x + y = 14$

e. Part d shows how we could get an outline of the region of points that represents possible numbers of cars and trucks. Are there some points in the region that seem more likely to give the maximum profit? Where are they? Why do you think they are the best coordinates? How can you represent the total profit if Otto makes $1.00 on each car and $2.00 on each truck? *certain vertices; profit = $x + 2y$*

f. What if Otto ended up with a profit of $8? Write an equation for the profit when it is $8.00. Use the graph of just one of your group members, and draw the graph of this equation on it. Do you think from looking at this graph that $8 is the maximum Otto could make? Why or why not? Try some other possibilities (such as $9, $10, $11, or $12). Write an equation and draw a graph for each. (Continue to use just one person's paper). Find the maximum and justify your answer for Otto. *$x + 2y = 8$; no, there are points above the line; the maximum is $11.*

It might be necessary to demonstrate part g of this problem. Have one of the groups demonstrate if possible.

g. What did all the profit lines you drew have in common? What was different? What were you trying to do? Ask your teacher for a transparency, place it on top of some graph paper, and draw just a y-axis and one line with the same slope as your profit lines. Now put the transparency on top of one of your group's graphs and align the y-axis. Slide the transparency up and down keeping the y-axes aligned to find the maximum profit. Try it on another group member's graph and explain why this method should work. *Same slope, y-intercepts are different. Slide the graph to find the maximum.*

An alternative way of dealing with parts f and g in LS-37 is to use the profit expression in part e and apply it to each vertex of the region to determine which vertex gives the maximum profit. While this produces the maximum, it does not lead to an explanation of why that particular vertex produces it.

LS-38. Use the method you developed in LS-37 to find Otto's maximum possible profit if he gets \$3.00 per car and \$2.00 per truck. *6 cars and 2 trucks for a profit of \$22*

EXTENSION AND PRACTICE

LS-39. Solve the following system algebraically, and explain what the solution tells you about the graphs of the two equations.

$$4x - 6y = 12$$
$$-2x + 3y = 7$$

No solution

LS-40. Marvelous Mark's Function Machines. Mark has set up a series of three function machines that he claims will surprise you:

a. Try a few numbers. Were you surprised? *input x, output x*

b. Caitlan claims she was not surprised, and she can show why the sequence of machines does what it does by simply dropping in a variable and writing out step by step what happens inside each machine. Try it (use something like *c* or *m*). Be sure to show all the steps.

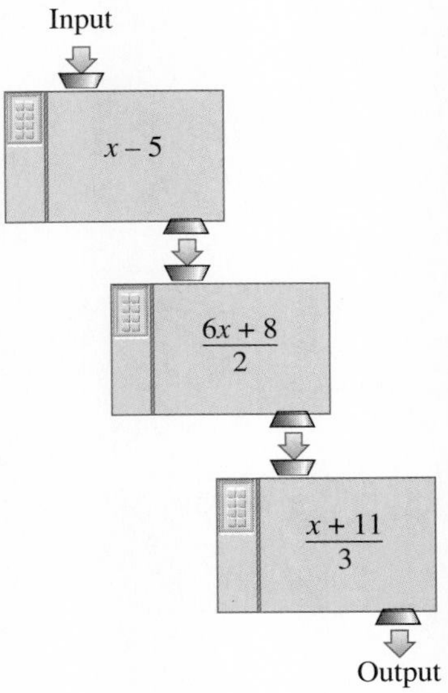

LS-41. Hamida inadvertently conducted an experiment by leaving her bologna sandwich in the science lab over the winter break. Much to her surprise, her sandwich (or what used to be a sandwich) was much larger than it was when she left it.

After inspecting it, her science instructor said Hamida had produced great quantities of a seldom-seen bacterial species, *Bolognicus sandwichae*. Based on a sample taken from the sandwich, they determined that there were approximately 72 million bacteria present. Hamida was greatly surprised by this number. Her biology teacher informed her that this is not too surprising since the number of this particular bacteria triples every 24 hours. Hamida thought the sandwich must have been loaded with bacteria initially since it had been made only 15 days ago.

Hamida is making plans to sue the meat company because she learned that the food industry standard for the most bacteria a sandwich could have at the time it was made was 100. Find out how many of the bacteria were present when the sandwich was made to determine whether Hamida has a case. *Only 5 bacteria were present.*

LS-42. Rewrite as an equivalent expression using fractional exponents.

a. $\sqrt[2]{5} =$ *$5^{1/2}$* **b.** $\sqrt[3]{9} =$ *$9^{1/3}$* **c.** $\sqrt[8]{17^x} =$ *$17^{x/8}$*

LS-43. Solve $2^x = 3$ for x by using guess and check. Be accurate to three decimal places. $x \approx 1.585$

LS-44. Solve for z:

$$\frac{2z+1}{5} - \frac{z}{10} = \frac{1-z}{4}$$

$1/11 \approx 0.0909$

LS-45. Graph $x = 4$:

a. on a number line.

b. in the xy-plane. *line parallel to the y-axis at x = 4.*

c. How do you think it would look in three-dimensional space? Try to sketch it. *plane parallel to yz-plane at x = 4*

d. How do the graphs differ in parts a, b, and c? *point, line, plane.*

LS-46. For every three turnips Kristin washes, Carlos eats two and one-half. How many has Kristin washed when Carlos has polished off a hearty meal of 18 turnips? *21.6, so she is in the middle of washing her 22nd turnip.*

LS-47. Solve each of the following:

a. $3x^2 - 5x - 11 = 0$

$x = \dfrac{5 \pm \sqrt{157}}{6}$

b. $\dfrac{3}{x-1} + \dfrac{x}{2} = 8$

$x = \dfrac{17 \pm \sqrt{201}}{2}$

c. $3^{x-2} = 81^{x+3}$

$x = -14/3$

d. $2x - y = 15$
$y = 3x + 7$

$(-22, -59)$

LS-48. Below are two lists. Within each list, briefly explain the differences in the graphs as you go from one equation to the next down the list.

a. $y = x$
$y < x$
$y \le x$

b. $y = x^2$
$y > x^2$
$y \ge x^2$

(a) First is a line; second is the same line, just dashed and shaded below; third is the same shading but now the line is included. (b) First is a parabola; second has the region above the parabola shaded with the parabola dashed; third is the same except the parabola is also included.

6.4 SANDY DANDY DUNE BUGGIES, INC.

LS-49 is another application of inequalities that may be used as a portfolio builder and is optional. (The whole day can be skipped.) You could use this problem as a group assessment, as a test of content, or as an assessment of how well the groups can work together when tackling a difficult problem. For this type of assessment you should circulate, listen, and be available to help with group questions and prompts. Students should have questions, and you should assist with prompts. That's how you will get to know how they are doing. Every group should complete the basic problem, after which you can assess how much help each group needed and what kind of questions they asked.

Remind students that our purpose in looking at these problems is to get an idea of how some of the algebraic and graphing techniques they have learned can be used in concert to solve complicated problems. We have selected artificially small, nice numbers so students won't have to deal with scaling. Computers can be programmed to handle such messiness and are used to resolve larger, more complicated situations using similar techniques. Students could work in groups and pairs. Each pair would be responsible for producing a report, but they would probably need their whole group for figuring out what to do.

If you choose to use this as a portfolio problem, students should do individual reports as well as write a description of what they learned and skills they used. Another option is to have them write a well-organized business letter to the company with their findings. Their letters should include their graphs, neatly reproduced, as a basis for their explanation.

LS-49. Sandy Dandy Dune Buggies, Inc. This company makes two models of off-road vehicles: the Sand Crab and the Surf Mobile. The company can get the basic parts to produce as many as 15 Sand Crabs and 12 Surf Mobiles per week, but two parts have to be special-ordered: a unique exhaust manifold clamp and a specially designed suspension joint. Each Sand Crab requires five manifold clamps and two suspension joints. Each Surf Mobile requires three manifold clamps and six suspension joints. For each week the maximum number of clamps available is 81 per week and the maximum number of joints is 78 per week.

It takes 20 hours to assemble one Sand Crab and 30 hours to assemble one Surf Mobile. The company has 12 employees who each work a maximum of 37.5 hours per week.

Sandy Dandy, Inc., has hired your group as consultants to advise them on how many Sand Crabs and how many Surf Mobiles they should make each week in order to maximize their profit. They know their profit margin on each Sand Crab is $500 and on each Surf Mobile $1000. Prepare a complete report including your graphs and an explanation to justify your conclusions.

Our goal in this problem is not simply to solve the problem. We want to see how it might be solved by using a particular method that relies on using your algebraic and graphing skills.

You may choose to work on this problem without further suggestions, or you can use parts a to c as a guide.

a. First you need to use the information in the problem to decide how many combinations of numbers of vehicles can be built. You could do this point by point, but it probably is more efficient to write five inequalities and graph them. Work in groups of four, but each *pair* should make a large, neat graph with equations written along the lines they represent.

Let x represent the number of Sand Crabs built in one week.

Let y represent the number of Surf Mobiles built in one week.

$0 \le x \le 15$; $0 \le y \le 12$; $5x + 3y \le 81$; $2x + 6y \le 78$; $20x + 30y \le 12(37.5)$

b. You can assume that $x \ge 0$ and $y \ge 0$. Why? What kind of polygon did you get? What would be useful to know about this polygon? Find them. *coordinates of the vertices*

c. What expression can you write for the profit? How can you use this expression and your graph to find the maximum possible profit and the number of Sand Crabs and Surf Mobiles that will produce it? Do it, and complete your report. *Vertices of the polygon are (0, 0), (0, 12), (3, 12), (6, 11), (12, 7), (15, 2), and (15, 0). Profit = 500x + 1000y; maximum profit of $14,000 when Sand Crabs = 6 and Surf Mobiles = 11.*

EXTENSION AND PRACTICE

You may want to assign only a few of these problems to allow students time to finish their work on the Sandy Dandy Dune Buggies problem. They should be able to complete the problem in class, but they may need time to finish their write-up.

LS-50. Graph the inequalities, and calculate the area bounded by them:

$$y \le 2x + 6$$
$$y \le -x + 3$$
$$y \ge -2$$

27

LS-51. Solve the following system:

$$2^{x+y} = 16$$
$$2^{2x+y} = \frac{1}{8}$$

$x = -7$, $y = 11$

LS-52. Paul states that $\sqrt{a+b}$ is equivalent to $\sqrt{a} + \sqrt{b}$. Joyce thinks that Paul is incorrect with his statement. Help Joyce show Paul that the two expressions are not equivalent.

LS-53. From each of the following pairs of domain and range "shadow" graphs, sketch two different graphs that satisfy the conditions of each pair of shadows. *Hint*: Draw a "box" that bounds the given domain and range. In part b, your "box" can have only two sides.

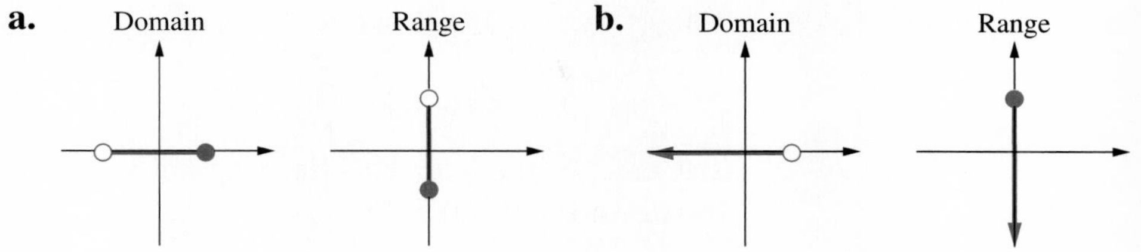

LS-54. You are standing 60 feet away from a five-story building in Los Angeles, looking up at its rooftop. In the distance you can see the billboard on top of your hotel, but the building is completely obscured by the five-story building in front of you. If your hotel is 32 stories tall and the average story is 10 feet, how far away from your hotel are you? ***about 384 ft as the crow flies***

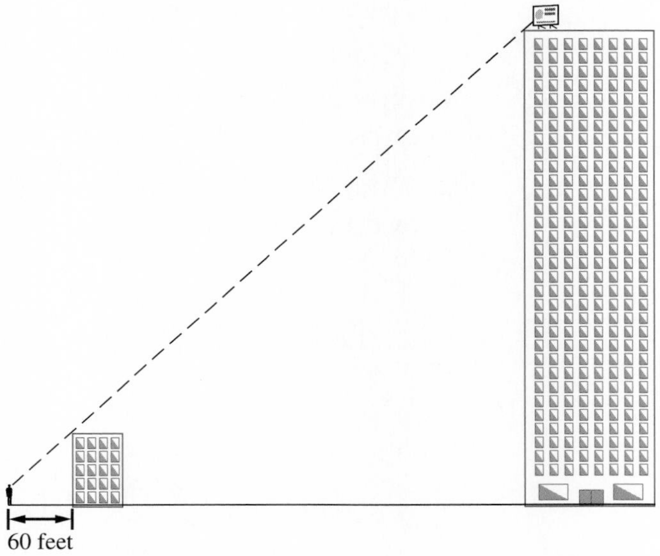

LS-55. Solve the following system of equations. What subproblems did you need to solve?

$$\begin{aligned} x + 2y &= 4 \\ 2x - y &= -7 \\ x + y + z &= -4 \end{aligned}$$

(–2, 3, –5)

LS-56. Consider the arithmetic sequence 2, $a - b$, $a + b$, 35, ... Find the values of a and b. *$a = 18.5$, $b = 5.5$*

LS-57. A square target 20 centimeters on a side contains 50 nonoverlapping circles. Each circle is 1 cm in diameter. Find the probability that a dart hitting the board at random hits one of the circles. ***$50(\pi/4)/400 \approx 9.8\%$***

LS-58. Think about the axes on the coordinate plane. What is the equation of the x-axis? What is the equation of the y-axis? ***$y = 0$, $x = 0$***

LS-59 is a preproblem for Chapter 8.

LS-59. A line intersects the graph of $y = x^2$ twice, at points whose x-coordinates are -4 and 2.

Draw a sketch of both graphs, and find the equation of the line. ***$y = -2x + 8$***

LS-60. An investment counselor advises a client that a safe plan is to invest 30 percent in bonds and 70 percent in a low-risk stock. The bonds currently have an interest rate of 7 percent, and the stock has a dividend rate of 9 percent. The client plans to invest a total of x dollars. In other words he doesn't know exactly how much he wants to invest.

a. Write an expression for the annual income that will come from the bond investment. ***bond: $0.07(0.3x)$ or $0.021x$***

b. Write an expression for the annual income that will come from the stock investment. ***stock: $0.09(0.7x)$ or $0.063x$***

c. Write an equation and solve it to find out how much the client needs to invest to have an annual income of \$5,000? ***$0.084x = 5000$; \$59,524***

6.5 THREE EQUATIONS, THREE VARIABLES

LS-61 is a good reference problem for solving systems in three variables. You might want to have students put it into their Tool Kits or write it up as a portfolio entry. They should work through the problem in their groups. If you don't want to spend a lot of time on systems of three equations with three unknowns, but would like to be able to do the application problems in Section 6.7, consider the note at the end of LS-61. This is a good point to mention using the graphing calculator to solve these systems. You will need to figure out the specifics for your calculator.

LS-61. Consider these three equations in three variables:

$$\begin{aligned} x + y - 2z &= 5 \\ 2x + y + z &= 0 \\ 3x - 2y + z &= 1 \end{aligned}$$

a. You already know everything you need to know to solve this system. This problem simply has a few more subproblems. As with two equations you need first to decide on one variable to *eliminate*.

b. Next, select any pair of equations from the given set of three equations. Use your knowledge of solving systems of equations with two variables to eliminate your chosen variable.

c. Then select a different pair of equations (choose one equation you have not used yet and one equation you've already used) and eliminate the same variable again.

d. You have modified the original system so that you now have a familiar situation: two equations with two variables. Rewrite and solve the new system.

e. You now know two of the three variables, but you're not finished. What is missing? Complete the solution. *$x = 1$, $y = 0$, $z = -2$*

f. Check your solution in each of the original equations.

NOTE: If solving systems of three equations with three variables causes you grief, you can learn to use your graphing calculator to solve them. Appendix D describes the general method for using the graphing calculator to solve systems of three equations using matrices. You will need to review your graphing calculator manual to learn how to use your specific calculator.

LS-62. What subproblems did you solve in the previous problem?
Answers will vary; eliminate z, eliminate x or y, solve, etc.

LS-63. Solve the following system of equations, and then check your solution in each equation. Be sure to keep your subproblems well organized.

$$x - 2y + 3z = 10$$
$$2x + y + z = 10$$
$$x + y + 2z = 14$$

$x = 1$, $y = 3$, $z = 5$

LS-64. Write a system of inequalities for the graph shown here.
$y \le -\frac{3}{4}x + 3$; $y \ge -\frac{3}{4}x - 3$; $x \le 3$; $x \ge -3$

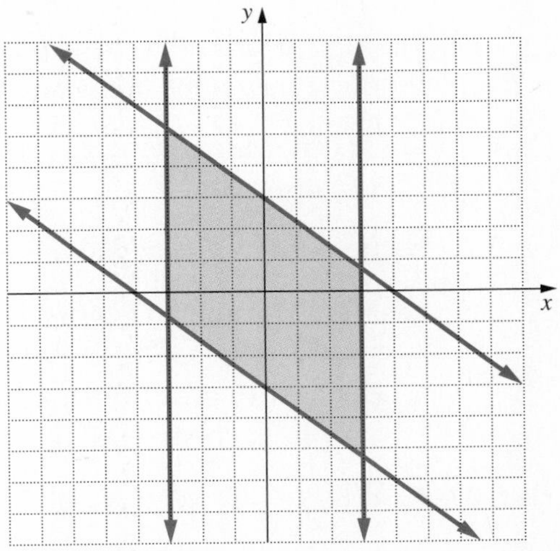

EXTENSION AND PRACTICE

LS-65. Find reasonable equations that will generate each graph. Each tick mark represents one unit. You may want to look at your Parent Graph Tool Kit.

a.

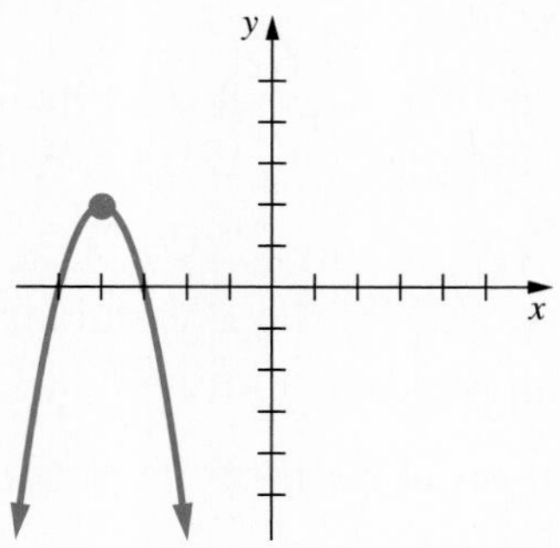

$y = -2(x+4)^2 + 2$

b.

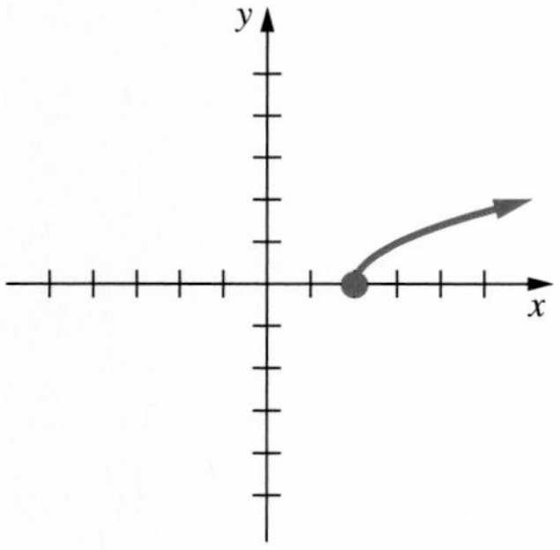

$y = \sqrt{x-2}$

c.

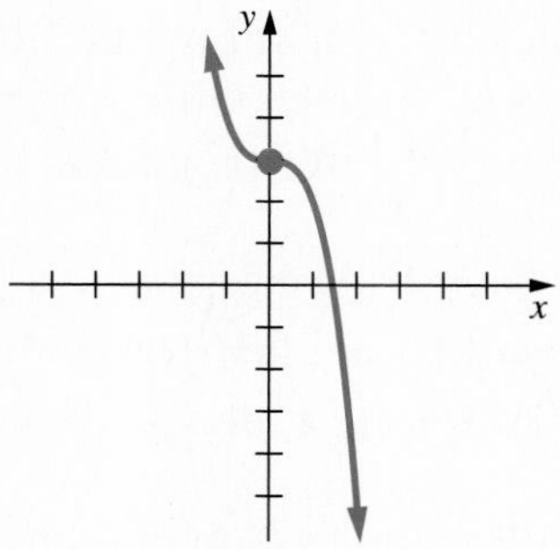

$y = -x^3 + 3$

LS-66. Solve each of the following for x:

a. $2x + x = b$ *$\frac{b}{3}$* **b.** $2ax + 3ax = b$ *$\frac{b}{5a}$* **c.** $x + ax = b$ *$\frac{b}{1+a}$*

LS-67. If an equation with three variables represents a plane, what would happen graphically if a system of three equations and three variables had no solution? *The planes don't intersect in a single point; two or three parallel lines are formed or all the planes are parallel.*

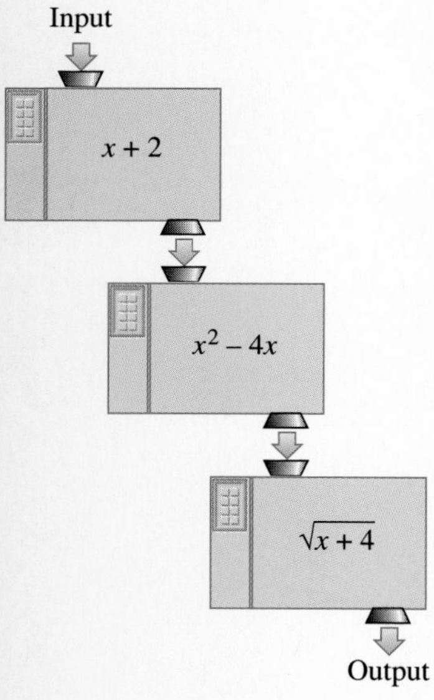

LS-68. Marcus has been busy. He claims to have created sequence of three function machines that always gives him the same number he started with.

a. Test out his machines. Do you think he is right? *No. Input equals output only if $x \geq 0$*

b. Be sure to test negative numbers. What happens for negative numbers?

c. Marcus wants to get his machine patented but he has to prove that it will always do what he says it will, at least for positive numbers. Show Marcus how to prove his machines work by dropping in a variable and writing out each step the machines must take.

d. Why do the negative numbers for inputs come out positive?

LS-69. Consider the following system of inequalities:

$$y \leq -2x + 3$$
$$y \geq x$$
$$x \geq -1$$

a. Graph this system.

b. Find the area of the shaded region. *6 sq. units*

In part f of LS-70, we have the students compose two functions. We have not officially introduced composition of functions, so don't expect mastery. We ask the question just to see whether students can figure it out.

LS-70. Given $f(x) = 2x^2 - 4$ and $g(x) = 5x + 3$, find the following:

a. $f(a) =$ *$2a^2 - 4$*

b. $f(3a) =$ *$18a^2 - 4$*

c. $f(a + b) =$ *$2a^2 + 4ab + 2b^2 - 4$*

d. $f(x + 7) =$ *$2x^2 + 28x + 94$*

e. $f(5x + 3) =$ *$50x^2 + 60x + 14$*

f. $g(f(x)) =$ *$10x^2 - 17$*

LS-71. If a movie ticket now averages $6.75 and has been increasing at an average rate of 15 percent per year, compute the cost:

a. in 8 years *$20.65*

b. 8 years ago *$2.21*

LS-72. Graph the solution of the following system of inequalities:

$$y > (x - 2)^2$$
$$y \leq (x - 2)^2 + 2$$

LS-73. You have a bag with 60 black jelly beans and 240 red ones.

a. If you draw one jelly bean out of the bag, find the probability that it is black. *1/5*

b. If you add 60 black jelly beans to the original bag and draw out a bean, what is the probability that it is black? *1/3*

c. How many black beans do you need to add to the original bag to double the original probability of drawing a black bean? *100*

d. Write an equation that represents the problem in part c. $\frac{x + 60}{x + 300} = 0.4$

LS-74. Graph the equation $y = x^2 + 4x - 2$, and use the graph to write the equation in vertex form. *$y = (x + 2)^2 - 6$*

LS-75. Sketch a graph showing the relationship between the average time a person spends exercising each week and the life span of the person. Be prepared to explain your graph to your group.

LS-76. Sketch a graph whose domain is from –2 through 6 and whose range is from 0 through 12.

6.6 BUILDING EQUATIONS: SOLVING FOR THE COEFFICIENTS

Students can use the graphing calculators to find the equation of the parabola. Instructions for finding the best-fit parabola are in Appendix D if you choose to use this approach.

LS-77. We already know how to find the equation of a line given two points. One way is to first find the slope of the line and then find the y-intercept. Another way is to solve a system of equations.

a. The fact that a line goes through two points (–3, 4) and (–2, –1) means those values for (x, y) make the equation $y = mx + b$ true. Use this information to write two equations in which m and b are the variables. *$4 = -3m + b$, $-1 = -2m + b$*

b. Use your equations from part a to find the equation of the line through the points (–3, 4) and (–2, –1). (That is, solve for m and b, not x and y.) Verify that these points make your equations true. *$m = -5$, $b = -11$, $y = -5x - 11$*

LS-78. A parabola with a vertical line of symmetry can be represented by the equation $y = ax^2 + bx + c$. It takes just two points to determine a line. How many points do you think it would take to determine such a parabola? Discuss this question with your group before doing parts a and b.

a. Use the idea you used in Problem LS-77 about the line to find an equation for a parabola that passes through the points (2, 3), (–1, 6), and (0, 3). You can start by writing three equations in which a, b, and c are the variables.

b. Solve the system of three equations and use the results to write the equation of a parabola. *$y = x^2 - 2x + 3$*

NOTE: As you might have guessed (or even figured out how by now!), it is possible to use your graphing calculator to find the equation of a parabola passing through three noncollinear points. See Appendix D for information on how to use your graphing calculator to do this. You will probably need to refer to your manual as well.

LS-79. Find the equation of the parabola through the points given:

a. (3, 10), (5, 36), and (–2, 15). *$y = 2x^2 - 3x + 1$*

b. (2, 2), (–4, 5), and (6, 0). *$y = -\frac{1}{2}x + 3$*

LS-80. What happened in part b of LS-79? (If you are not sure, plot the points.) Why did this occur? *The result is the equation of a line. These three points are collinear, which yields $a = 0$. You must have three noncollinear points to get a parabola.*

LS-81. Would it be possible to write an equation for a parabola for any three noncollinear points? Explain your thinking. *Yes*

EXTENSION AND PRACTICE

LS-82. Write the system of inequalities that will give you the graph shown here. *$y \le -x + 4$; $y > \frac{1}{3}x$*

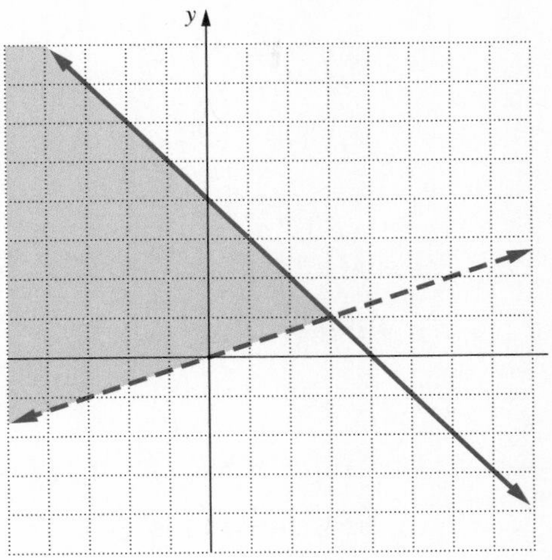

LS-83. Arturo was supposed to solve the following system of equations:

$$5x + y + 2z = 6$$
$$3x - 6y - 9z = -48$$
$$x - 2y + z = 12$$

Arturo decided to solve this system by eliminating x. Starting with the second and third equations, he realized he could divide both sides of the second by -3 to get:

$$-x + 2y + 3z = 16$$

He then combined this result with the third equation to get:

$$-x + 2y + 3z = 16$$
$$x - 2y + z = 12$$
$$4z = 28$$
$$z = 7$$

"Wow! Two with one shot!" said Arturo. But then he didn't know what to do next. What should he do to find x and y? Do it. *(–1, –3, 7)*

LS-84. Solve for x:

$$1 - \frac{b}{x} = a$$

$x = \frac{b}{1 - a}$

LS-85. Compare and contrast the graphs of $y = (x - 3)^2$ and $y = x^2 - 3$.

LS-86. A stick two feet long is accidentally broken into two pieces. (A diagram is essential in making sense of this problem.)

a. What is the probability that each of the pieces is at least 9 inches long? ***1/4***

b. What is the probability that each of the pieces is at least x inches long? $1-\frac{x}{12}$

LS-87. Solve for x, y, and z:

$$2^x(3^y)(5^z)=(2^3)(3^{x-2})(5^{2x-3y})$$

$x = 3$, $y = 1$, $z = 3$

LS-88. Consider the functions $f(x) = 2x^2 - 4$ and $g(x) = 5x + 3$.

a. Find $g(-2)$. *−7*

b. Find $f(-7)$. *94*

c. Evaluate $f(g(-2))$by substituting the result for $g(-2)$ into f. ***94***

d. How do you think you would calculate $f(g(1))$? Find the value. ***124***

For LS-89 the answer section reminds students that angle A equals angle C and that vertical angles are equal, but you should not assign LS-89 if you think it will create too much anxiety for your students. It is not essential to the development, but simply another opportunity to practice solving systems of equations.

LS-89.

a. Find the measure $\angle CPM$. ***60°***

b. List any subproblems that were necessary to solve this problem.

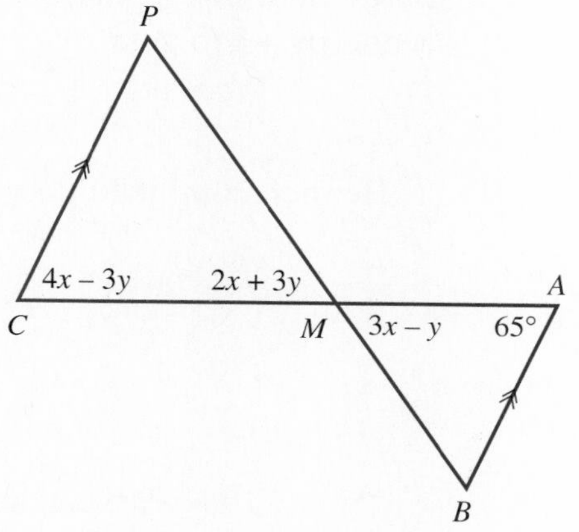

LS-90. Describe how the graphs of $y + 3 = -2(x + 1)^2$ and $y = x^2$ are different.

LS-91. Explain two different ways to find where the graphs of $y = (x - 4)^2 - 7$ and $3x + 2y = 12$ cross.

LS-92. Davis has been using his graphing calculator so much, he has actually rubbed the writing off the keys! He cannot remember the order of the four operation keys: ÷ "divide," × "times," + "plus," and – "minus." If he guesses which one might be the "+" key, what is the probability he is right? ***1/4***

6.7 APPLICATIONS OF PARABOLAS, SUMMARY

The goal of this section is for students to consolidate the ideas of the chapter. The next seven problems are all very similar, and our intent is that each group do just one of them. The first four problems (LS-93 through LS-96) can be done by hand, but the rest should not be attempted without using a graphing calculator, either for matrices or curve fitting. Remind students that these problems involve real data, and so are not going to produce "nice" equations.

The first four problems can be used as an assessment of the students' ability to solve systems of three equations. Our suggestion is to assign a different problem to each group and have each group prepare a poster to be displayed to the class. These posters will be their assessment for the chapter. The posters should be well organized and visually appealing, but most importantly, they should clearly display the fact that the students understand completely the problem and its solution. Large newsprint and large felt-tip pens are a good medium for in-class poster making that should not be too elaborate. These problems should be completed within 20 minutes.

The last three applications (LS-97, LS-98, and LS-99) use real data, and the equations are very messy. Do not use them unless students are going to be using their calculators to find the equation. These applications have no value in assessing the solving of systems, but they can be used to assess the student's ability to interpret models.

Since each group is assigned just one application, the students should have time to work on their summaries in class.

LS-93. A spaceship is approaching a star and is caught in its gravitational pull. When the ship's engines are fired, the ship will slow down, momentarily stop, and then hopefully pick up speed, move away from the star, and not be pulled in by the star's gravitational field. The engines were engaged when the ship was 750,000 miles away from the star. After one minute the ship was 635,000 miles away, and after two minutes the ship was 530,000 miles away.

a. Find three points from the information given above. Let x represent the time since the engines were fired and let y represent the distance (in thousands of miles) from the star. *(0, 750), (1, 635), (2, 530)*

b. Plot the points you found in part a. We want the distance to reach a minimum and then increase again over time. What kind of model follows this pattern?

c. Find the equation of the parabola that fits the three points you found in part a. $y = 5x^2 - 120x + 750$

d. If the ship comes within 50,000 miles of the star, the shields will fail and the ship will burn up. Use your equation to determine whether the spaceship has failed to escape the gravity of the star. ***After 10 min the ship is 50,000 miles away; therefore, the ship burns up.***

LS-94. Melvin has contracted an infection and gone to the doctor for help. The doctor has the nurse take a blood sample and finds 900 bacteria per cc. ("cc" stands for cubic centimeter; 1 cc is equal to 1 milliliter.) Melvin gets a shot of a strong antibiotic from the doctor. The bacteria will continue to grow for a period of time, reach a peak, and then decrease as the medication succeeds in overcoming the infection. After ten days, the infection has grown to 1600 bacteria per cc. After 15 days it has grown to 1875.

a. What are the three data points? *(0, 900), (10, 1600), (15, 1875)*

b. Make a rough sketch that will show the number of bacteria per cc over time.

c. Find the equation of the parabola that contains the three data points. $v = -t^2 + 80t + 900$

d. Based on the equation, when will Melvin be cured? *after 90 days*

e. Based on the equation, how long had Melvin been infected before he went to the doctor? *10 days.*

LS-95. Gertrude has a job that is 35 miles from her home. She needs to be at work by 8:15 A.M. Gertrude wants to maximize her sleep time by leaving as late as possible but still get to work on time. From experimentation, Gertrude discovers that if she leaves at 7:10, it will take her 40 minutes to get to work. If she leaves at 7:30, it will take her 60 minutes to get to work. If she leaves at 7:40, it will take her 50 minutes to get to work. Since her commute time increases and then decreases, Gertrude decides to use a parabola to model her commute. Assume the time it takes to get to work depends on the number of minutes after 7:00 that Gertrude leaves.

a. Let x = the number of minutes after 7:00 that Gertrude leaves.
Let y = the number of minutes it takes Gertrude to get to work.

b. One ordered pair is (10, 40). What are the other two ordered pairs given in the problem? *(30, 60), (40, 50)*

c. Find the quadratic equation that fits these three points. Round your coefficients to three decimal places. $y = -0.067x^2 + 3.667x + 10$, *or* $y = -\frac{1}{15}x^2 + \frac{11}{3}x + 10$

d. Use your equation to find how long it would take to get to work if Gertrude left at 7:20 *56.5 min*

e. According to your equation, how long would it take to get to work if she left at 7:58? Does this make sense? Why? *−2.7 min; no, she cannot arrive before she leaves.*

LS-96. Buckets O' Buckets manufactures and sells children's buckets. The marketing manager noticed that when they priced a bucket at $1.50, approximately 65 buckets were sold each month, and with each 10¢ increase in the price, three fewer buckets were sold.

a. Figure out how much money they made from selling buckets at \$1.00 each, \$1.50 each, and \$2.00 each. *\$80, \$97.50, \$100*

b. Use the data points you found in part a to write the equation of the parabola that contains the three points. $y = -30x^2 + 110x$

c. At about what price is it best to sell the buckets? Explain. *If they can only increase the cost in 10¢ increments, the best price is \$1.80, which makes \$100.80 (\$1.83 makes \$100.83).*

d. How many buckets should they sell to take in \$95 each month? *Selling 44 buckets at \$2.20 will bring in \$96.80. Selling 68 buckets at \$1.40 will bring in \$95.20—but why not save all those buckets?*

*The following problems will lead to equations containing **very** messy numbers. They should **not** be done by hand. You can use the quadratic regression feature on your graphing calculator to find the equation of the parabola passing through the three points or you can use matrices to solve the 3×3 system of equations. Appendix D provides an introduction to both methods.*

LS-97. According to the *Information Please, Almanac*, the median age for females for first marriages in 1930 was 21.3 years of age. In 1950 the median age dropped to 20.3, and by 1990 the median age was up to 23.9 years of age.

a. What are the three data points? *(30, 21.3); (50, 20.3); (90, 23.9)*

b. Make a rough sketch that will show the median age for first marriages over time.

c. Find the equation of the parabola that contains the three data points. $y \approx 0.002333x^2 - 0.2366666x + 26.3$

d. Based on the equation you found in part c, when will the median age for first marriages be over 30 years of age? *The year 2015*

LS-98. The real estate market can change quickly and drastically. In 1988 a three-bedroom, two-bath house in Davis, California, sold for \$113,520. By 1991 the same house sold for \$176,020. Shortly after that, the housing market crashed, and in 1996 the same house was worth \$132,720.

a. What are the three data points? *(88, 113.52), (91, 176.02), (96, 132.72)*

b. Make a rough sketch that will show the market value of the house over time.

c. Find the equation of the parabola that contains the three data points. $y \approx -3.687x^2 + 680.75x - 31242.64$

d. Based on the equation you found for part c, when could the house have been sold for \$113,520 again? *in late 1996, about October*

e. Based on the equation, when is the house worth \$0? Does this make sense? Explain. *in 1985 and in 1999*

LS-99. According to an article in *U.S. News and World Report* (for June 3, 1996), there were 39,643 deaths from automobile accidents in 1937, 54,633 deaths in 1970, and 40,676 deaths in 1994.

a. What are the three data points? *(37, 39643), (70, 54633), (94, 40676)*

b. Make a rough sketch that shows the number of auto-related deaths over time.

c. Find the equation of the parabola that contains the three data points. $y \approx -18.17165072x^2 + 2398.609051x - 24{,}228.54506$

d. Based on the equation, has there ever been, or will there ever be, a year in which there are no auto-related deaths? Does this make sense? *in about 1911 and 2021.*

e. How accurate is this model? What happened in 1996 that might affect the number of auto-related deaths? *The model is not that good since it predicts no accidents in the years 1927 or 2078. In 1996, the maximum speed limit was raised. This might affect the number of deaths.*

LS-100. *Summary* Write and solve a problem to represent each of the following ideas. Make your solution clear enough to convince your instructor that you understand each of these.

a. Graphing systems of inequalities

b. Solving a system of three equations and three variables

c. Find the equation of a parabola given three points

EXTENSION AND PRACTICE

LS-101. Write an explanation for someone just coming into the class about how to use subproblems to solve a system of three linear equations with three variables. Your explanation should be so clear that your instructor will be convinced that you know what to do and will decide never to ask you to solve another one.

LS-102. *Self-Evaluation*

a. In this chapter you have looked at solving linear equations (and at when they don't have a solution) for systems with both two and three variables. Write a paragraph to explain the relationship between the *geometry* of the graphs and the *algebra* of solving the equations.

b. What were the most difficult parts of this chapter? List sample problems and discuss the hard parts.

c. What problem did you like best, and what did you like about it?

d. How has your role in your group changed throughout this course? How has your role in your group affected your learning? Explain.

LS-103. Tool Kit Check For each of the topics listed in LS-100, make sure you have a Tool Kit entry that is helpful to you. Add anything else you might need.

In LS-104, you may want to tell students that a part of your evaluation will be based on their selection of a reasonably challenging polygon in part b, but remind them to keep it reasonable enough that they can do a thorough and accurate job of writing the inequalities. You also may want them all to be different!

LS-104. On graph paper draw a polygon and label its vertices. Then write the set of inequalities for which the intersection will be the area inside the polygon you drew. You can do an easy one, such as a triangle, or you can consider this a challenge and see how interesting a polygon you can represent.

LS-105. Portfolio: Growth over Time — Problem No. 2 On a separate piece of paper (so you can hand it in separately or add it to your portfolio) explain *everything* that you now know about:

$$f(x) = 2^x - 3.$$

LS-106. Graph the following system and shade the solution:

$$y \geq x^2 - 4$$
$$y < \frac{1}{3}x + 1$$

LS-107. Solve the following system:

$$\begin{aligned} 5x - 4y - 6z &= -19 \\ -2x + 2y + z &= 5 \\ 3x - 6y - 5z &= -16 \end{aligned}$$

−1, 1/2, 2

LS-108.

a. Find the equation of the parabola that passes through (2, 3), (−1, 6), and (0, 3). *$y = x^2 - 2x + 3$*

b. Find the vertex of the parabola. *(1, 2)*

LS-109. Portfolio Assignment

a. Find what you believe to be your two best pieces of work for this chapter. Explain why you are particularly proud of these assignments. Be sure to include a restatement of the original problem or a copy of it.

b. Select two problems that you are still having difficulty with. Copy each problem and solve as much of it as you can. Then explain what part of each problem you don't understand.

LS-110. Consider the pattern shown below:

a. Draw the figure for $n = 4$.

b. Write down the number of dots in each figure from $n = 1$ to $n = 4$? Is this sequence arithmetic? Is this sequence geometric? *1, 3, 6, 10; no; no*

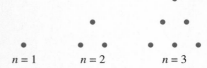

c. Find the equation where, given n, you can determine the number of dots in the figure n. If you get stuck, try graphing the points (n, d) where d is the number of dots. *$0.5x^2 + 0.5x$*

LS-111. Three red rods are 2 centimeters longer than two blue rods. Three blue rods are 2 cm longer than four red rods. How long is each rod? *red = 10 cm, blue = 14 cm*

LS-112. Solve the following system:

$$x + 3y = 16$$
$$x - 2y = 31$$

Now, rewrite the system by replacing x with x^2. What effect will this have on the solution to the system? Solve the new system. *(25, –3); new system (5, –3), (–5, –3)*

LS-113. Consider the function $f(x) = 2x^2 - 7x - 15$.

a. Calculate $f(0)$, $f(-1.5)$, and $f(5)$? *–15, 0, 0*

b. What is important about the results you found in part a? Explain. *–15 is the y-intercept. Since the other two numbers gave an output of 0, they are the x-intercepts.*

c. Sketch a graph of this function.

LS-114. To pick their wedding day, Emma and Sammy throw a dart at the calendar that is open to the months of January and February. Emma is hoping the dart will land on Valentine's Day, while Sammy is hoping the dart will not land on Super Bowl Sunday.

a. What is the probability Valentine's Day will be the day? *in a nonleap year, 1/59; in a leap year, 1/60*

b. What is the probability that Sammy will miss the Super Bowl to be at his wedding? *Nonleap year, 1/59; in a leap year, 1/60*

c. Who is more likely to be pleased with the choice of the wedding day? Explain. *Sammy*

CHAPTER SEVEN

THE CASE OF THE COOLING CORPSE

Logarithms and Other Inverses

IN THIS CHAPTER YOU WILL HAVE THE OPPORTUNITY TO:

- undo all the functions you have worked with so far. You will learn about inverse functions, which reverse the order in which their original functions operate;
- learn about a new function, the "undoing" or inverse function for an exponential function;
- appreciate the usefulness of logarithms for solving some equations and other problems.

AGAIN, A GRAPHING CALCULATOR WILL BE A VALUABLE TOOL, AND YOU WILL CONTINUE TO NEED AT LEAST A SCIENTIFIC CALCULATOR FOR HOMEWORK.

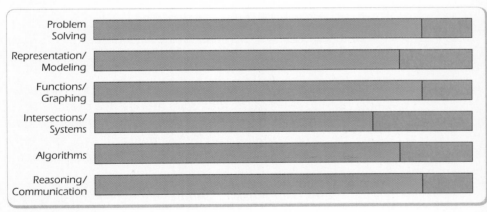

MATERIALS	CHAPTER CONTENTS		PROBLEM SETS
	7.1	Undoing Machines	CC-1 – CC-18
	7.2	Inverses and Interchanges	CC-19 – CC-32
	7.3	More on Inverse Relations	CC-33 – CC-43
	7.4	Interchanging an Exponential Function	CC-44 – CC-57
	7.5	Introducing Logarithms	CC-58 – CC-72
	7.6	Investigating Logarithms	CC-73 – CC-86
	7.7	Translating the Graphs of Logarithmic Functions	CC-87 – CC-101
	7.8	Using Logarithms to Solve Problems	CC-102 – CC-113
	7.9	Writing Equations for Exponential Functions	CC-114 – CC-126
	7.10	Solving the Mystery of the Cooling Corpse	CC-127 – CC-138

There are two goals for this chapter. First, we want students to use concrete examples to develop the concept of an inverse as an "undoing" function. We start by emphasizing what a function machine does to the inputs and what is needed to undo those operations. Then the question "How do you undo $y = 2^x$?" will arise naturally. Attempting to answer this question generates the need for logarithms. We spend some time on composition of functions but only to emphasize further the undoing aspect of inverses, not to make students proficient at composing functions. Finally, when students are comfortable with logs, we come to our second goal: we want students to use logs to solve equations, and in particular, to use them to find equations for exponential functions that model given sets of data.

7.1 UNDOING MACHINES

In this section we want the students to feel comfortable "undoing" equations. This is how we algebraically develop the idea of an inverse. We want students to verbalize what operations must be performed to undo what the machine does, so don't push for equations until the students are ready. They should be able to undo fairly complicated equations if they really understand what is going on.

Start the day with some "brain games" similar to "What's my rule?" For example: "I'm thinking of a number; when I add 5 to it and multiply the result by 2, I get 14. What's my number?" Do several of these examples, enough so that everyone can play this game. Then have students ***justify*** *how they know what the number is. Require them to be precise in their verbal descriptions! A good answer would be "I know your number is 2 because I can reverse, or undo, the process: 14 divided by 2 is 7; and 7 minus 5 is 2."*

Move on to more complicated rules, those involving squares, square roots, and even higher powers. The following is a possible list of equations to use.

$3x + 5 = 17$	$x = 4$
$2x^2 = 50$	$x \pm 5$
$\frac{1}{2}x - 1 = 4$	$x = 10$
$3x^3 - 2 = 22$	$x = 2$
$2 + \sqrt{x - 4} = 5$	$x = 13$

Encourage students to say out loud what a function is doing to x. In CC-1, for example, f is a function that multiplies any number by 2 and then adds 1. Similarly, in the problems that involve fractions, have students interpret the fractions as division.

There are several analogies that can give students a familiar frame of reference for the idea of undoing. A Jelly Maker puts jelly into a jar and then puts on the lid. You undo this by taking off the lid and then taking out the jelly. Another example would be wrapping a present: First you put the gift in the box. Then you close the box, wrap it, and put ribbon around it. To open a present, you take off the ribbon, remove the wrapping paper, open the box, and then remove the gift. You undo each step, following the reverse order. A demonstration of this can be very effective if you get the students involved in identifying the undoing steps.

Some students will follow this process with no hints, while others will simply guess and check a rule. Either process is OK to start with, but be sure there is a discussion of both methods. At some point fairly early on, we want students to begin thinking, "Reverse the steps, in reverse order."

Consider the following problem:

> Suppose you put $5000.00 in a bank at 3 percent interest compounded annually. How many years would it take for you to double your money?

You know from your work with multipliers that the equation you need to solve is:

$$5000(1.03)^x = 10{,}000$$

Think about how to solve this equation. We could divide by 5000 to obtain $(1.03)^x = 2$. At this point, solving the equation $(1.03)^x = 2$ might seem like a simple problem. After all, you saw similar equations back in Chapter 4. Try solving it now. Is it easy? Did you use guess and check? One of the goals of this chapter is to provide you with more of the algebraic tools you need to solve bigger problems, as well as problems similar to this one. Along the way, you should increase your understanding of functions. *It will take a little over 23 years to double the money.*

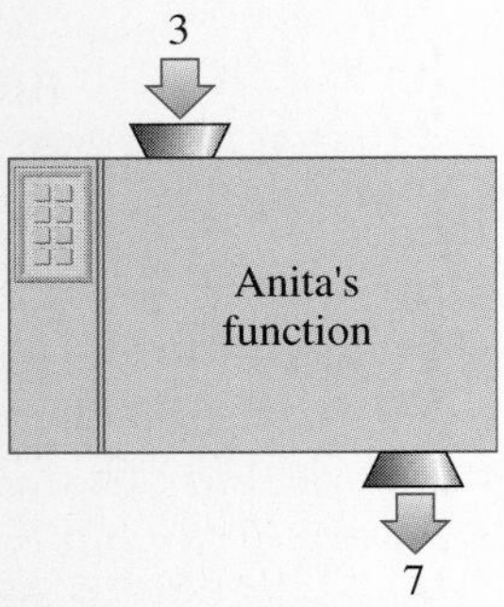

CC-1. A picture of Anita's function machine is shown here. When she put a 3 into the machine, the machine put out a 7. When she put in a 4, the machine gave her a 9, and when she put in a −3, out came −5.

a. Explain in words what this machine does to a number. *It multiplies the input by two and then adds 1.*

b. Suppose it is possible to put the machine into reverse, causing it to pull 7 back up into it. What do you think will come out the top? Explain. *We hope they will see that 3 would come out the top and, if that's true, that the machine must "undo" itself—work backward—to get this result.*

c. Suppose Anita wants to build another machine that will **undo** the effects of her first function machine. That is, she wants a machine that will take 7 and turn it back into 3, take 9 and turn it back into 4, and so on. Write a rule that she should program into this new machine to make it do this. *Subtract 1, and then divide by 2.*

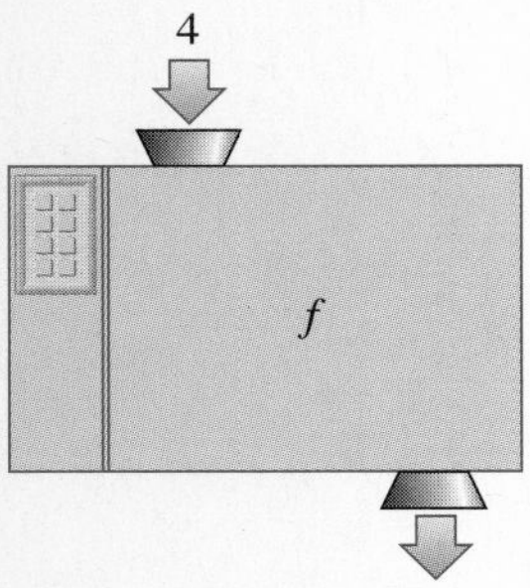

CC-2. The function machine f shown here has the rule

$$f(x) = 5x + 2$$

a. What is $f(4)$? *22*

b. If the machine is put in reverse, what number should be pulled up into the machine in order to have a 4 come out the top? *22*

c. Keiko wants to build a new machine that will undo what the function machine f does. What must Keiko's machine do to 17 to undo f and come out with 3? Write your rule in function notation and call it $g(x)$. *Subtract 2 and divide by 5;* $g(x) = \frac{x-2}{5}$

You could break the steps of the function down, using a flowchart. This method may make it easier for some students to see the steps and figure out what is needed to reverse them.

CC-3. Find the rule that undoes each function below. Write each rule in function notation, but be sure to use a name that is different from the given function name. For example, if the function is named $f(x)$, you can't use $f(x)$ for the undoing machine too. You'll need to use some other letter, such as $g(x)$.

a. $f(x) = 3x - 2$ $y = \frac{x+2}{3}$ **b.** $h(x) = \frac{x+1}{5}$ $y = 5x - 1$

c. $p(x) = 2(x + 3)$ $y = \frac{x}{2} - 3$ **d.** $q(x) = \frac{x}{2} - 3$ $y = 2(x + 3)$

CC-4. The formal mathematical name for an undoing function is **inverse**. Record this in your Tool Kit! Find the inverse for each of the following functions. Then graph each function and its inverse on the same set of coordinate axes. In other words, use one set of axes for part a, a new set of axes for part b, and so on. This is *not* a good time to divide up the graphing among your group members. Make sure each of you does the work for each part. Check with each other as you go.

a. $f(x) = 2x + 4$ *$y = \frac{x-4}{2}$* **b.** $f(x) = -\frac{2}{3}x$ *$y = \frac{-3}{2}x$*

c. $y = \frac{1}{3}x + 2$ *$y = 3x - 6$* **d.** $y = x^3 + 1$ *$g(x) = \sqrt[3]{x-1}$*

CC-5. When you have completed all the pairs of graphs in CC-4, look for patterns in the graphs. What relationships do you see between the graph of a function and the graph of its inverse? Do the pairs of graphs have a line of symmetry? Justify your answer. ***Yes, the line $y = x$ is a line of symmetry for each pair.***

*In CC-6 we graph only half of the parabola in order to keep the function one-to-one so it will have an inverse. We don't deal with the one-to-one issue right now (we'll get to it later) because we want to focus students' attention on the **interchange** idea.*

CC-6. Use a full sheet of graph paper. Put the axes in the center of the page, and label each mark on each axis as one unit. Use pencil (the softer the lead the better).

a. Graph $y = \left(\frac{x}{2}\right)^2$ over the domain $0 \le x \le 8$. How could you describe this graph? Label the graph with its equation. Trace over the graph with the pencil until the graph is heavy and dark. Crayon works even better than pencil. ***half of a parabola***

b. On the same graph paper, graph $y = x$ using a ballpoint pen (use a straightedge, and press down hard).

c. What is the equation of the inverse of the half parabola in part a? *$y = 2\sqrt{x}$*

d. Using a ruler as an edge, fold the paper along the line $y = x$, with the graphs on the *inside* of the fold. Put the folded paper on your desktop. You will be able to see the darkened graph of the half parabola through the paper. Now press the paper with a hard object, such as the back of your fingernail or the top of a pen, so that you are rubbing the graph from the back of the paper. The purpose is to make a "carbon copy" of the graph of the parabola, reflected across the line. Then open the paper and fill in the picture if it is not completely copied. Write your observations on the graph. ***The carbon copy is the reflection across the line $y = x$. The equation of the inverse is the equation of the reflection.***

e. Find three points (x, y) that satisfy the equation you wrote in part c. Find each point on the carbon-copy graph. Each *should* be a point on this graph. Explain why.

EXTENSION AND PRACTICE

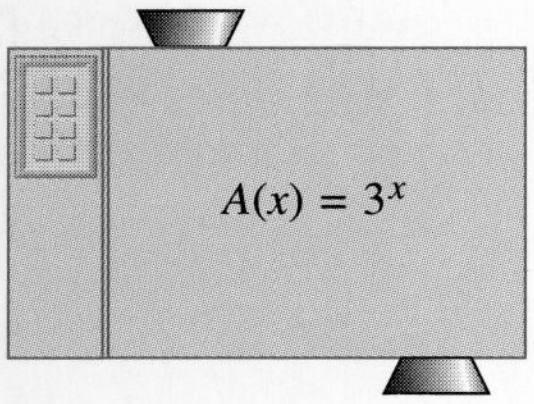

CC-7. Antonio's function machine is shown at the left.

a. What is $A(2)$? *9*

b. If 81 came out, what number was dropped in? *4*

c. If 8 came out, what number was dropped in? Be accurate to two decimal places. *$x \approx 1.89$*

CC-8. Diane claims that $f(x) = \frac{3}{x}$ is its own undo rule. Is her conjecture correct? Show how you know. *yes; use two function machines, back to back*

CC-9. If $10^x = 10^y$, what is true about x and y? Justify your answer. *equal*

CC-10. Sketch the solution of the following system of inequalities:

$$y > x^2 - 5$$
$$y < -(x - 1)^2 + 7$$

If the students do not see the line of symmetry in the next problem, don't panic. We will come back to this idea.

CC-11. Graph $y = \frac{1}{2}x - 3$ and its undoing function on the same set of axes.

a. What is the equation of the undoing function? *$y = 2(x + 3)$*

b. Does the pair of graphs have a line of symmetry? If so, what? *yes, $y = x$.*

CC-12. Solve the equation $x^2 = 91$. *$x \approx \pm 9.5394$*

CC-13. Solve the equation $3 = 8^x$ for x, accurate to two decimal places. *$x \approx 0.53$*

CC-14 might be a good opportunity for students to learn how to solve a 3-by-3 system on the graphing calculator, if they haven't already. Appendix D provides an introduction.

CC-14. Solve the following system:

$$x + y + z = 2$$
$$2x - y + z = -1$$
$$3x - 2y + 5z = 16$$

(–3, 0, 5)

CC-15. Dana's mother gave her $175 on her 16th birthday. "But you must put it in the bank and leave it there until your 18th birthday," she told Dana. Dana already had $237.54 in her account, which pays 3.25 percent annual interest, compounded quarterly. How much money will she have on her 18th birthday if she makes *no* deposits and *no* withdrawals before then. Justify your answer. ***If she adds nothing else to the account, and the money just sits there earning interest, she will have $440.13 on her 18th birthday.***

CC-16. If $2^{x+4} = 2^{3x-1}$, what is x? ***$x = 2.5$***

CC-17. Jennifer was given a parabola $y = x^2 - 4x - 2$ in standard form and had to rewrite it in graphing form but didn't know how. Explain to her the way to do it, and show the graphing form for her equation. *$y = (x - 2)^2 - 6$*

The project that follows pulls together a lot of the work of the course. It should be introduced now so that students can turn it in toward the end of the quarter or semester after completing this chapter. Doing the report will require students to use the big ideas of several chapters in the context of a problem they have chosen. The project also could be used as a possible assessment problem or included as part of the final exam.

CC-18. Report on Growth Read the following project description now and come to class prepared with any questions and ideas.

REPORT ON GROWTH

We have been learning about several different families of functions such as lines, parabolas, exponential functions, and others. All of these can be used to describe something in the real world. In this project, you will learn more about how mathematical ideas are used to model reality. This project has four parts.

a. ***Project Proposal:*** The first step is to choose something that has been growing or shrinking steadily over time. Choose something you are really interested in studying; cancer rates, world record times for some sport, crime rates, AIDS, teenage pregnancy rates, the number of lawyers in the United States, or the population of your family's country of origin are some possibilities. You are free to choose one of these or a topic of your own.

   Once you have chosen your topic, you must turn in a proposal. In this proposal, state your topic, explain why you chose it, and describe exactly where you plan to collect your data. *Don't* say, "at the library". Describe in detail exactly where: Who are you going to call or ask? In what book will you look? Be specific, as this is your bibliography for your project. Note that your data must show an increase or decrease for at least *three* times (days, hours, months, years—whichever is relevant for the topic).

b. ***Mathematical Modeling:*** Your job in this part is to construct different mathematical models for your data, and use the models to make predictions about the future. Using your three or more data points, find the equation of a line, a parabola, and an exponential function that best fit the data. For extra credit, you may use other families of functions to fit your data. Make a large graph (using a full sheet of paper), clearly labeled, for each. Show *completely* how you derived your equations (this is one of the most important parts).

c. ***Analysis:*** Finally, tell what your equations predict for the future and explain what the consequences of your predictions are. Which model do you think is best, and why? What factors would make your model less reliable? Is there some point at which the model is obviously wrong? Your explanations must be thorough and complete.

d. ***Report Presentation:*** The report should be typed, neatly presented, and complete. Your instructor will give you a timeline and assign point values for parts a through d.

Topic	Due Date	% Value
Proposal		
Modeling		
Analysis		
Presentation		

This project is to be done *individually*. Please note all due dates. Be sure to include *at least* everything mentioned above.

7.2 INVERSES AND INTERCHANGES

*CC-19 ties the **undo** idea to the **interchange** concept. Group discussion is important as students work on this problem.*

CC-19. Ngan and Nai have found a useful pattern between functions and their inverses. They didn't remember any shortcuts for graphing lines, so they made a table to graph $f(x) = 2x + 4$ and its inverse. That's when they made their discovery.

a. Make and complete the two tables below, and see if you can find their pattern. *x- and y- values are interchanged.*

x	$f(x)$
0	4
1	*6*
2	*8*
−1	*2*
−2	*0*

x	$g(x)$
4	0
6	*1*
8	*2*
2	*−1*
0	*−2*

b. If (10, 24) is a point on the graph of the function, what point do you automatically know will be on the inverse graph? What if (3.5, 11) is on the function's graph? What if (a, b) is on the graph? What about (x, y)? *(24, 10), (11, 3.5), (b, a), (y, x)*

c. What is the inverse of $f(x) = 2x + 4$? Write it as $i(x) =$ ____. $i(x) = \left(\frac{x-4}{2}\right)$

d. Discuss the significance of this discovery with your group before going on to the next problem. Record important conclusions in your Tool Kit.

CC-20. Kalani began to think that the method used in CC-19 to *graph* the inverse of a function might help him find the *equation* of the inverse. He said to his friend Macario, "If you can just interchange x and y to find points on the graph of the inverse, why not just switch x and y in the equation itself to find the equation of the inverse?"

"Well, I think I see what you are saying," Macario said. "A function and its inverse are reflected across the line $y = x$. Also, in the table of values, the values for the independent and dependent variables are just switched. But what makes you think we can do the same thing with the equation?"

"Think about it: an equation represents *all* the points on the line, so if you can interchange all of the x and y values, why not interchange the x and y variables?"

a. Try this method: Interchange the x- and y-variables in the equation $y = 3x - 1$. Then solve the new equation for y. $y = \frac{x+1}{3}$

b. Did this method work? Show how you know. *Yes; check by substituting numbers into the equations.*

c. The inverse of a function could be called the ***x-y* interchange** of the function. Give a good reason why someone might want to call it that. *We interchange the x and y variables.*

CC-21. Find the inverse of each function below. Write your answer in function notation. Remember to use a new name for the new function!

a. $f(x) = 8x + 6$ $\quad g(x) = \frac{x-6}{8}$

b. $f(x) = 3x^5$ $\quad g(x) = \sqrt[5]{\frac{x}{3}}$

By inspecting responses to Uyregor's problem as students are working in their groups on CC-22, you will get a good idea whether or not they are understanding the concept that if f(x) and g(x) are inverses then f(g(x)) = g(f(x)) = x.

CC-22. Uyregor thinks his instructor is sadistic! He gave Uyregor the function $f(x) = \frac{3}{5}x + 9$ and told him to find the inverse of this function and call it $g(x)$. And that's the easy part. Once he has done that, he is supposed to find $f(1), f(2), f(3), \ldots, f(50)$! But wait—that's not all. Then he is supposed to substitute each of those results into $g(x)$. Since you have mastered this concept, you can help Uyregor. Explain to him why his instructor is not so mean after all and how he should already know all of the final results. *He will get the numbers 1, 2, 3, . . . 50*

CC-23. The graphs of three functions are sketched below. For each function, sketch the graph of the inverse. Then state the domain and range for each function and for its inverse. The dotted line is the graph of $y = x$.

a.

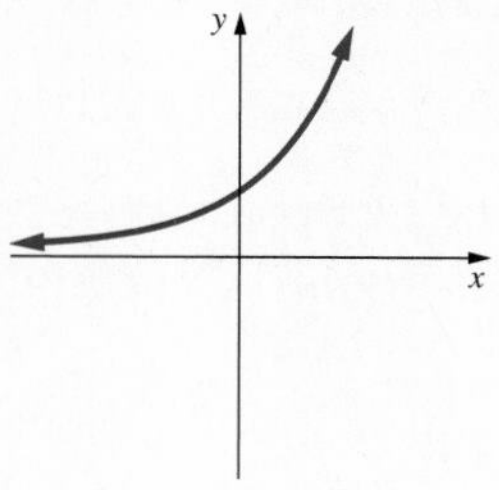

f: D, reals; R, $y > 0$
f^{-1}: D, $x > 0$; R, reals

b.

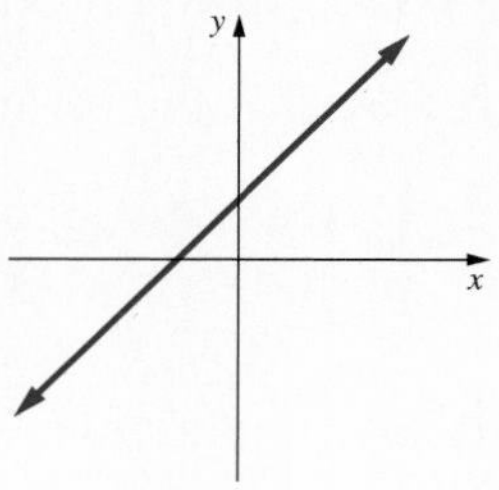

f: D, reals; R, reals
f^{-1}: D, reals; R, reals

c.

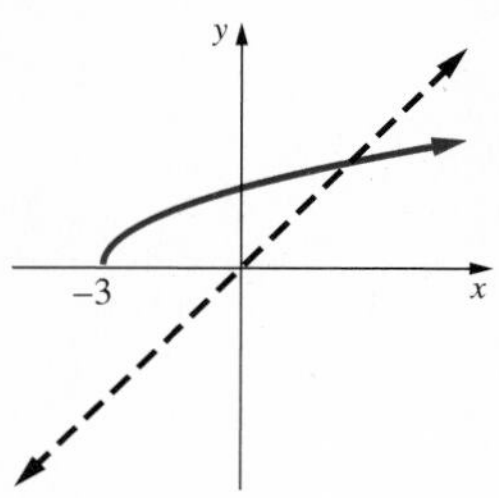

f: D, $x \geq -3$; R, $y \geq 0$
f^{-1}: D, $x \geq 0$; R, $y \geq -3$

EXTENSION AND PRACTICE

CC-24. Trejo says that if you know the x- and y-intercepts and the domain and range of an equation, then you automatically know the x- and y-intercepts, domain, and range of the inverse. Hilary disagrees. She says you know the intercepts for the inverse, but that is all you know for sure. Who is correct? Justify your answer. ***Trejo is correct.***

Next we introduce the concept of composition, not to make students proficient at composing two or more functions, but rather to reinforce the undoing aspect of inverse functions. We start with the concrete function machines, and students should have a strong grasp of the ideas before moving to the abstract. Students should be able to do these for homework.

CC-25. Lacey and Richens each have their own personal function machines. Lacey's machine, $L(x)$, squares the input and then subtracts 1. Richens's function, $R(x)$, adds 2 to the input, and then multiplies by 3.

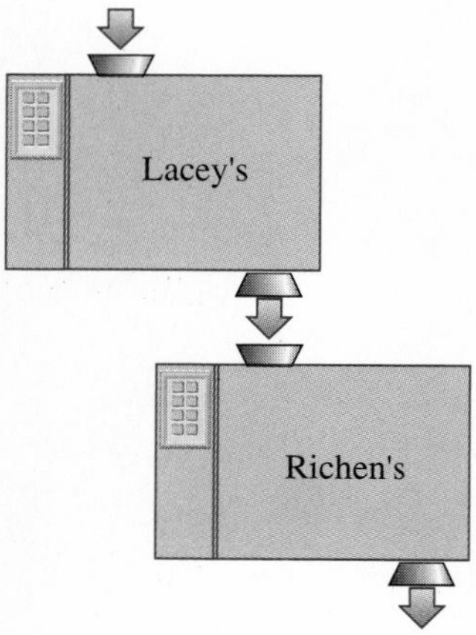

a. Write the equations that represent $L(x)$ and $R(x)$. ***$L(x) = x^2 - 1$; $R(x) = 3(x + 2)$***

b. Lacey and Richens decide to connect their two machines, so that Lacey's output becomes Richens's input. Eventually, what is the output if 3 is the initial input? ***30***

c. What if the order of the machines were changed? Would the final output change? Justify your answer. ***Order does matter—show by substituting a number for x.***

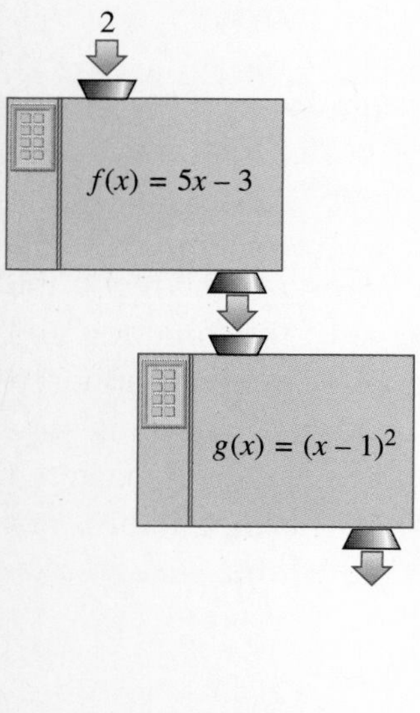

CC-26. Two function machines, $f(x) = 5x - 3$ and $g(x) = (x - 1)^2$, are shown here. Suppose the result for $f(2)$ is dropped into the $g(x)$ machine. This is written as $g(f(2))$. What is this output? ***36***

CC-27. Using the same function machines as in CC-26, find $f(g(2))$. Be careful! The result will be different from the last one because the order in which you use the machines has been switched! With $f(g(2))$, first you will find $g(2)$ and then you will substitute that answer into the f machine. *2*

CC-28. When we push two (or more) function machines together, we say we have a new function, which is called the **composition** of the two functions. The composition can be written different ways, as $f(g(x))$ or sometimes as $f \circ g(x)$. Does it seem to matter in what order you use the functions? That is, does $f(g(x)) = g(f(x))$? Explain. *Yes, order matters. Not necessarily equal.*

CC-29. Rebecca thinks she has found a quick way to graph an inverse of a function. She figures that if you can interchange x and y to find the inverse, then she will interchange the x- and y-axes by flipping the paper over so that when she looks through the back, the x-axis is vertical and the y-axis is horizontal, as shown here: Copy the graph below on the right on a separate sheet and try her technique. What do you think?

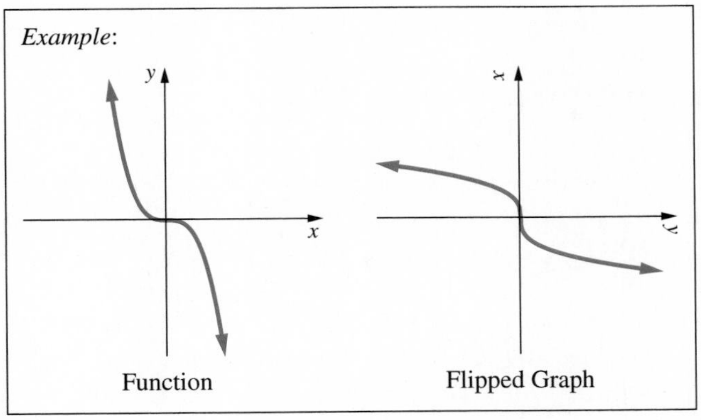

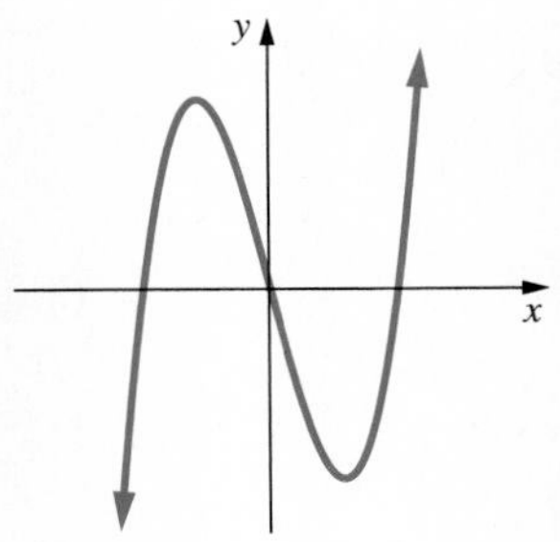

CC-30. Nancy just hit the \$10,000 jackpot playing the giant slot machine in Reno. She would like to invest the money for a down payment on a new house. The local banker offers her a one-year renewable Certificate of Deposit (CD) earning 6.78 percent interest compounded quarterly. For how many years will Nancy need to renew her CD in order to earn enough money to make a 10 percent down payment on a \$200,000 house? (*Hint*: $y = k(m^x)$) *Use the equation* $20{,}000 = 10{,}000(1.01695)^x$; $x \approx 10.5$ *yr*

CC-31. Samy climbs a 10-ft ladder to reach the roof of his house. The roof is 12 ft above the ground. The base of the ladder must be at least 1.5 feet from the base of the house. If he feels safe standing on the top step of the ladder, how far will he have to step up to get onto the roof? Draw a sketch. $\approx$*2.11 ft*

CC-32. Later, after Samy had recovered from his fall off the roof—a fall that broke his ladder into two pieces—he needed to reach the top of a window he was washing. He needed a ladder that was at least 4 feet long. At the same time, his neighbor, Elisa, needed to borrow a ladder, and she *also* needed one that was at least 4 feet long. What is the probability that Samy's fall, which broke the ladder into two random pieces, caused the ladder to break in a spot that would give Samy and Elisa each a ladder at least 4 feet long? *1/5*

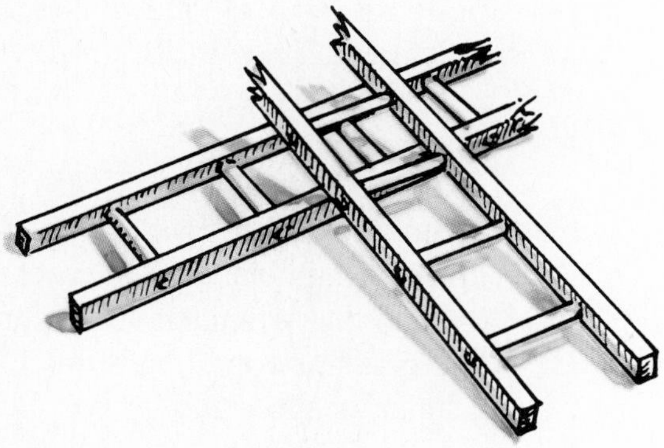

7.3 MORE ON INVERSE FUNCTIONS

CC-33 deals with the issue of functions that don't have inverse functions, but for which students could find inverse relations. We will not go into a major blitz of definitions, but through the problems, we do want students to figure out how to restrict domains in order to get functions that have inverse functions. This is an opportunity for students to refine their ideas about what a function is and about domains and ranges, but not a time to introduce new definitions.

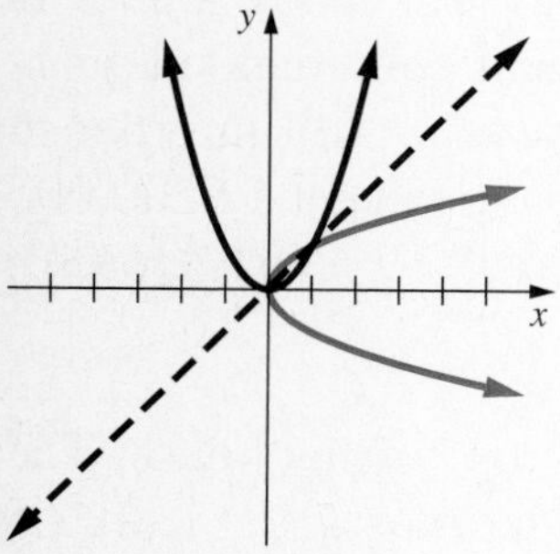

CC-33. Danielle was having trouble finding the undo function (the inverse) for $g(x) = x^2$. So she graphed the function $y = x^2$ and used the carbon-copy method from CC-29 to get the interchanged graph at left.

a. "Aha," she said, "now I know the interchange function." What equation was she thinking of? $x = y^2$

b. "But wait!" said Regis. "The equation $x = y^2$ is not a function." "Picky, picky, picky," Danielle replied, "it works, doesn't it?" What do you think? Why can't $y = x^2$ have an inverse *function*?

c. Think of another function that does not have an inverse function. Discuss this with your group. ***$y = x^4$ and $y = x^3 - 4x$ are examples. You may want to get the whole class to pool their examples. Examples are important at this point, not definitions.***

CC-34. As you saw in CC-33, some functions do not have inverse functions. There may be an inverse equation that works, but it is not a function. Sometimes when this happens, we "cheat." In order to get an inverse function we use only a part of the original function. CC-33 is an example.

For each function below, graph the original function and its inverse function. Then find the equation of the inverse function.

a. $f(x) = x^2$ with domain $x \geq 0$ $\quad$ *$y = \sqrt{x}$*

b. $g(x) = (x - 2)^2$ with domain $x \geq 2$ $\quad$ *$y = \sqrt{x} + 2$*

CC-35. Graph each function on a separate pair of axes. Clearly label the equation of the graph.

1. $y = -\frac{2}{3}x + 6$

2. $y = \frac{1}{2}(x + 4)^2 + 1$; domain: $x \geq -4$

3. $y = \frac{1}{x}$

a. For each of these functions, state the x-intercepts, y-intercept, domain, and range. *(9, 0), (0, 6), reals, reals; no x-intercept, (0, 9), $x \geq -4$, $y \geq 1$; no intercepts, $x \neq 0$, $y \neq 0$*

b. Graph the inverse of each function by using the carbon-copy method or by interchanging points. Clearly label the inverse.

c. State the x-intercepts, y-intercept, domain, and range for each of the inverse graphs. *(6, 0), (0, 9), reals, reals; (9, 0), no y-intercept, $x \geq 1$, $y \geq -4$; no intercepts, $x \neq 0$, $y \neq 0$*

d. Find the "undoing" equation for each function, and write it next to the graph of the inverse function. *$y = -\frac{3}{2}x + 9$; $y = -4 + \sqrt{2(x - 1)}$; $y = \frac{1}{x}$*

EXTENSION AND PRACTICE

CC-36. Amanda's favorite function is $f(x) = 1 + \sqrt{x + 5}$. She has built a function machine that performs these operations on the input values. Her brother Eric always tries to mess up Amanda's stuff, so he created the inverse of $f(x)$, called it $e(x)$, and programmed it into a machine.

a. What is the equation of $e(x)$, Eric's function, which is the inverse of Amanda's? (Remember, it must *undo* Amanda's machine.)
$e(x) = (x - 1)^2 - 5$

b. What happens if the two machines are pushed together? What is $e(f(-4))$? Explain why this happens. *One machine undoes the other, so $e(f(-4)) = -4$.*

c. If $f(x)$ and $e(x)$ are graphed on the same set of axes, what would be true about the two graphs? *They would be reflections of each other over the line $y = x$.*

d. Draw the two graphs on the same set of axes. Be sure to notice the restricted domain and range of Amanda's function.

CC-37 is a good assessment problem. You can see how many students are understanding how to undo equations with a correct reversal of order of operations.

CC-37. If $g(x) = 4x + 7$, find the equation of a machine $f(x)$ so that $f(x)$ is the inverse of $g(x)$. Explain completely how you got your function and show that it works on at least three different numbers. *$f(x) = \frac{x-7}{4}$*

CC-38. Solve the following system:

$$\begin{aligned} a - \ \ b + 2c &= 2 \\ a + 2b - \ \ c &= 1 \\ 2a + \ \ b + \ \ c &= 4 \end{aligned}$$

What does the solution tell you about the relationship among the graphs? *No solution; planes don't intersect in a single point—parallel lines are formed.*

CC-39. Sketch the solution to this system of inequalities:

$$y \geq (x+5)^2 - 6$$
$$y \leq (x+4)^2 - 1$$

CC-40 is a preproblem to the Case of the Cooling Corpse problem in Section 7.10. Be sure to assign it. The data for this problem were gathered using a CBL. You may want to make this problem a lab and use your own CBL to gather data.

CC-40. The following table shows the change in temperature of a cup of coffee as it sits on the kitchen table cooling. Sketch the graph.

Time (sec.)	Temperature (°C)
0	93.7
10	93.3
20	92.9
30	92.0
40	91.2
50	90.8
60	90.0
70	89.2

a. What are the independent and dependent variables? *Temperature depends on time.*

b. What are the domain and range? *D: 0 to ∞; R: room temp. to starting temp., presumably boiling*

c. Is there an asymptote? Explain. *theoretically yes, actually no*

CC-41. Solve each equation for x. Be accurate to three decimal places.

a. $x^2 = 27$ *$x \approx \pm 5.196$* **b.** $2^x = 27$ *$x \approx 4.755$*

CC-42. Brian keeps getting negative exponents and fractional exponents confused. Help him by explaining the difference between $2^{1/2}$ and 2^{-1}.

CC-43. Eniki has a sequence of numbers given by the formula $t(n) = 4(5^n)$.

a. Starting with $n = 0$, what are the first three terms of Eniki's sequence? *4, 20, 100*

b. Chelita thinks the number 312,500 is a term in Eniki's sequence. Is she right? Justify your answer by either giving an appropriate value for n or explaining why it isn't. *$n = 7$*

c. Elisa thinks the number 94,500 is a term in Eniki's sequence. Is she right? Explain. *no*

7.4 INTERCHANGING AN EXPONENTIAL FUNCTION

CC-44 is the crucial problem for introducing logarithms. Use it to start the class as a whole-class activity. Copy the table on the chalkboard, horizontally as shown. We use this table and the need to solve the equation $2^x = 3$ without using guess and check as the motivation for introducing logarithms. After completing the Silent Board Game, you may be able to help students understand the thinking behind the process by reminding them of the square-root function. If you have not yet done the Silent Board Game with $y = \sqrt{x}$, you may want to use that example first to lead up to this one. Be sure to include values that do not have real solutions.

CC-44. The Silent Board Game Do this individually and silently (don't discuss it with your group). Figure out the rule for this function. The table is written on the board. As you figure out a number that fits, go up to the board and fill in the number (silently). If your number is not correct, the instructor will erase it. Only one number from each person, please. Copy the table below onto a full sheet of graph paper. *You may need to slow down the process; otherwise the quicker students will finish the table before everyone gets a chance to think about it.*

x	8	$\frac{1}{2}$	32	1	16	4	3	64	2	0	0.25	−1	$\sqrt{2}$	0.2	$\frac{1}{8}$
$g(x)$	3	−1	*5*	0	*4*	*2*	*	6	*1*	**	*−2*	**	*$\frac{1}{2}$*	*	*−3*

** = guess and check on calculator*
*** = impossible*

CC-45. When the Silent Board Game table in CC-44 is complete, discuss with your group how you found the answers, and write a brief description of the method. *Class discussion will be useful. At this point you may want to do the table for log, base 3, in CC-46 with the whole class as a further example.*

a. Write your rule for the function in symbols. *$x = 2^y$, or better $x = 2^{g(x)}$*

b. $x = 1024$ is not in the table. What is $g(1024)$? *10*

CC-46. Copy the following table, and fill in the missing numbers:

a.

x	$f(x)$
81	4
$\frac{1}{3}$	*−1*
3	*1*
27	*3*
1	0
−1	*none*
0	*none*
9	2
$\frac{1}{9}$	*−2*
$\sqrt{3}$	*$\frac{1}{2}$*

b. Write an equation for the function the table represents. *$x = 3^{f(x)}$*

CC-47. Using a full sheet of graph paper, put the axes in the center of the page, and label each mark on each axis as one unit.

a. Draw the graph of the function from the Silent Board Game in CC-44 by plotting the table of values. On the same axes, in a contrasting color, graph $y = 2^x$; and label it with its equation.

b. What would happen if you reflected the graph $y = 2^x$ across the line $y = x$? Reflect it by the carbon-copy method if you would like to check it. On the graph, answer this question, and write any other observations you can make about the two graphs. *inverse*

CC-48 We will call the function, $x = 2^y$, **the inverse exponential function, base 2**. Label the graph in part b of CC-47 with this name. Then describe this function as completely as possible. More precisely, investigate it. Write all the important information on the graph paper. *Observations should include domain, range, asymptote, smooth, continuous, etc.*

CC-49 can be a struggle for students. The problems in parts a through i correspond with the values in the table of the Silent Board Game in CC-44, which will allow all of the students to figure out the various answers. Try to get them to take it a step further by getting them to think, "what power of 2 equals 32," as in part a below. Reminding them of what they think when doing square roots also may help with this problem.

CC-49. Suppose we label the inverse exponential function, base 2, as $g(x)$. Refer back to the Silent Board Game table in CC-44 and your graph to evaluate $g(x)$ below. What is y and/or x in each of the following cases?

a. $g(32) = ?$ *5* **b.** $g\left(\frac{1}{2}\right) = ?$ *−1* **c.** $g(4) = ?$ *2*

d. $g(0) = ?$ *does not exist* **e.** $g(?) = 3$ *8* **f.** $g(?) = \frac{1}{2}$ *$\sqrt{2}$*

g. $g\left(\frac{1}{16}\right) = ?$ *−4* **h.** $g(?) = 0$ *1* **i.** $g(8) = ?$ *3*

EXTENSION AND PRACTICE

CC-50. Sketch the graph of $y + 3 = 2^x$.

a. What is the domain and range of this function? *x: all numbers; $y > -3$*

b. Does this function have a line of symmetry? If so, what? *no*

c. What are the x- and y-intercepts? *(0, −2), (1.585, 0)*

d. Change the equation so that the graph of the new equation has no x-intercepts. *$y = 2^x + a$, where $a \geq 0$*

CC-51. What is the equation of the inverse of $f(x) = \sqrt{5x + 10}$? Make a graph of both the function and its inverse on the same set of axes.

$g(x) = \left(\frac{x^2 - 10}{5}\right)$; be sure the domain and range are properly restricted.

CC-52. Write the equation of an increasing exponential function that has a horizontal asymptote at $y = 15$. *Answers could vary—something like $y = 2^x + 15$.*

In CC-53, we hope that students will begin to realize that, even though the first equation can be simplified through some number crunching, the second equation is essentially the same as the first. This is not a big deal, however.

CC-53. On Jason's math homework last night, he had to solve for x in several equations. He did well on all of the problems involving numbers, but he just didn't believe that methods he used for equations such as

$$5 \cdot 4 + 2(x^3) = 8 + 7$$

would work on equations such as

$$ay + bx^3 = c + 7$$

Show Jason how he can use the same basic steps he used for the first equation to solve the second equation for x.

CC-54. What is the difference between the graphs of these two functions?

$$y + 3 = (x - 1)^2 \quad \text{and} \quad x + 3 = (y - 1)^2$$

Explain completely enough so that someone who doesn't know how to graph either equation could graph them after reading your description. Are the two graphs inverses of each other? Are they both functions? Justify your answer. *Yes, they are inverses since the x- and y-variables are simply interchanged, and the graphs are reflections of each other over the line $y = x$. The second is not a function.*

CC-55. Consider the sequence given in the following table:

n	0	1	2	3	4
$t(n)$	4	2	2	4	8

a. Plot the points and draw in staircases to help you decide what type of sequence it is. *It is neither arithmetic nor geometric. Some students may think of calling it quadratic or a sequence of squares.*

b. Write the equation that represents this sequence. If you have trouble getting started, look back to Chapter 6, Section 6.6. *$t(n) = n^2 - 3n + 4$*

CC-56. Which is larger, $n - 1$ or $n - 2$? Is your answer the same regardless of the value you chose for n? Explain. *$n - 1$; yes, $n - 1$ is always one more than $n - 2$.*

CC-57. Solve each equation for x. Be accurate to two decimal places.

a. $x^3 = 243$ *$x \approx 6.24$* **b.** $3^x = 243$ *$x = 5$*

7.5 INTRODUCING LOGARITHMS

Still more practice with the inverse exponential function, base 2. If students begin to ask about the "log" key on the calculators, it would be well worth their time to ***discover*** *what base the calculator works in, rather than simply telling them. If you don't want to give them hints, tell them to read CC-76.*

CC-58. If $x = 2^y$, how can you solve for y? Discuss this with your group, and be prepared to report to the class what you think. *Be prepared for answers like "take the yth root" or other ideas. It shows they are trying.*

CC-59. When mathematicians can't solve a problem or don't have the tools to do a particular problem, do you know what they usually do? They invent a tool to do what they want. In the case of $x = 2^y$ they invented a new operation, the inverse exponential function, base 2. Taadaa! At one point in history they said, "From now on, the inverse exponential function, base 2, will be known as **logarithm, base 2**." Why did they choose "logarithm" when thousands of other names were possible? That's an extra credit problem to check out with your instructor. You graphed this function back in CC-48 (we called the graph the inverse exponential function, base 2). Find your graph, and label it "logarithm, base 2." Add the graph of $y = \log_2(x)$ to your Parent Graph Tool Kit.

Logarithm, base 2, is usually abbreviated "log, base 2" and written "$\log_2$." When we see this symbol, we read it aloud as "log, base 2." The comma means we pause a bit when we say it. Since it is a function, we can use function notation, as in $g(x) = \log_2(x)$. Notice that the base number, 2 in this case, is always written a little lower, as a subscript. Use your graph of $g(x)$ to find each of the missing values:

a. $\log_2(32) = ?$ **b.** $\log_2\left(\frac{1}{2}\right) = ?$ **c.** $\log_2(4) = ?$

d. $\log_2(0) = ?$ **e.** $\log_2(?) = 3$ **f.** $\log_2(?) = \frac{1}{2}$

g. $\log_2\left(\frac{1}{16}\right) = ?$ **h.** $\log_2(?) = 0$ **i.** $\log_2(8) = ?$

CC-60. How do your answers in the previous problem compare with your answers in CC-49? *exactly the same*

Let groups work on making connections for the inverse exponential function, base 2, expressed as $x = 2^y$ and logarithm, base 2, expressed as $y = \log_2 x$. Then follow with a whole-class discussion. If you don't have a discussion, then CC-61 might be very difficult for your students. Since this is a very important problem, you want (need) them to be successful.

CC-61. Let $y = \log_2(x)$. Rewrite this equation so that it is solved for x. Remember how we defined $y = \log_2(x)$. Put a large box around both equations. Do the two equations look the same? Do the two equations mean the same thing? Are they equivalent? How do you know? Think about it; don't rush it, this is very important. *$x = 2^y$, no, yes, yes*

CC-62. Since a logarithm is the interchange (or inverse) of an exponential function, each logarithmic function has a particular base. Note that we write the base *below* the line as a subscript. For example, we write $\log_2(x)$. This looks a bit confusing, since the x is above the 2. It almost looks like 2^x, but it isn't. When you write it, make it very clear. Every log equation can be written as an exponential equation and vice versa, as you saw in CC-61. Copy each equation shown below. Then rewrite each equation in the other form.

a. $y = 5^x$
$x = \log_5 y$

b. $y = \log_7(x)$
$x = 7^y$

c. $8^x = y$
$x = \log_8 y$

d. $A^K = C$
$K = \log_A C$

e. $K = \log_A(C)$
$C = A^K$

f. $\log_{1/2}(K) = N$
$K = (1/2)^N$

EXTENSION AND PRACTICE

CC-63. On Wednesdays at Tara's Taqueria four tacos are the same price as three burritos. Last Wednesday the Lunch Bunch ordered five tacos and six burritos, and their total bill was $8.58 (with no tax or drinks included). Nobody in the Lunch Bunch can remember the cost of one of Tara's tacos. Help them figure it out. *$0.66*

CC-64. Dolores has two big bags of hard candies that she is planning to put into small packages for the Ides of March picnic. One bag contains 450 assorted fruit-flavored candies; the other contains 500 coffee-toffees. She figures the children will like the fruit candies better and the adults will prefer the coffee-toffees. So she plans to make packets containing two coffee-toffees and six fruit candies for the children and packets of five coffee-toffees and three fruit candies for the adults. Now she is trying to figure out how many child packets and how many adult packets she can make. Help her out by writing two inequalities and drawing their graphs. Let x be the number of packets for children and y the number of packets for adults. If 60 children and 70 adults go to the picnic, will there be enough packets of candy? *$2x + 5y \le 500$, $6x + 3y \le 450$; no, since $6(60) + 3(70) > 450$.*

CC-65. Shakeah's grandfather is always complaining that back when he was a kid, he used to be able to buy his girlfriend dinner for only $1.50.

a. Assuming he wasn't just a cheap date, why would this be true? *inflation*

b. If the same dinner that Shakeah's grandfather purchased for $1.50 sixty years ago now costs $25.25, and the increasing prices over time form a geometric sequence, write an equation that will give you the costs at different times. *$t(n) = 1.5(1.048)^n$*

CC-66. Make a sketch of a graph that is a decreasing exponential function with the x-axis as the horizontal asymptote.

Now make a similar sketch, this time with the horizontal asymptote as the line $y = 5$.

CC-67. If $f(x) = x^4$ and $g(x) = 3(x + 2)$, find each value:

a. $f(2)$ *16* **b.** $g(2)$ *12*

c. $f(g(2))$ *$12^4 = 20{,}736$* **d.** $g(f(2))$ *54*

e. If $f(x) = 81$, what is x? *3*

CC-68. Suppose $h(x) = 3^x$ and $k(x) = \log_3 x$. What is $h(k(x))$? What about $k(h(x))$? Start with a few numbers to convince yourself. Explain completely why this is true. *$h(k(x)) = k(h(x)) = x$*

CC-69. Write-a-Text Factory At the Write-a-Text Factory, workers spend grueling hours at computer terminals trying to be creative writing textbooks. One day, during a brief bout of boredom, Karna and Carlos decided to play a trick on Darell as he worked on his document. They implanted a strange code into his computer so that as soon as he had typed 60,000 characters 10 percent of his document, starting from the beginning, would be deleted every hour. Just as the clock struck 5:00 P.M. and he was anxiously waiting for the whistle to blow telling him he could go home, Darell typed in his 60,000th character! He left, not to return until 8:00 A.M. the next day.

a. How many characters were left when he returned the next morning? *≈12,353*

b. If Darell starts typing 3600 characters per hour when he arrives, will he increase the size of his document faster than the mutant code can decrease it, or vice versa? ***He will increase the size.***

CC-70. A triangle is formed by a line that has a slope of 2, and the x- and y-axes. The area of the triangle is 30 square units. Find the equation of the line. A diagram would be helpful. $y = 2x \pm 2\sqrt{30}$

CC-71. Eeew! Ever eat a maggot? Guess again! The FDA publishes a list, the Food Defect Action Levels list, which indicates limits for "natural or unavoidable" substances in processed food (*Time* magazine, October 1990). So in 100 grams of mushrooms, for instance, the government allows 20 maggots! The average rich and chunky spaghetti sauce has 350 grams of mushrooms. How many maggots is that? *70*

CC-72. If $x = 3^y$, what do you think you would write to solve for y? Explain. $y = \log_3 x$

7.6 INVESTIGATING LOGARITHMS

Although the problems in this section include a medium-sized problem and a rather big function investigation, the purpose of the section is primarily consolidation. The graphing calculator will not be of much help in the function investigation. Students should gain in their understanding of what a logarithm is through the practice required to make tables without the calculator. CC-73 is optional. Don't do it if students have already discovered what base the calculators use with logarithms.

CC-73. There is a log key on your calculator. On most scientific calculators, you enter the number first, then press the log key. On the graphing calculator, however, it is entered just as you would write it or say it (press "log" first, then the number). But notice that the base is omitted. Unfortunately, you do not get the option of putting in any base you want. Instead, the calculator's log has a fixed base. Your mission is to figure out this base. As a hint, try entering "log 2," "log 3," and so on. Keep track of your results. Use complete sentences to write an explanation of the problem and your solution. *log 10 gives the simplest way to find the base.*

CC-74. Investigate $y = \log_b x$. Be complete, and be sure to include several appropriate graphs using different values for b. Do not expect to use your graphing calculator.

CC-75. When computing with logarithms using the calculator, what base must we use?

> Since the calculator's base for common logarithms is always 10, and since the calculator is so useful in computations, it saves time to agree that when we write "log x" *without writing a base*, we mean $\log_{10} x$.
>
> Log is a function, and like any function its inputs should be in parentheses like $f(x)$. But since logs are so useful and written so frequently, most people omit the parentheses, and just write log x (or $\log_N x$, if the base is not 10). We will omit the parentheses, except where it might be confusing, such as when the inputs are fractions or complex expressions, for example, $\log_2\left(\frac{1}{32}\right)$ and $\log(3x + 5)$.

CC-76. Copy and solve these equations for x. If possible, obtain a numerical answer without a calculator.

a. $\log_5 25 = x$
2

b. $\log_4\left(\frac{1}{4}\right) = x$
−1

c. $3 = \log_x 343$
7

d. $\log_6 0 = x$
n.a.

e. $3 = \log_5 x$
125

f. $\log_9 x = \frac{1}{2}$
3

g. $x = \log_{64} 8$
1/2

h. $\log_{11} x = 0$
1

i. $x = \log_{10} 0.01$
−2

CC-77 is extremely important. Take the time to make it a group quiz or an activity in which the students make a poster to demonstrate that they understand.

CC-77. Your friend is not quite sure how to rewrite exponential equations as log equations, and vice versa. Write out an explanation for your friend. It might be easier to use equations, symbols, and letters of the alphabet in addition to sentences. What you are really doing here is developing a definition of logarithm. It may help to rewrite the equations from CC-76 as exponential equations to see a pattern. *Hopefully students will come up with "$\log_K D = E$ means $K^E = D$."*

EXTENSION AND PRACTICE

CC-78. Late last night, as Agent 008 negotiated his way back to headquarters after a long day, he saw the strangest glowing light. It came closer and closer until finally he could see that it was some kind of spaceship! He was frozen in his tracks. It landed only 15 feet from him, and a hatch slowly opened. Four little creatures came out carrying all sorts of equipment, from calculators to what appeared to be laser beams. They did not seem to notice that Agent 008 was there—that is, until he sneezed! Suddenly, the creatures turned around, looking very startled. They dashed into the spaceship, closed the hatch, and rocketed into the night. Could he believe what he had seen? Maybe it was just a dream? After a few minutes of standing there dazed and confused, he started walking on, his eyes glazed over. He was just coming to his senses when he stepped on something strange. He picked it up and to his surprise it was one of the creatures' calculators. What a prize! He started playing with it as he walked on. "Boy, headquarters is going to love this!" he thought. It appeared to have a "log" button, and as he played with it he noticed something interesting: log 10 did not equal 1 as it did on his own calculator. With this calculator, $\log 10 \approx 0.926628408$! He tried some more calculations: $\log 100 \approx 1.853256816$, and $\log 1000 \approx 2.779885224$. This was most peculiar. Obviously, the creatures did not work in base 10!

a. What base do the space creatures work in? Explain how you got your answer. You may want to rewrite the problem as $\log_b 10 = 0.926628408$ and try to figure out the value of b. ***Base 12***

b. How many fingers do you think these space creatures have? Explain. ***base 12 implies 12 fingers!***

CC-79. We are trying to get a more accurate answer to the question given at the beginning of this chapter: How long it will take you to double \$5000 invested at 3 percent compounded annually? Solve the following equation for x. Your answer must be accurate to three decimal places this time.

$$2 = 1.03^x$$

$x \approx 23.450$

CC-80. If $10^{3x} = 10^{x-8}$, solve for x. Show that your solution works by checking your answer. ***$x = -4$***

CC-81. For $f(x) = 3 + \sqrt{2x - 1}$, answer each of the following questions:

a. What are the domain and range of $f(x)$? ***$1/2 \le x < \infty$, $3 \le y < \infty$***

b. What is $f(x)$'s inverse, or interchange? Call it $g(x)$. ***$g(x) = \dfrac{(x-3)^2 + 1}{2}$***

c. What are the domain and range of $g(x)$? ***$3 < x < \infty$, $1/2 \le y < \infty$***

d. What is $f(g(6))$? ***6***

e. What is $g(f(6))$? What do you notice? Why does this happen? ***6; f(g(6)) and g(f(6)) are the same.***

CC-82. Which of the following statements are true? If a statement is false, explain why.

a. $\frac{x+3}{5} = \frac{x}{5} + \frac{3}{5}$ *true*

b. $\frac{5}{x+3} = \frac{5}{x} + \frac{5}{3}$ *false: if $x = 2$, then $1 \neq \frac{5}{2} + \frac{5}{3}$*

CC-83. Solve for x. Check that your solution works.

a. $5 - \frac{8}{x+4} = \frac{2x}{x+4}$

no solution, $x \neq -4$

b. $\frac{1}{x} = \frac{x}{1-x}$

$\frac{-1 \pm \sqrt{5}}{2}$ or 0.61803 and −1.61803

CC-84. While working on his math homework, Pietro came across this problem:

> "If $F(x) = (3)^{-x} - 1$, find $F(2)$, $F(3)$, $F(4)$, and $F(5)$. Then explain what happens to the values of $F(x)$ as bigger and bigger numbers are substituted in for x."

He didn't have any idea how to do the problem.

a. Help him out by calculating $F(2)$, $F(3)$, $F(4)$ and $F(5)$. *−0.889, −0.963, −0.988, −0.996*

b. Plot the points and draw the graph of $F(x)$.

c. Use your graph to explain to Pietro what does happen to $F(x)$ as x gets larger and larger. *The values get closer to −1.*

CC-85. Solve for m: $m^5 = 50$. *$m \approx 2.1867$*

CC-86. Darts hit each of these dartboards at random. What is the probability that a dart will *not* land in the shaded area?

a.

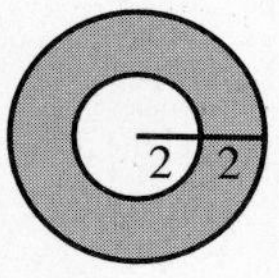

1/4

b.

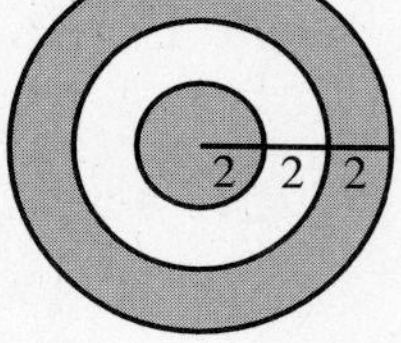

1/3

7.7 TRANSLATING THE GRAPHS OF LOGARITHMIC FUNCTIONS

While the concepts of inverse and logarithm settle into your mind, we explore the translated graphs of log functions. If your Tool Kit is current, you can refer to your notes from the Chapter 5 investigations on translations of graphs.

CC-87. Using a full sheet of graph paper, make a fairly accurate graph of $f(x) = \log x$. Clearly write the equation on the graph.

CC-88. On the same set of axes used in CC-87, but in another color, graph $g(x) = 5 + \log x$. How is this graph different from the first? Explain. You may use the graphing calculator, but if you refer to your notes from Chapter 5 in your Tool Kit, you can probably do these graphs more efficiently without it. *shifted up 5 units*

CC-89. On the same set of axes used in CC-87, but in yet another color, graph $h(x) = 5 \log x$. How is it different from the graph of $f(x) = \log x$? Explain. *stretched up*

CC-90. Now graph $j(x) = \log(x + 5)$ in a fourth color on the CC-87 axes. How is this graph different from the graph of $f(x) = \log x$? Explain. *shifted 5 to the left*

CC-91. Last one! Graph $k(x) = \log(x - 5)$. How is it different? Explain. *shifted 5 to the right*

CC-92. Prove that you are the expert log grapher. Explain completely, so that even a friend who has not studied logarithms can follow, how to graph $y = 2 + 7 \log(x + 4)$.

EXTENSION AND PRACTICE

CC-93. Copy the following equations, and solve for x without a calculator:

a. $\log_x 25 = 1$
25

b. $x = \log_3 9$
2

c. $3 = \log_7 x$
343

d. $\log_3 x = \frac{1}{2}$
$\sqrt{3}$

e. $3 = \log_x 27$
3

f. $\log_{10} 10{,}000 = x$
4

Be sure to assign CC-94. It prepares students for the work in Section 7.8.

CC-94. Calculator Investigation

a. Compute the decimal values for the following pairs of logarithmic expressions:

$4 \log 2$ and $\log 2^4$ $\quad$ $3 \log 4$ and $\log 4^3$ $\quad$ $2 \log 3$ and $\log 9$

b. What do you notice in all three examples? Explain completely.
$a \log k = \log(k^a)$

c. Try $\log 1000^5$ and $5 \log 1000$ and $\log (10^3)^5$.

d. Make up three examples of your own. Use unusual numbers in one of them.

e. Make a note of this pattern. You will use it again soon.

CC-95 is just a restatement of an earlier problem, and we are asking students to make their previous answer more accurate. We will use it to start Section 7.8 as a means to motivate the use of logarithms.

CC-95. Oh no, not this problem again:

$$2 = 1.03^x$$

How long will it take to double your money? This time your solution must be correct to *four* decimal places. ***$x \approx 23.4498$***

CC-96. Solve the following systems of equations. Look for a shortcut based on good algebraic reasoning. If you use a shortcut, be sure to explain your thinking.

a. $13x + 17y = 0$
$8x - 10y = 0$
(0, 0)

b. $435x + 334y = 0$
$1.78x - 37.7y = 0$
(0, 0)

CC-97. Is it true that $\log_3 2 = \log_2 3$? Justify your answer. ***No; $\log_3 2 < 1$ and $\log_2 3 > 1$.***

CC-98. Consider the general form of an exponential function, $y = km^x$.

a. Solve for k. $k = \dfrac{y}{m^x}$

b. Solve for m. ***m is the xth root of y/k.***

CC-99 is a preproblem for solving the Case of the Cooling Corpse (CC-127) in Section 7.10. Don't skip it.

CC-99. Graph the following two functions on the same set of axes:

$$y = 3(2^x)$$
$$y = 3(2^x) + 10$$

a. How do the two graphs compare? ***The second is the same as the first shifted up 10.***

b. Suppose the first equation is $y = k \cdot m^x$, and the graph is shifted up b units. What is the new equation? $y = k \cdot m^x + b$

CC-100. Write the equations of two different curves that cross the line $y = 10$ but do not cross the line $y = 11$. Make sketches of each.
$y = a(x + b)^2 + c$ with $a < 0$ and $10 < c < 11$

CC-101. Given the function $y = 3(x + 2)^2 - 7$, how could you restrict the domain to give "half" of the graph?

a. Find the equation for the inverse function for your "half a function."

$y = \sqrt{\frac{x+7}{3}} - 2$ assuming domain is restricted to $x \geq -2$.

b. What are the domain and range for the inverse function? *$x \geq -7$; $y \geq -2$*

7.8 USING LOGARITHMS TO SOLVE PROBLEMS

In this section we finally get to a most useful aspect of logs. Start by taking a quick survey to find out what solutions students got to $2 = 1.03^x$ from CC-95. Then ask if anyone has found a shortcut or noticed anything that can make the problem easier to do. Assuming no one has, start planting the seed with comments like "Wouldn't it be nice if someone could come up with a method besides guess and check?"

The whole class will need some guidance in this section as you bring together the two ideas needed to solve an exponential equation. Read through the section to decide how you will balance group work with guidance for the whole class.

CC-102. Discuss with your group how you solved the double-your-money problem, $2 = 1.03^x$, accurately to four decimal places. Lay out the subproblems or key ideas. *Put bounds on x; guess and check.*

Have students work together on these problems. Then you may want to go over the problem, asking the groups what answers they came up with and emphasizing the fact that the relation $\log(X^a) = a \log X$ will be very useful.

CC-103. Aren't you getting tired of all this guess and check? Don't you *wish* there were an easier way of doing this sort of problem? Well, there is, and you know most of the pieces already. You just need to put some ideas together! Converting this equation from exponential form to log form gives $x = \log_{1.03} 2$, but our calculators use log10, so this does not help.

Look back to CC-94 and make sure you see the pattern. (If, $\log 4^3 \approx 1.806$ and $3 \log 4 \approx 1.806$, for example, what can you conclude?) This pattern is a step toward solving our familiar problem. Use this pattern to write each of the following in a different form.

a. $\log 2^5 = ?$ *$5 \log 2$*

b. $\log 10^5 = ?$ *$5 \log 10$*

c. $\log 10^x = ?$ *$x \log 10$*

d. $\log 3^x = ?$ *$x \log 3$*

e. $\log X^a = ?$ *$a \log X$*

Be sure to add a statement describing this property of logarithms to your Tool Kit, along with at least three numerical examples.

You may want to lead the whole class through the development in CC-104.

CC-104. How can we use the idea from CC-103 to solve our old friend, $2 = 1.03^x$?

When two expressions are equal, we know that adding the same number to each or multiplying by the same number gives equal results. What if we took the logarithm, base 10, of each side of the equation:

$$\log 2 = \log 1.03^x$$

Now solve for x, accurate to eight decimal places. ***$x \approx$ 23.44977225***

Now have students do CC-105 for practice. Be sure they all understand that they can take the log, base 10, of both sides of an equation, and then use the relation $\log N^m = m \log N$ to create a simple (if messy) numerical equation they know how to solve.

CC-105. Use the ideas from CC-103 and 104 to solve the following equation. Be accurate to three decimal places.

$$5 = 1.04^x \qquad \textit{41.035}$$

CC-106. Solve each of the following equations to the nearest thousandth:

a. $25^x = 145$ *1.546*
b. $1.28^x = 4.552$ *6.139*
c. $240(0.95^x) = 100$ *17.068*

EXTENSION AND PRACTICE

CC-107. Write three different-looking but equivalent expressions for each of the following expressions. For example: $\log 7^{3/2}$ can be written as

$$\frac{3}{2}\log 7$$

$$\frac{1}{2}\log 7^3$$

$$3\log\sqrt{7}, \text{ etc.}$$

a. $\log 8^{2/3}$ **b.** $-2 \log 5$ **c.** $\log (na)^{ba}$

CC-108. Margee thinks she can use logs to solve $56 = x^8$ since logs seem to make exponents disappear. Unfortunately, Margee is wrong. Explain the difference between equations like $2 = 1.03^x$, where you can use logs, and equations like $56 = x^8$ where you don't need logs. ***take the eighth root.***

CC-109 leads toward solving with a calculator equations in the form $x^a = b$ for x.

CC-109. Investigate:

a. What can we multiply 8 by to get 1? *1/8*

b. What can we multiply a by to get 1? *1/a*

c. By using the rules of exponents, find a way to solve $m^8 = 40$. (Obtain the answer as a decimal approximation using your calculator. Check your result by raising it to the eighth power.) *m = 1.586 . . .*

d. Now solve $n^6 = 300$. *±2.587; it's not likely students will think of the negative answer, but it appears in their answer section, so maybe they'll ask.*

e. Find a rule for solving $x^a = b$ for x with a calculator. *Descriptions will vary; $x = b^{1/a}$.*

CC-110. What is the equation of the line of symmetry of the graph of $y = (x - 17)^2$? Justify your answer. *x = 17*

CC-111. Solve each system of equations:

a.
$$\begin{aligned} 8x - 3y + 4z &= 1 \\ y + z &= 12 - x \\ -4x &= z - 2y + 12 \end{aligned}$$
x = −3, y = 5, z = 10

b.
$$\begin{aligned} 3x + y - 2z &= 6 \\ x + 2y + z &= 7 \\ 6x + 2y - 4z &= 12 \end{aligned}$$
infinitely many solutions

c. Remember that the graph of each equation in parts a and b is a plane. What does the result in part b tell you about the planes? *They intersect in a line.*

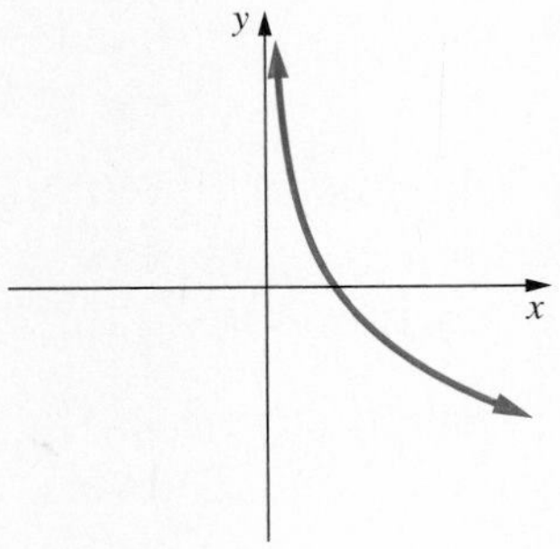

CC-112. At the left is a graph of $y = \log_b x$. Find a reasonable value for b. *0 < b < 1*

CC-113. Describe the transformation and sketch the graph of $y = \log_3 (x + 4)$.

7.9 WRITING EQUATIONS FOR EXPONENTIAL FUNCTIONS

After learning a method for finding the equation of an exponential function given its asymptote and two of its points, students will work on some applications. You may need to guide the whole class at times, but wait until they have had time to do some work first so they can pinpoint their questions.

CC-114. As you learned in Chapter 3, an exponential function has the general form

$$y = km^x$$

a. Does this type of equation have an asymptote? If so, what? *Yes, the x-axis*

b. A particular exponential function goes through the points (2, 36) and (1, 12) and has the x-axis as a horizontal asymptote. Substitute these values for x and y into $y = km^x$ to create two equations with two unknowns, k and m. *$36 = km^2$, $12 = km^1$*

c. Solve one of your equations for k or m. Substitute that result into the remaining equation. Now you should be able to figure out one of the variables. When you know one you should be able to find the other. *$m = 3$, $k = 4$*

d. What is the exponential function that passes through (2, 36) and (1, 12) and has the x-axis as an asymptote? *$y = 4 \cdot 3^x$*

CC-115. I'd like to have $40,000 in 8 years (hey, I'm greedy), and I only have $1000 now.

a. What interest rate compounded yearly do I need to satisfy my greed? To help me solve this, set up an exponential equation, which should be in the form $y = km^x$.

Let y = the amount of money.
Let x = the number of years.
Let m = the multiplier.

Since $1000 = km^0$, $k = 1000$. Substitute 1000 for k in the other equation. $40{,}000 = 1000m^8$. The multiplier is 1.586. Therefore, the rate of increase is 0.586 . . . , and the interest is 58.6%—a little too greedy, I guess!

b. Try another scenario: I start with $7800 and I want to have $18,400 twenty years from now. What interest rate do I need (compounded yearly)? *4.4%*

c. Which of these two scenarios do you think is more likely to happen? Justify your response. *Students better say the second!*

CC-116. Suppose an exponential function contains the points (3, 12.5) and (4, 11.25).

a. Is the function increasing or decreasing? Justify your answer. *decreasing*

b. This function does not have the x-axis as a horizontal asymptote. Its horizontal asymptote is the line $y = 10$. Make a sketch of this graph showing the horizontal asymptote.

c. If this function has the equation $y = km^x + b$, what would be the value of b? Explain. *10, a shift up from the general case*

d. Substitute the known points into the equation $y = km^x + 10$, and solve the system for k and m. *$k = 20$, $m = 0.5$*

e. What is the equation of the function? *$y = 20(0.5^x) + 10$*

CC-117. Joyce has been using her online computer service too much. Her average monthly bill has been $325.00. She needs to buy a new car, so she has decided to decrease her monthly online use until she uses only the basic 10 or fewer hours per month. The minimum monthly charge is $9.95 for up to 10 hours. Joyce would then have an extra $315.05 per month to help pay for her car. The online service charges an extra $2.95 per hour for each hour of use after the first 10 hours. Two months later her bill is $192.76. Three months later her bill is $147.06.

a. As the months go by, Joyce's bill changes. This change can be represented by an exponential function. Is this function increasing or decreasing? Justify your answer. *decreasing*

b. This exponential function does not have the x-axis as a horizontal asymptote. Make a sketch of this graph showing the horizontal asymptote.

c. If this function has the equation $y = km^x + b$, what is the value of b? Explain. *9.95, a shift up from the general case.*

d. Using your graph write the equation of the function. Your equation does not have to be exact; you are making an approximation.
$y = 325(0.75^x) + 9.95$

e. If she continues this pattern, how many months will it take Joyce to decrease her bill to $10.00? *about 30.5 months*

EXTENSION AND PRACTICE

CC-118. Solve each equation to the nearest thousandth.

a. $5.825^{x-3} = 120$
5.717

b. $18(1.2^{2x-1}) = 900$
11.228

CC-119. Hollywood Glitter The economy has worsened to the point that the merchants in downtown Hollywood can't afford to replace the lightbulbs when they burn out. On average about 13 percent of the lightbulbs burn out every month. Assuming there are now about one million outside lights in Hollywood stores, how long will it take until there are only 100,000 bulbs lit? until there is only one bulb lit? *16.5 months; 99.2 months*

CC-120, CC-121, and CC-122 lead to the relationships

$$\log xy = \log x + \log y \qquad \text{and} \qquad \log \frac{x}{y} = \log x - \log y.$$

If this is not a "must cover" topic, you may want to assign problems from the Section 7.10's Extension and Practice set instead.

CC-120. Using your calculator, compare each expression in Column A with its paired expression in Column B. Use one of these symbols to show the relationship: >, < , or =

Column A	Column B
log 30	log 5 + log 6
log 27	log 9 + log 3
log 24	log 2 + log 12
log 132	log 12 + log 11

a. Write a conjecture (a statement you think is true) based on the pattern you noticed.

b. What is another way of expressing log 65?

c. Make up three more examples of your own, and check them.

CC-121. Using your calculator, compare the expressions in Columns C and D.

Column C	Column D
log 30	log 300 – log 10
log 27	log 81 – log 3
log 8	log 24 – log 3
log 12	log 60 – log 5

a. Write a conjecture based on the pattern you noticed.

b. What is another way of expressing log 13?

c. Make up three more examples of your own, and check them.

CC-122. (Last time!) Using your calculator, compare the expressions in Columns E and F.

Column E	Column F
$\log(3 \cdot 4)$	log 3 + log 4
$\log\left(\frac{72}{8}\right)$	log 72 – log 8

a. What is another way of expressing log $(a \cdot b)$?

b. What is another way of expressing log $\left(\frac{a}{b}\right)$?

CC-123. Solve each system of equations:

a. $2x + y = 1$
$-3y = 10x$
(–3/4, 5/2)

b. $\frac{x+y}{5} = 1$
$\frac{2x+3y}{3} = 1$
(12, –7)

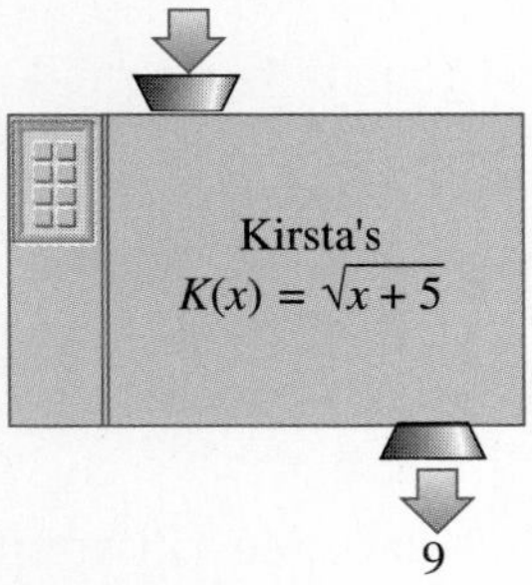

CC-124. Kirsta was working with her function machine, but when she turned her back, her friend Caleb dropped a number in. She didn't see what was dropped in, but she did see what fell out: 9.

a. What operations must she perform on 9 to undo what her machine did? Use these to find out what Caleb dropped in. *Square 9 and subtract 5; he dropped in a 76.*

b. Write a rule for a machine that will undo Kirsta's machine. Call it $c(x)$. *$c(x) = x^2 - 5$*

In CC-125 students may find it difficult to accurately describe the characteristic asked for in part c. A follow-up class discussion may be needed.

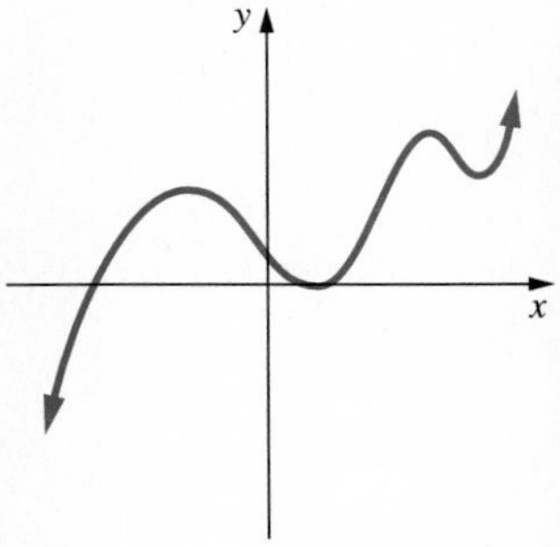

CC-125. Is the graph shown here a function? Explain. *Yes*

a. Make a sketch of the inverse of this graph. Is the inverse a function? Justify your answer. *no*

b. Must the inverse of a function be a function? Explain. *not necessarily*

c. Describe what is characteristic about functions that do have inverse functions. (You might want to add this characteristic to your Tool Kit.)

d. Could the inverse of a nonfunction be a function? Explain or give an example.

CC-126. Fifty grams of uranium are sealed in a box. The half-life of uranium is 1000 years.

a. How much uranium is left after 10,000 years? *≈ 0.0488 g*

b. How long will it take to reduce to one percent of the original amount? *≈ 6640 yr*

c. How long will it take until all of the original mass of uranium is gone. Support your answer. *never!*

7.10 SOLVING THE MYSTERY OF THE COOLING CORPSE

CC-127 is an application problem for logarithms.

CC-127. The Case of the Cooling Corpse The coroner's office was kept at a cool 17°C. Agent 008 kept pacing back and forth trying to keep warm as he waited for any new information. For over three hours now, Dr. Dedman had been performing an autopsy on the Sideroad Slasher's latest victim, and Agent 008 could see that the temperature of the room and the deafening silence were beginning to irritate even Dr. Dedman. The Slasher was creating more work than Dr. Dedman cared to investigate.

"Dr. Dedman, don't you need to take a break?" Agent 008 queried. "You've been examining this dead body for hours! Even if there were any clues, you probably wouldn't see them at this point."

"I don't know," Dr. Dedman replied. "I just have this feeling something is not quite right. Somehow the Slasher slipped up with this one and left a clue. We just have to find it."

"Well, I have to check in with HQ," 008 stated. "Do you mind if I step out for a couple of hours?"

"No, that's fine." Dr. Dedman responded. "Maybe I'll have something by the time you return."

"Yeah, right," 008 thought to himself. These small-town people always want to be the hero and solve everything. They just don't realize how big this case really is. The Slasher has left a trail of dead bodies through five states!

Agent 008 left, closing the door quietly. As he walked down the hall, he could hear the doctor's voice fading away while describing the victim's gruesome appearance into the tape recorder.

The hallway from the coroner's office to the elevator was long and dark. This was the only way to Dr. Dedman's office. Didn't this frighten most people? Well, it didn't seem to bother old Ajax Boraxo, who was busy mopping the floor, thought 008. Most of the others wouldn't notice, he reminded himself.

He stopped briefly to use the restroom and bumped into one of the deputy coroners.

"Dedman still at it?"

"Sure is, Dr. Quincy. He's totally obsessed. He's dead sure there's a clue."

As usual, when leaving the courthouse, 008 had to sign out. Why couldn't these small towns move into the 20th century? *Everyone* used surveillance cameras these days.

"How's it going down there, 008?" Sergeant Foust asked. Foust spent most of his shifts monitoring the front door, forcing all visitors to sign in while he recorded the time next to the signature. Agent 008 wondered if Foust longed for a more exciting aspect of law enforcement. He thought if he were doing Foust's job, he would get a little stir-crazy sitting behind a desk most of the day. Why would someone become a cop to do this?

"Dr. Dedman is convinced he'll find something soon. We'll see!" Agent 008 responded. He noticed the time: Ten minutes to 2:00. Would he make it to HQ before the chief left?

"Well, good luck!" Foust shouted as 008 headed out the door.

Agent 008 sighed deeply when he returned to the courthouse. Foust gave his usual greeting: "Would the secret guest please sign in!" he would say, handing a pen to 008 as he walked through the door. Sign in again? 5:05 P.M. Agent 008 had not planned to be gone so long, but he'd been caught up in what the staff at HQ had discovered about that calculator he'd found. At that moment he saw a good reason for having anyone who comes in or out of the courthouse sign in: He knew by quickly scanning the list that Dr. Dedman had not left. In fact, the old guy must still be working on the case.

As he approached the coroner's office, he had a strange feeling something was wrong. He couldn't hear or see Dr. Dedman. When he slowly opened the door, the sight he saw inside stopped him in his tracks. Evidently, Dr. Dedman had become the *newest* victim of the Slasher. But wait! The other body, the one the doctor had been working on, was gone! Immediately, the security desk, with its annoying sign-in sheet came to mind. Yes, there were lots of names on that list, but if he could determine the time of Dr. Dedman's death, he might be able to just scan the roster to find the murderer! Quickly, he grabbed the thermometer to measure the doctor's body temperature. Then he turned around and hit the security buzzer. The bells were deafening. He knew the building would be sealed off instantly and security would be there within seconds.

"My God!" Foust cried as he rushed in. "How did this happen? Who could have done this? I spoke to the doctor less than an hour ago. What a travesty!"

As the security officers crowded into the room, Agent 008 explained what he knew, which was almost nothing. He stopped long enough to check the doctor's body temperature: 27°C—10°C below normal. He figured the doctor had been dead at least an hour. Then he remembered the tape recorder. Dr. Dedman had been taping his observations. That was standard procedure. Agent 008 and the security officers began looking everywhere for that blasted thing. Yes, the Slasher must have realized that the doctor had been taping and taken that as well. Exactly an hour had passed during the search, and Agent 008 noticed that the thermometer still remained in Dr. Dedman's side. The thermometer clearly read 24°C. Agent 008 knew he could now determine the time of death precisely. He looked at the security sign-in sheet:

Coroner's Office—Please Sign In

Name	Time In	Time Out
Rufin Vonboskin	12:08	2:47
Lizzy Borden	12:22	1:38
Chuck Manson	12:30	2:45
Hanibal Lechter	12:51	1:25
Ajax Boraxo	1:00	2:30
D. C. Quincy	1:10	2:45
Agent 008	1:30	1:50
Ronda Ripley	1:43	2:10
Jeff Domer	2:08	2:48
Stacy Stiletto	2:14	2:51
Blade Butcher	2:20	2:43
Pierce Slaughter	3:48	4:18
Gashes Wound	3:52	5:00
Slippery Eel	3:57	4:45
Candy Carcass	4:08	4:23
Milly Maniacal	4:17	4:39
D.C. Quincy	4:26	4:50
Fred Cruger	4:35	
Danny Demented	4:48	4:57
Larry Laceration	5:04	
Agent 008	5:05	
Security	5:12	

a. Make a sketch showing the relationship between body temperature and time. What type of function is it? Justify your answer. *Exponential*

b. What is the asymptote for this relationship? Explain. *Room temperature*

c. Use your data and the equation $y = km^x + b$ to find the equation that represents the temperature of the body at a certain time. $y = 10(0.7^x) + 17$

d. When did Dr. Dedman die? *Set* $y = 37°C$, *normal body temperature; solving gives* $x \approx -1.94$ *hr, so about 1.94 hr before 5:12 P.M., or a little after 3:16 P.M.*

e. Who is the murderer? *Since no one is logged into the building between 2:51 and 3:48, it must be Foust, who lied about when he last saw the doctor alive. Could he be the Slasher, or is he a "copycat criminal"?*

EXTENSION AND PRACTICE

CC-128. *Summary* In this chapter you have worked with inverses of functions and more specifically logarithms in many contexts. Your responses to the following situations should convince your fellow group members and your instructor that you can apply what you have learned.

a. Explain how to find the inverse of $f(x) = 3x + 5$.

b. Explain how to solve $3 = 1.05^x$ using logarithms.

c. Make up a real situation that can be represented by an equation that can be solved using logarithms.

d. Create an exam problem using the equation and real situation you made up in part c. Be sure to show the solution to your problem.

CC-129. *Self-Evaluation*

a. Throughout the year so far, you have been working with different groups. What are the types of contributions you have made? What role have you taken in your group? Would you like these to change or stay the same? Explain.

b. What are the important topics of this chapter?

c. For each of the topics you chose in part b, find a problem that represents that topic and that you enjoyed doing. Write out the problem (or outline the key pieces of information of the problem) along with the complete solution.

d. Find a problem that you still cannot solve. Write out the problem and as much of the solution as you can. What is it you need to know to finish the problem?

e. How does this chapter seem to connect to any other topics in mathematics that you have studied? Explain.

CC-130. Tool Kit Check Now would be a good time to review your Tool Kit and make sure you have included any new ideas you learned in this chapter. Include examples and explanations that will help you understand and remember what each of the following means.

- inverse
- composition of functions
- $\log x = \log_{10} x$
- $\log_N x$, where N is not base 10
- $\log_K D = E$ means $K^E = D$
- $\log X^a = a \cdot \log X$

CC-131. A general rule used by car dealers is that the trade-in value of a car decreases by 30 percent each year.

a. Explain how the phrase "decreases by 30 percent each year" tells you that the trade-in value varies *exponentially* with time. In other words: What tells you that the trade-in value can be represented by an exponential function? *Decreasing by 30% means we multiplying by 0.7 each time; "multiplier" implies "exponential."*

b. Suppose the initial value of the car is \$23,500. Write an equation expressing the trade-in value of your car as a function of the number of years from the present. $y = 23{,}500(0.7^x)$

c. How much is the car worth in three years? *\$8060.50*

d. In how many years will the trade-in value be \$6000? $\approx$ *3.83 yr*

e. If a car is 2.7 years old and its trade-in value was 23,500, what was its value when it was new? $\approx$ *\$61,560.64*

CC-132. Portfolio: Growth-over-Time Problem No. 1 On a separate sheet of paper, explain *everything* you now know about

$$f(x) = x^2 - 4 \qquad \text{and} \qquad g(x) = \sqrt{x+4}.$$

CC-133. Growth-over-Time Reflection Look at your three responses to the growth-over-time problem. Write an evaluation of your growth based on those responses. Consider the following while writing your answer:

a. What new concepts did you include the second time you did the problem? In what ways was your response better than your first attempt?

b. How was your final version different from the first two? What new concepts did you include?

c. Did you omit anything in the final version that you used in one of the earlier problems? Why did you omit that item?

d. Draw some bars like the ones here, and shade them to represent the amount that you knew when you did the problem at each stage. (A blank means you knew nothing about the problem, and completely shaded means you knew and answered everything that could be answered.)

First attempt:

Second attempt:

Third attempt:

e. Is there anything you would add to your most recent version? What?

CC-134.

a. Explain completely how to get a good sketch of the graph of $y = (x + 6)^2 - 7$.

b. Explain how to change the graph of $y = (x + 6)^2 - 7$ into the graph of $y < (x + 6)^2 - 7$.

c. Restrict the domain of the parabola in part a to $x \geq -6$, and graph the inverse function.

And for those who want a challenge, . . .

d. What would be the inverse function if we had restricted the domain of the function in part a to $x \leq -6$?

CC-135. Solve for x without using a calculator:

a. $x = \log_{25} 5$
1/2

b. $\log_x 1 = 0$
$x > 0$

c. $23 = \log_{10} x$
1.0×10^{23}

CC-136. Using your calculator, solve for a positive number to the nearest thousandth:

a. $x^6 = 125$
2.236

b. $x^{3.8} = 240$
4.230

c. $x^{-4} = 100$
0.316

d. $(x + 2)^3 = 65$
2.021

e. $4(x - 2)^{12.5} = 2486$
3.673

f. $n^3 = 49$
$n \approx 3.659$

CC-137. Find the inverse of each function. Write your answer in function notation.

a. $p(x) = 3(x^3 + 6)$

$$g(x) = \sqrt[3]{\frac{x}{3} - 6}$$

b. $k(x) = 3x^3 + 6$

$$f(x) = \sqrt[3]{\frac{x - 6}{3}}$$

c. $h(x) = \dfrac{x+1}{x-1}$

$$f(x) = \frac{x+1}{x-1}$$

d. $y = \dfrac{2}{3 - x}$

$$g(x) = \frac{3x - 2}{x}$$

CC-138. A two-bedroom house in Vacaville is worth \$110,000. It appreciates at a rate of 2.5 percent each year.

a. What will it be worth in 10 years? $\approx$ ***\$140,809.30***

b. When will it be worth \$200,000? $\approx$ ***24.2 yr***

c. In Davis, houses are depreciating at a rate of 5 percent each year. If a house is worth \$182,500 now, how much would it be worth 2 years from now? $\approx$ ***\$164,706.25***

CHAPTER EIGHT

AT THE COUNTY FAIR

Polynomials and General Systems

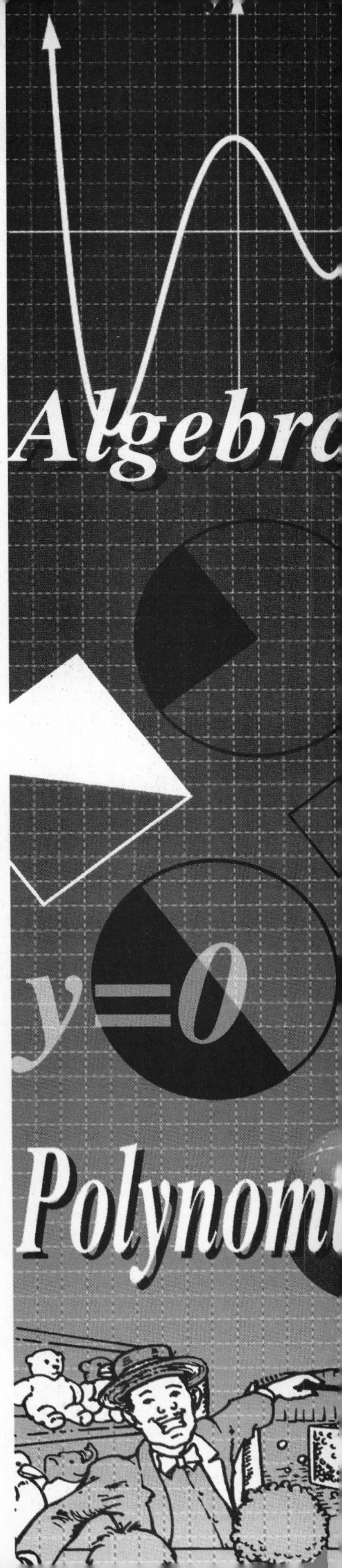

IN THIS CHAPTER YOU WILL HAVE THE OPPORTUNITY TO:

- explore a new category of equations and their graphs: polynomials;
- find intersections of nonlinear functions and other equations, both graphically and algebraically;
- discover that some systems of equations cannot be solved algebraically. Their solutions can be approximated only by graphing and estimating;
- learn about other equations that appear unsolvable at first glance but can be solved when we create a whole new set of numbers;
- expand your thinking in order to work with numbers having real and imaginary parts;
- summarize the important ideas from the course.

KEEP A GRAPHING CALCULATOR HANDY—YOU WILL BE SURPRISED AT HOW USEFUL, EVEN NECESSARY, IT WILL BE IN SOLVING MANY OF THESE PROBLEMS!

Problem Solving	
Representation/ Modeling	
Functions/ Graphing	
Intersections/ Systems	
Algorithms	
Reasoning/ Communication	

MATERIALS	CHAPTER CONTENTS		PROBLEM SETS
	GRAPHS OF POLYNOMIAL FUNCTIONS		
Graphing calculators, Resource Pages	**8.1**	Polynomial Functions Investigation	CF-1 – CF-13
	8.2	Observations about Polynomial Functions	CF-14 – CF-28
	8.3	Finding the Equations of Polynomial Functions	CF-29 – CF-40
	INTERSECTIONS OF NONLINEAR FUNCTIONS		
Graphing calculators	**8.4**	Nonlinear Systems of Equations	CF-41 – CF-55
Graphing calculators	**8.5**	More Nonlinear Systems	CF-56 – CF-65
	COMPLEX NUMBERS		
	8.6	Nonintersecting Systems: Introduction to Complex Numbers	CF-66 – CF-80
	8.7	Complex Numbers as Solutions of Systems of Equations	CF-81 – CF-95
	CULMINATING ACTIVITIES		
Paper, rulers, scissors	**8.8**	The Game Tank Lab	CF-96 – CF-103
Graphing calculators Tool Kit Clean-up Page	**8.9**	Chapter Summary	CF-104 – CF-114
	8.10	Course Summary	CF-115 – CF-135

This chapter has four main parts. You may want to select from among these parts those topics you feel your class needs to address.

GRAPHS OF POLYNOMIAL FUNCTIONS (Sections 8.1–8.3)*: We go back to looking at where a curve $y = P(x)$ intersects the very important line $y = 0$. We investigate the graphs of polynomial functions by looking at the roots of the equation formed when $P(x) = 0$.*

INTERSECTIONS OF NONLINEAR FUNCTIONS (Sections 8.4–8.5)*: We examine equations and pairs of equations that cannot be solved algebraically but require graphing to solve. We will look at the intersections that occur between parabolas, cubics, lines, and exponential functions.*

COMPLEX NUMBERS (Sections 8.6–8.7)*: We continue to develop the concepts introduced in the lessons on polynomials (Sections 8.1 to 8.3) into an investigation of a new set of numbers created by mathematicians as solutions to problems when there are no real numbers to make polynomials equal zero. We look at conjugates and at complex solutions and what they represent in terms of the graphs (namely, nonintersections).*

CULMINATING ACTIVITIES (Sections 8.8–8.10)*: These include (1) The "Game Tank," a lab in which students take on the role of a carnival game-booth designer who wants to maximize the volume of an open tank; (2) a Chapter Summary, Tool Kit Review, and Growth-Over-Time problem; (3) a summary of the course as a whole; and (4) the video,* ***Polynomials,*** *by Project Mathematics! (We recommend showing the video toward the end of the unit because it covers a lot of material rather quickly, and therefore serves better in helping students consolidate their ideas than as an introduction. It is available from the Caltech Bookstore, 1-51 Caltech, Pasadena, CA 91125; phone: 818-395-6161 as well as from the NCTM, 1906 Association Drive, Reston, VA 22091, phone: 703-620-9840.)*

8.1 POLYNOMIAL FUNCTIONS INVESTIGATION

The Game Tank problem (CF-1) is given here only as an introduction. Students should read it now if you plan to assign it at the end of the chapter.

CF-1. The Game Tank The Mathamericaland Carnival Company wants to premiere a special game at this year's county fair. The game will consist of a tank filled with Ping-Pong balls. Most are ordinary white ones, but there are limited numbers of orange, red, blue, and green prize Ping-Pong balls. The blue ones have a prize value of $2.00, the reds $5.00, the orange ones $20.00, and the green Ping-Pong ball (there's only one!) is worth $1000.00. There are 1000 blue Ping-Pong balls, 500 reds, and 100 orange ones. Fair-goers will pay $1.00 for the opportunity to crawl around in the tank for a set amount of time, blindfolded, trying to find one of the colored Ping-Pong balls.

The owner of the company thinks she will make the biggest profit if the tank has maximum volume. In order to cut down on the labor costs of building this new game, she hires you to find out what shape to make the tank.

In this section students start with introductory problems about polynomials and work on an investigation. They need graphing calculators to work on this investigation. Use the Resource Pages entitled "Graphing Polynomial Functions." Students will be using graphing calculators to identify characteristics and properties of polynomials. CF-2 is an example that you will want to model for the class so that students will understand the expectations of the activity. Students may need more than one day of class time to complete the investigation.

Be sure to do all of CF-2 and CF-4 yourself before the class session, so you know what to expect on the graphing calculator.

We suggest doing CF-2 with the whole class to serve as a model for CF-4. Note: While we introduce the term ***root****, we don't introduce the terminology* ***zero of a polynomial*** *because it tends to confuse rather than clarify at this point.*

CF-2. The Polynomial Functions Investigations (CF-2 through CF-4) We are going to be investigating polynomial functions, but first we will do one example as a class activity. There are Resource Pages for these problems at the end of the text.

The example we will use is

$$P_1(x) = (x - 2)(x + 5)^2$$

a. Graph the function on the graphing calculator using the standard viewing window. *The first grid on the Resource Page is for this graph.*

b. As you can tell, we do not see the best view of the function with the standard window, so we want to use the zoom feature of the graphing calculator to obtain a better view of the graph. What should we be looking for when we zoom? *where the graph curves; the parts of the graph that are "cut out"*

Start by zooming out. You may have to do this more than once to get the complete graph. Now use the box feature to select the portion of the graph which you would like to view. Sketch this graph on the second set of axes. Label the important points of the graph such as the intercepts.

c. The roots are the solutions of the equation $0 = (x - 2)(x + 5)^2$. Find the roots. roots: $x = -5$ *and* $x = 2$

d. On the number line provided on the Resource Page, mark the roots with open circles and then shade the regions where the function outputs are positive (that is, where the graph is above the x-axis).

e. Describe the shape of the graph and its relation to the equation. Make sure that you include the degree and the intercepts as part of your description. Pay close attention to the way the function curves. ***In this example, the function turns twice (or goes three directions; it goes up, then down, and then up again). Sketching direction arrows will help the students see this property.***

CF-3. In the Parabola Investigation in Chapter 5, you discovered how to make a parabola "sit" on the x-axis, and you also looked at ways of making parabolas intersect the x-axis in two specific places. These x-intercepts for the graph of the function are often called the **roots of the equation of the function**. Sometimes roots can be found by factoring. Add **roots** to your Tool Kit.

CF-4. Use the same approach that you used for $P_1(x)$ to investigate the following five polynomials:

a. $P_2(x) = (x - 1)(x + 1)(x - 3)$

b. $P_3(x) = 0.2x(x + 1)(x - 3)(x + 4)$

c. $P_4(x) = (x + 3)^2(x + 1)(x - 1)$

d. $P_5(x) = -0.1x(x + 4)^3$

e. $P_6(x) = x^4 - x^2$

EXTENSION AND PRACTICE

CF-5. Find the roots of the equation of each function:

a. $y = x^2 - 6x + 8$
$x = 2, 4$

b. $f(x) = x^2 - 6x + 9$
$x = 3$

c. $y = x^3 - 4x$
$x = -2, 0, 2$

CF-6. To sketch the following graphs, make the recommended table and plot the points. Do not use a graphing calculator.

a. Graph $y = (x - 1)^2(x + 1)$. Use a table of values from $x = -2$ to $x = 2$. What do you think is the parent for this equation?

b. Now graph $y = (x - 1)^2(x + 1)^2$. Again, make a table from $x = -2$ to $x = 2$. What do you think is the parent for this equation?

c. Finally, sketch a graph of $y = x^3 - 4x$. Use a table of values from $x = -3$ to $x = 3$.

CF-7. Algebraic expressions, like those in CF-6, that include at most addition, subtraction, and multiplication are called **polynomials**.

These are polynomial functions	These are not polynomial functions
a. $f(x) = 8x^5 + x^2 + 6.5x^4 + 6$	**d.** $y = 2^x + 8$
b. $y = \frac{3}{5}x^6 + 19x^2$	**e.** $f(x) = 9 + \sqrt{x} - 3$
c. $P(x) = 7(x - 3)(x + 5)^2$	**f.** $y = x^2 + \dfrac{1}{x^2 + 5}$

The general expression for a polynomial with one variable is often written in textbooks as:

$$(a_n)x^n + (a_{n-1})x^{n-1} + \cdots + (a_1)x^1 + (a_0)$$

where n is a positive integer that represents the highest power of the x terms, and $a_n, a_{n-1}, \ldots, a_1, a_0$ usually represent specific numbers.

In $7x^4 - 5x^3 + 3x^2 + 7x + 8$, the highest power of x is 4, so $n = 4$, and

$$a_n = a_4 = 7 \qquad a_{n-1} = a_3 = -5 \qquad a_{n-2} = a_2 = 3 \qquad a_1 = 7 \qquad a_0 = 8$$

Question: Why are the expressions in parts d, e, and f *not* polynomial functions?

CF-8. In CF-7, the polynomial function in part a has degree 5, the polynomial function in part b has degree 6, and the polynomial function in part c has degree 3. Figure out the meaning of the term **degree of a polynomial function**, and explain how you can tell that the polynomial in part c has degree 3. Discuss this with your group, and include this definition in your Tool Kit along with your own definition and examples of **polynomial functions**.

CF-9. Describe the possible numbers of intersections for each of the following pairs of graphs. Sketch a graph for each possibility. For example, in part b, a parabola could intersect a line twice, once, or not at all. Your solution to each part should include all of the possibilities and a sketched example of each one.

a. two different lines *0 or 1*

b. a line and a parabola *0, 1 or 2*

c. two different parabolas *0, 1, 2, 3, or 4*

d. a parabola and a circle *0, 1, 2, 3. or 4 (1 and 3 require the parabola to be tangent to the circle.)*

CF-10. Solve the following system:

$$y = x^2 - 5$$
$$y = x + 1$$

(–2, –1); (3, 4)

CF-11. Answer the following questions for CF-10:

a. What method did you use?

b. Can this problem be solved algebraically? Explain.

CF-12. Which of the following are polynomial functions? For each one that is not, give a brief justification as to why it is not.

a. $y = 3x^3 + 2x^2 + x$

b. $y = (x - 1)^2(x - 2)^2$

c. $y = x^2 + 2^x$ *not*

d. $y = 3x - 1$

e. $y = (x - 2)^2 - 1$

f. $y = \sqrt{(x-2)^2 - 1}$ *not*

g. $y = \frac{1}{x^2} + \frac{1}{x} + \frac{1}{2}$ *not*

h. $y = \frac{1}{2}x + \frac{1}{3}$

CF-13. Describe the difference between the graphs of $y = x^3 - x$ and $y = x^3 - x + 5$. *The second graph is the first graph shifted up 5 units.*

8.2 OBSERVATIONS ABOUT POLYNOMIAL FUNCTIONS

If students have not finished their work on the Resource Pages for the previous section, they should complete them before answering the following questions.

CF-14. Look back at the work you did on the Polynomial Functions Investigation in CF-2 and CF-4. Then answer the following questions:

a. What is the maximum number of roots a polynomial of degree 3 can have? Sketch an example. *3*

b. What is the maximum number of roots a polynomial of degree n can have? *n*

c. Can a polynomial of degree n have fewer than n roots? Under what conditions? *Yes, if any of the minimum value "turns" are above the x-axis, or any of the maximum value "turns" are below the x-axis*

d. For each function, give the minimum degree its polynomial equation could have.

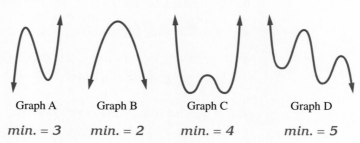

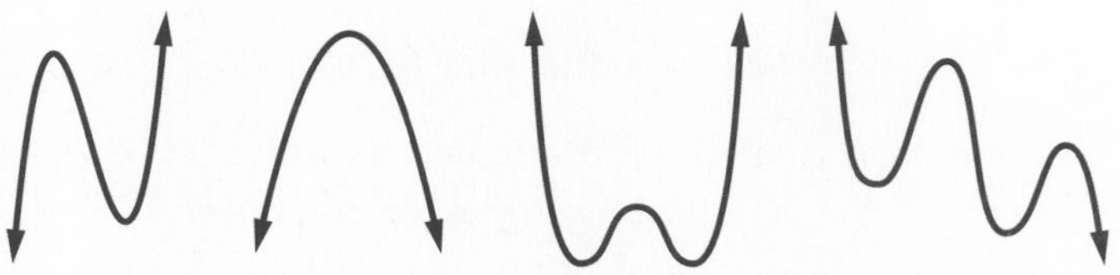

Graph A — *min. = 3*

Graph B — *min. = 2*

Graph C — *min. = 4*

Graph D — *min. = 5*

Negative orientation *refers to the stretch factor being negative. The resulting graphs all have f(x) diverging to $-\infty$ as x approaches $+\infty$. Or, in terms the students will understand, "The right arrow goes down!"*

e. Which of the graphs in part d have a negative orientation? (They are opposites of their parent graphs.) Explain how you determine the orientation of a graph. *b and d*

CF-15. In CF-14, part d, we illustrated how to find the minimum degree. It may be possible that these graphs have a higher degree.

Look at the graphs of $y = x^2$ and $y = x^4$ on your graphing calculator.

a. How are these graphs similar? How are they different? *They have the same general shape, but $y = x^4$ is flatter around the origin and rises more quickly.*

b. Could Graph B in CF-14d have degree 4? *yes*

c. Could Graph B have degree 5? Explain. *No, it would have to turn again.*

CF-16. In the first example from the Polynomial Functions Investigation in CF-2, $(x + 5)^2$ is a factor. This factor produces what is called a **double root**.

a. What effect does this have on the graph? *It causes the graph to be tangent, (i.e., to "bounce off") at the root.*

b. In $P_5(x)$ there is a triple root. What does the equation have that lets us know it has a triple root? *a factor with a power of 3*

c. What does this triple root do to the graph? *It gives the graph a turn similar to that of $y = x^3$ at the root.*

CF-17. Each number line shows where the output values of a polynomial function are positive, in other words, where the graph of the function is above the x-axis. Sketch a possible graph of a function to fit the description shown by each number line. (*Note*: Each number line represents a different polynomial.)

a.
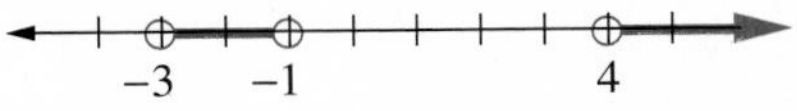

b.
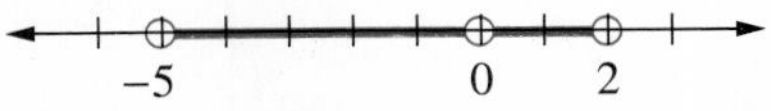

c.
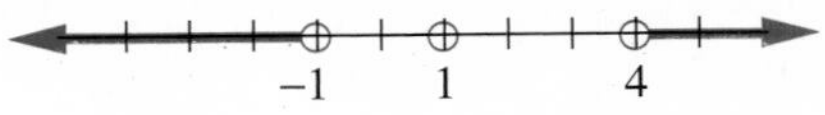

d.
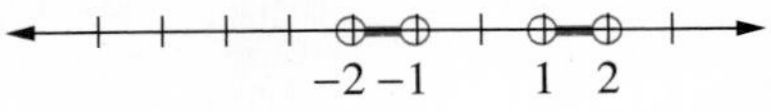

CF-18. Sketch rough graphs of the following functions:

a. $P(x) = -x(x + 1)(x - 3)$

b. $P(x) = (x - 1)^2(x + 2)(x - 4)$

c. $P(x) = (x + 2)^3(x - 4)$

CF-19. Where does the graph $y = (x + 3)^2 - 5$ cross the x-axis? *at $x = -3 \pm \sqrt{5}$*

EXTENSION AND PRACTICE

CF-20. Sketch a graph of $y = x^2 - 7$.

a. How many x-intercepts does the graph of this function have? 2

b. What are the roots of its equation? $\sqrt{7}$ and $-\sqrt{7}$

CF-21. Solve $x^2 + 2x - 5 = 0$. $x = -1 \pm \sqrt{6}$

a. How many x-intercepts does the graph of $y = x^2 + 2x - 5$ have? 2

b. What are they? $(-1 + \sqrt{6}, 0)$ and $(-1 - \sqrt{6}, 0)$

c. Approximately where does the graph of $y = x^2 + 2x - 5$ cross the x-axis? at $x \approx 1.45$ and $x \approx -3.45$

CF-22.

a. Are parabolas polynomial functions? Explain.

b. Are lines polynomial functions? Explain.

c. Are cubics polynomial functions? exponentials? square roots? Explain why or why not for each.

CF-23. If you were to graph the function $f(x) = (x - 74)^2(x + 29)$, where would the graph of $f(x)$ intersect the x-axis? at (74, 0), a double root, and at (–29, 0)

CF-24. What is the degree of each of the following polynomial functions?

a. $P(x) = 0.08x^2 + 28x$ 2

b. $y = 8x^2 - \frac{1}{7}x^5 + 9$ 5

c. $f(x) = 5(x + 3)(x - 2)(x + 7)$ 3

d. $y = (x - 3)^2(x + 1)(x^3 + 1)$ 6

CF-25. Find x if $2^{P(x)} = 4$, where $P(x) = x^2 - 4x - 3$. -1 or 5

CF-26. Start with the graph of $y = 3^x$, and write a new equation that will shift the graph as described:

a. down 4 units $y = (3^x) - 4$

b. to the right 7 units $y = 3^{(x-7)}$

CF-27. Explain the relationship between the solutions of $3x = x^2 + 5x$ and the coordinates of the points where the graphs of $y = 3x$ and $y = x^2 + 5x$ intersect.

CF-28. Judy claims that since $(xy)^4$ is x^4y^4, it must be true that $(x + y)^4$ is $x^4 + y^4$. What do you think? Explain your reasoning. The first statement is correct; the second is false.

8.3 FINDING THE EQUATIONS OF POLYNOMIAL FUNCTIONS

The problems in this section deal primarily with looking at graphs of polynomial functions, using the x-intercepts to get a general equation, then using the y-intercept to find the stretch factor for the specific equation. We expect students to be able to find the nonspecific equation with little or no difficulty. Finding the stretch factor **a** *and the specific equation, which first comes up in CF-31 part b, is more difficult. When they get stuck on this problem, you will probably need to suggest to the groups that they use substitution. But do wait until they ask.*

CF-29. Find a reasonable equation for each polynomial function.

a.

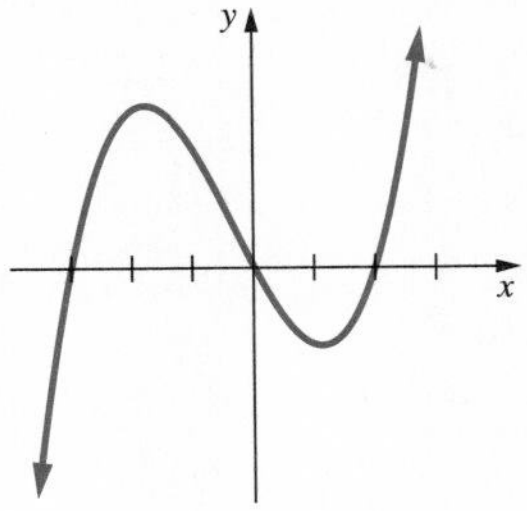

$y = x(x + 3)(x - 2)$

b.

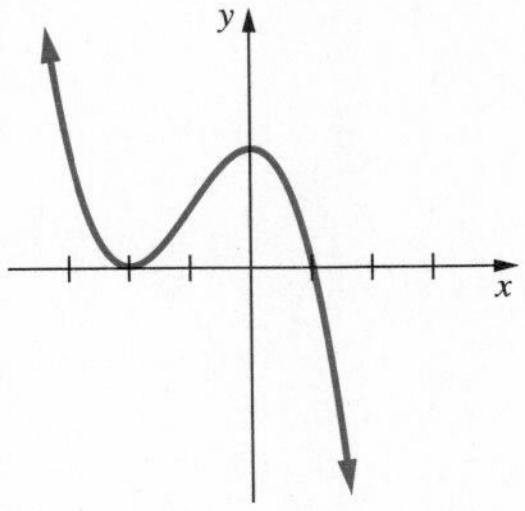

$y = -(x + 2)^2(x - 1)$

c.

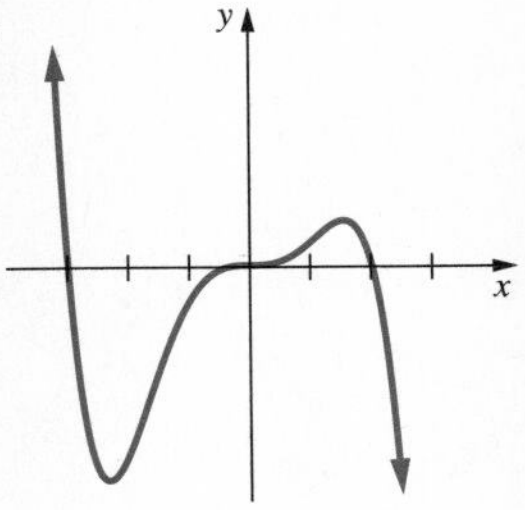

$y = -x^3(x + 3)(x - 2)$

CF-30. What is the difference between the graphs of the functions $y = x^2(x - 3)(x + 1)$ and $y = 3x^2(x - 3)(x + 1)$? ***The second graph is a vertical stretch of the first.***

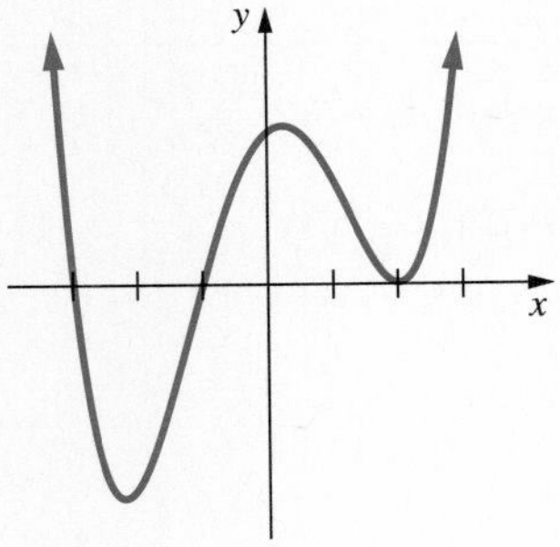

CF-31. Consider the graph shown here.

a. Find an equation for the graph. *$y = (x + 3)(x + 1)(x - 2)^2$*

b. If you multiply the polynomial by 2, how would that alter the graph? ***stretches the graph vertically***

c. The polynomial for this graph can be written as $P(x) = a(x + 3)(x + 1)(x - 2)^2$. Find the value of a if you know that the graph goes through the point (1, 16). *$a = 2$*

CF-32. The County Fair Coaster Ride The Mathamericaland Carnival Company has decided to build a new roller-coaster to use at this year's county fair. The new roller-coaster will have the special feature that part of the ride will be underground. The company will use polynomial functions as models to construct various sections of the track. Part of the design is shown here:

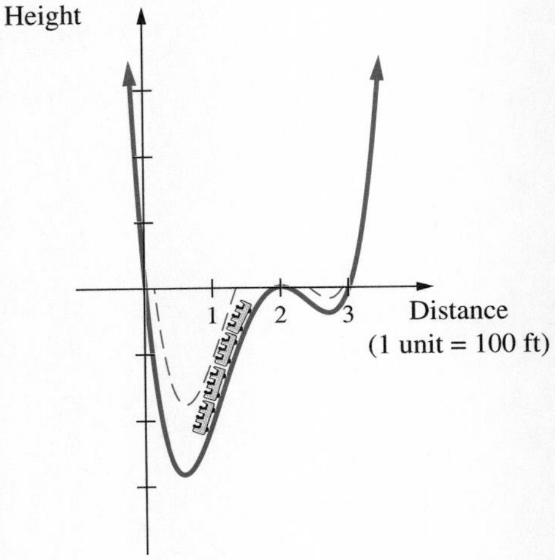

The numbers along the x-axis are in hundreds of feet. At 250 feet, the track is to be 20 feet below the surface. This will give the point (2.5, −0.2).

a. What degree polynomial is demonstrated by the graph? *min. degree: 4th*

b. What are the roots? *0, 2, and 3; (2 is a double root).*

c. Find the equation of the polynomial that will generate that curve. $y = 0.64x(x-2)^2(x-3)$

d. Find the deepest point of the tunnel. (*Note*: It is not likely that anyone would build a roller-coaster that goes this far underground; however, it is true that roller-coasters are designed based on pieces of several different polynomial functions.) *about 181 ft*

CF-33. Write a specific equation for each graph:

a.

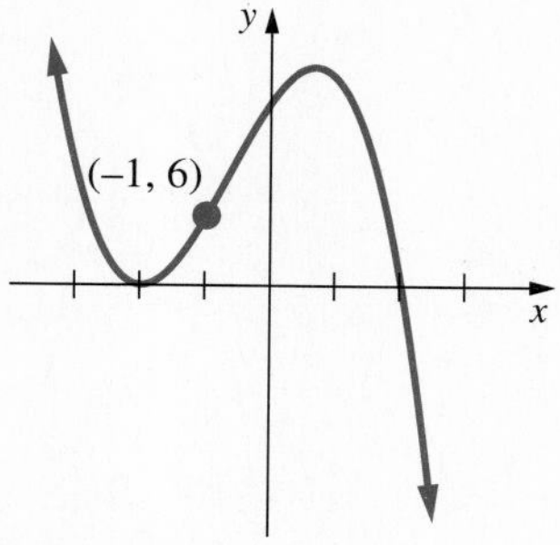

$y = -2(x+2)^2(x-2)$

b.

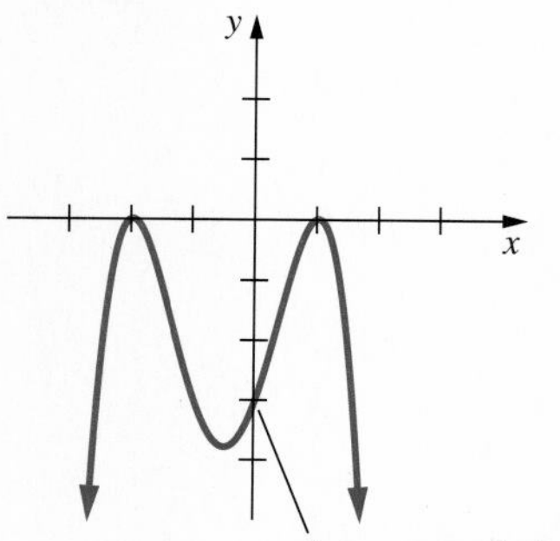

$y = -\frac{3}{4}(x+2)^2(x-1)^2$

EXTENSION AND PRACTICE

CF-34. Consider the function $y = x^3 - 9x$.

a. What are the roots of the function? (Factoring will help!) *3, 0, –3*

b. Sketch a graph of the function.

CF-35. Consider the functions

$$y = \frac{1}{2} \quad \text{and} \quad y = \frac{16}{x^2 - 4}.$$

Find the coordinates where their graphs intersect. *when* $(x^2 - 4) = 32$, *or* $x = \pm 6$, *and* $y = 1/2$

CF-36. Remember, a function $g(x)$ that undoes what another function $f(x)$ does is called its inverse. In the last chapter, for example, we saw that $\log_2 x$ is the inverse of 2^x. We often use a special notation for this relationship and say that $g(x)$ is actually $f^{-1}(x)$ (read "*f*-inverse of *x*"). In other words, f^{-1} undoes whatever f does to x.

a. If $f(x) = 2x - 3$, then $f^{-1}(x) = ?$ $\frac{x+3}{2}$

b. If $h(x) = (x-3)^2 + 2$, then $h^{-1}(x) = ?$ $\sqrt{(x-2)}+3$

CF-37. Another useful tool in finding an inverse function is to make a table showing the effect of doing each operation in the function. Consider the following table for $f(x) = 2\sqrt{x-1} + 3$:

	Operations			
	1st	**2nd**	**3rd**	**4th**
What *f* does to *x*:	subtract 1	square root	multiply by 2	add 3

Since the inverse must undo these operations, in the opposite order, the table for $f^{-1}(x)$ would look like this:

	Operations			
	1st	**2nd**	**3rd**	**4th**
What f^{-1} does to *x*:	subtract 3	divide by 2	square	add 1

a. Copy and complete the following table for $g^{-1}(x)$ if $g(x) = \frac{1}{3}(x+1)^2 - 2$:

	Operations			
	1st	**2nd**	**3rd**	**4th**
What *g* does to *x*:	add 1	square	divide by 3	subtract 2
What g^{-1} does to *x*:	*add 2*	*multiply by 3*	*square root*	*subtract 1*

b. Write the equations for $f^{-1}(x)$ and $g^{-1}(x)$. $f^{-1}(x) = \left(\frac{x-3}{2}\right)^2 + 1$; $g^{-1}(x) = \sqrt{3(x+2)} = 1$

CF-38. A device used for measuring distance on a playground or football field is a handle attached to a rolling wheel that clicks with each full rotation. Brian needs to measure the distance across his backyard. He sets the device at the start of a click and walks across his backyard counting the clicks. He counts a total of 25 clicks. The radius of the wheel is 1 foot. What is the distance across Brian's backyard? *50π ft ≈ 157 ft*

CF-39. Consider the equation $x^2 = 2^x$.

a. How many solutions does the equation have? *3*

b. What are the solutions? *$x = 2$, $x = 4$, and $x \approx -0.767$*

CF-40. In one of the games at the county fair, people pay to shoot a paint pistol at the target to the right. The center circle has a radius of 1 inch. Each concentric circle has a radius 1 inch larger than the preceding circle. Assuming the paint pellet hits the target randomly, what is the probability it hits anywhere within:

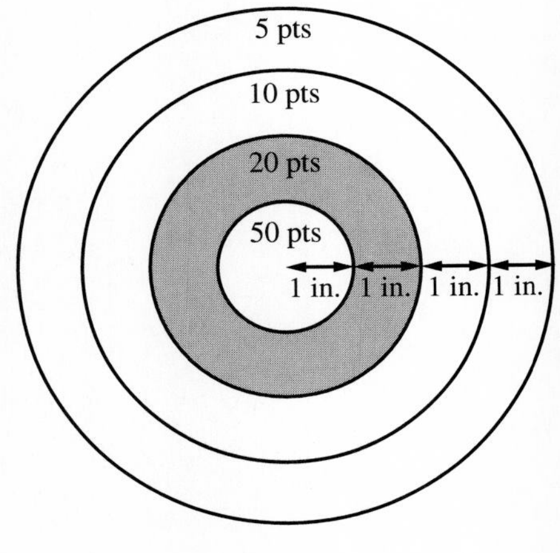

a. the 50 point ring? *1/16*

b. the 20 point ring? *3/16*

8.4 NONLINEAR SYSTEMS OF EQUATIONS

Problems in this section involve intersections of exponentials, polynomials, logarithms, and so on. The graphing calculator is a must for some of these problems, although a guess-and-check approach (interpolation) is tediously possible. Have the students work carefully with their groups to try to find reasonable answers for each of the following problems.

Students are likely to get stuck on the first equation (CF-41). It does not have an algebraic solution and can only be solved by guess and check or by graphing. It is important to have graphing calculators available so students can discover the benefits of solving this equation by graphing. If students get really stuck, ask leading questions such as, "What kind of function is $y = 2^x$?" and "What does $y = x + 3$ look like?" Allow time for this problem. Students have done the reverse process of going from two graphs to a single equation. Let someone get the inspiration to work backward from a single equation to two graphs. Follow this problem with a class discussion about the methods the students attempted. Many students will be surprised to discover that there is no algebraic method for solving this problem (and never will be).

Be sure to do these problems yourself in advance on the graphing calculator. *Students may need additional time to complete the later problems, or you may not want to assign all of them.*

CF-41. Consider the equation $2^x = x + 3$.

a. Discuss in your group different methods you could use to solve this equation. ***Write two equations and use the zoom and/or trace features on the calculator, or use guess and check (interpolation).***

b. Solve the equation. ***The solutions are: ≈ 2.44 and ≈ −2.86.***

c. Be sure you found two solutions. How can you be certain there are no more than two? ***Students will probably describe their graphs in reference to possible intersections.***

CF-42. When solving systems of equations, it is important to recall the significance of the solution. What does the solution represent in terms of

a. the equations? ***the value that satisfied the equations simultaneously***

b. the graphs of the equations? ***the point of intersection of the lines or planes***

CF-43. Consider the following system:

$$y = 2^x + 1$$
$$y = x + 6$$

a. What kind of equation is $y = 2^x + 1$? ***exponential***

b. What kind of equation is $y = x + 6$? ***linear***

c. In how many points could these graphs intersect? Must they intersect? (This will tell us how many solutions we can expect.) ***0, 1, or 2***

d. Solve the system by any method that your group discussed for CF-41. ***(3, 9); (−4.97, 1.03)***

CF-44. Kenya needs to graph $y = \log_2 x$ and wants to use her graphing calculator. She figures she can change the equation into exponential form and then solve that equation for y. She did the following steps:

$y = \log_2 x$	
$2^y = x$	She changed to exponential form,
$\log 2^y = \log x$	took the log (base 10) of both sides,
$y \cdot \log 2 = \log x$	used the exponent rule,
$y = \dfrac{\log x}{\log 2}$	and divided by log 2.

a. Will this work for any base? Show the method for $y = \log_7 x$. *$y = \dfrac{\log x}{\log 7}$*

b. How would you change $y = \log_3 (x + 5)$ to be able to graph it on a graphing calculator?
$y = \dfrac{\log (x+5)}{\log 3}$

c. How would you change $y = \log_2 x - 3$ to be able to graph it on a graphing calculator?
$y = \dfrac{\log x}{\log 2} - 3$

CF-45. Find the points where the graphs of $y = \log_2 (x - 1)$ and $y = x^3 - 4x$ intersect. *(2, 0); ≈(1.1187, −3.075)*

CF-46. Solve $\log_2 (x - 1) = x^3 - 4x$. *$x = 2$ or $x \approx 1.1187$*

CF-47. The systems you have been solving so far in this section are nonlinear systems.

a. What does "nonlinear" mean?

Some nonlinear systems may be *impossible* to solve algebraically; others may just be extremely difficult.

b. What other methods do we have available to solve these systems? ***guess and check, graphing***

EXTENSION AND PRACTICE

CF-48.

a. Sketch a graph of the equation $y = x^2 + 4$. Where does the graph cross the x-axis? ***nowhere***

b. Solve the equation $x^2 + 4 = 0$, and explain how this relates to the answer you found in part b. ***It has no real solution; therefore the graph cannot cross the x-axis.***

c. What occurred when you tried to solve the equation in part b that told you the equation has no real solution? ***got a square root of a negative***

CF-49. The beginning of a sequence of pentagonal numbers is shown.

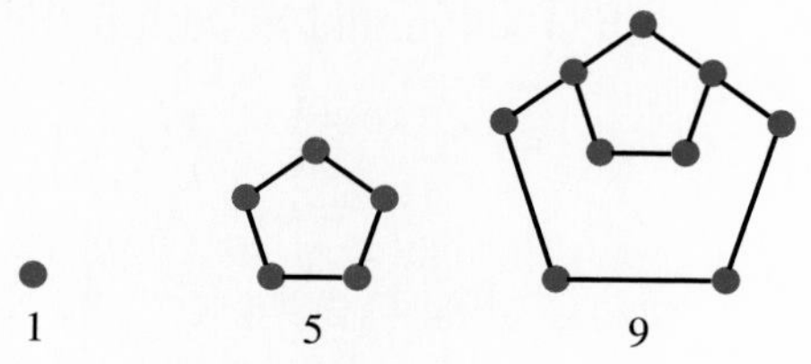

a. Find the next three pentagonal numbers. *13, 17, 21*

b. What kind of a sequence is formed by the pentagonal numbers? *arithmetic*

c. Write an expression that represents the *n*th pentagonal number. *$4n + 1$*

CF-50. Sketch the graph of each of the following on the same set of axes:

a. $y = 2^x$ **b.** $y = 2^x + 5$ **c.** $y = 2^x - 5$

CF-51. Determine whether $x = -2$ is a solution to the equation $x^4 - 4x = 8x^2 - 40$. *Actually, it is not!*

CF-52. Solve. (Remember to use the "fraction busters" reviewed in Appendix A.)

a. $\frac{3x}{x+2} + \frac{7}{x-2} = 3$ *$x = -26$*

b. $\frac{x-7}{x-5} + \frac{6}{x} = 1$ *$x = \frac{15}{2}$*

CF-53. An arithmetic sequence starts out with the terms $-23, -19, -15, \ldots$.

a. What is the rule? *$4n - 23$*

b. How many times must the initial value be put through the function machine so that the result is greater than 10,000? *2506*

CF-54. Multiply:

a. $(2x - 1)(3x + 1)$
$6x^2 - x - 1$

b. $(x - 3)^2$
$x^2 - 6x + 9$

CF-55. James was putting up the sign at the County Fair Theater for the movie "Elvis Returns from Mars"—a sure draw for this year's fair! He got all of the letters he needed for the sign and put them in a box. Then he reached into the box and pulled out a letter at random.

a. What is the probability that he got the first letter he needed when he reached into the box? *2/20 = 1/10*

b. Once he put the first letter up, what is the probability that he got the second letter he needed when he reached into the box? *1/19*

8.5 MORE NONLINEAR SYSTEMS

CF-56, CF-57, and CF-58 continue the development of methods for solving various systems of lines, parabolas, exponentials, and cubics. Students should discuss whether to use the graphing calculator as they do each problem. Although they may be inclined to ignore the algebraic approach, CF-56 lends itself to that solution. CF-57 will require the calculator, but let them discover that on their own. CF-58 can be done either way. With only three problems, students should have enough time to explore and investigate with and without a graphing calculator.

CF-56. Sketch the parabola $y = (x + 4)^2$ and the line $y = x + 4$ on the same set of axes.

a. Find and clearly label the point(s) of intersection of the parabola and line. ***(–4, 0) and (–3, 1)***

b. Did you solve part a algebraically? If not, do so now, and remember that the key to solving systems is to eliminate a variable.

CF-57 does not have an algebraic solution and can be solved only by guess and check or graphing. Our intent is for the students to realize just that, and then solve (approximately) by graphing and by checking their solutions with calculators. Note the size window they use, and let them discover when they need to adjust it. The answers will be approximations based on guessing and checking. We do not intend that they learn to use interpolation here.

CF-57 and CF-58 will take a significant amount of time. Try them yourself so that you can judge how much time may be needed. You may want to have groups split them, with each pair in the group doing one. For each problem select a pair of students to report to the whole class.

CF-57. Sketch the exponential function $y = 2^x$ and the parabola $y = 1.5x^2 - 4$.

a. How many points of intersection are there for the two curves? Are you sure? ***3***

b. Set up a new window on your calculator with $-6 \le x \le 6$, and $-10 \le y \le 50$. Now what do you think?

c. Find the coordinates of these points. ***(–1.695, 0.3089), (2,579, 5.973), (5.180, 36.243)***

d. Try to solve this system algebraically. Explain what happens. ***It can only be solved by graphing or by guess and check.***

CF-58. Solve the following system of equations, and explain your method. Be sure to sketch the graphs on your paper.

$$y = x^3 - 9x$$
$$y = -2x + 3$$

(–2.398, 7.795), (–0.441, 3.882), (2.838, –2.677)

EXTENSION AND PRACTICE

CF-59. Verify that the graphs of the equations $x^2 + y^3 = 17$ and $x^4 - 4y^2 - 8xy = 17$ intersect at (3, 2).

a. Give the equation for the vertical line through this point. *$x = 3$*

b. Give the equation for the horizontal line through this point. *$y = 2$*

CF-60. Each year the Strongberg Construction company builds twenty more houses than in the previous year. Last year they built 180 homes. Business for Acme Homes is increasing also—by 15 percent per year, and they built 80 homes last year.

a. Assuming you don't have a graphing calculator with you now, what can you do, using just a scientific calculator, to figure out the year in which both construction companies will build the same number of homes?

b. Write an equation for each construction company, and find the year in which both construction companies will build the same number of homes. *Strongberg: $y = 20x + 180$; Acme: $y = 80(1.15^x)$. The are equal in about 11.8 yr.*

CF-61. Julia says that

$$\frac{a}{x+b} = \frac{a}{x} + \frac{a}{b}.$$

Jason is not sure; he thinks that the two expressions are not equivalent. Who is correct? Justify your answer. *Jason*

CF-62. The area of ΔABC is 24 square inches. If $\overline{AB}$ is perpendicular to $\overline{BC}$ and $\overline{BC} = 8$ inches, find $\overline{AC}$. *$\overline{AC} = 10$ in.*

CF-63. Solve the equation $y^2 + 2y = 1$. *$-1 \pm \sqrt{2}$*

*CF-64 could be done using the sum formula for an arithmetic series. Using this method is **not** the intent. Students can figure it out by counting as they add, using a calculator.*

CF-64. A contractor working for the county failed to complete the new County Fair Pavilion within a specified time. His contract compelled him to forfeit $10,000 a day for the first 10 days of extra time required, and for each additional day, beginning with the 11th, the forfeit increased by $1000 a day. If he lost a total of $255,000, how many days did he overrun the stipulated completion date? What kind of sequence did his fines form starting after the 10th day? *20 days, arithmetic*

CF-65. Each of the dartboards shown here is a target at the County Fair dart-throwing game. What is the probability of hitting the *shaded* region of each target? Assume that you always hit the board, but where you hit on the board is random.

a.

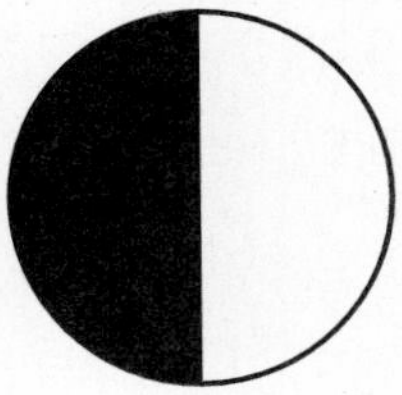

1/2

b.

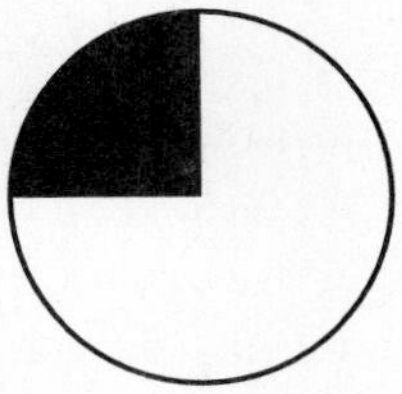

1/4

c.

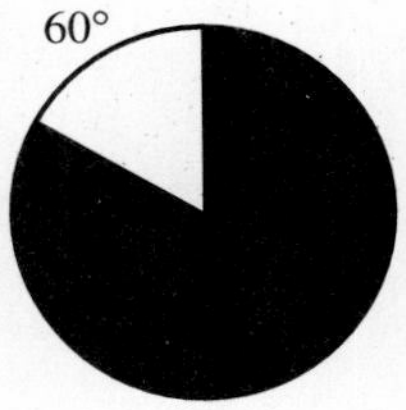

5/6

d.

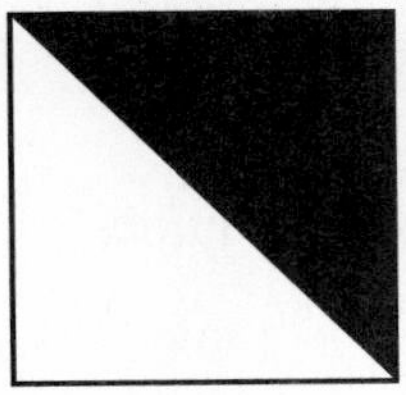

1/2

e.

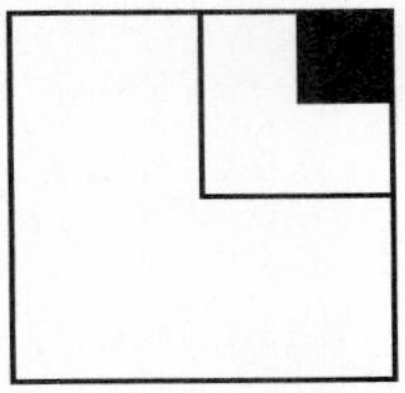

1/16

f.

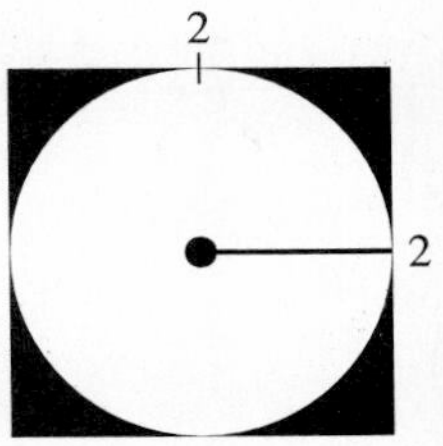

1 – π/4

8.6 NONINTERSECTING SYSTEMS: INTRODUCTION TO COMPLEX NUMBERS

In this section the focus is on the need for complex numbers and the development of their notation. In the next section, the students will see the relationship between graphs not intersecting and complex solutions to their systems of equations.

CF-66. Consider the graphs of the equations $y = x^2$ and $y = 2x - 5$.

a. Use algebra to find the intersection of the graphs.

b. Sketch the graphs and label the intersection. *The two graphs do not intersect.*

CF-67. Discuss CF-66 with your group. What happened when you tried to solve the system algebraically? *Likely answers would be "we had no solution" or "it could not be solved." Ask students why it could not be solved. What algebraic problem tells them that it cannot be solved? The key here is to get them to articulate the problem of taking the square root of a negative number.*

CF-68. Solve $x^2 = 2$.

County Fair Information Booth

In Ancient Greece, people believed that all numbers could be written as fractions (one integer over another). Many individuals realized later that some numbers (such as the solution to $x^2 = 2$) could not be written as fractions and challenged the accepted beliefs. Some of these people were exiled or killed over these challenges. As you can see, the Greeks took their mathematics very seriously. Eventually, they could not avoid the situation and invented a new way of thinking. They even invented new numbers, square roots. The equation $x^2 = 2$ has no rational solutions. Fractions won't work, but square roots will. There are decimal approximations that can be found on your calculator, but the exact answer, which is an irrational number, can only be represented in radical form.

a. How do you undo squaring a number? *Take the square root.*

b. When you solved $x^2 = 2$, how many solutions did you get? *two*

c. How many x-intercepts will $y = x^2 - 2$ have? *two*

d. Write your solutions both as radicals and as decimal approximations. $\pm\sqrt{2}$, ± 1.4142

CC-69. Mathematicians throughout history have challenged the notion that some equations were not solvable and have invented new ways of dealing with these situations. Recently you have come across a similar situation where a solution to a problem could not be found because of an existing mathematical belief.

a. What happens if we try to solve $x^2 = -1$? We need to take the square root of a negative number.

County Fair Information Booth Part 2

Instead of simply settling on no solution, mathematicians decided to invent a new number that would be the solution. "Let's call it **imaginary**," they said, having no trouble assuming all of their previous inventions were real. They decided to call this new number i.

b. If we let $i = \sqrt{-1}$, and we define $i^2 = -1$, then what would be the value of

$$\sqrt{-16} = \sqrt{(16)(-1)} \quad \text{and} \quad \sqrt{16\left(i^2\right)} = ?$$

4i

CF-70. Use the definition of i to show that the following statements are true:

a. $\sqrt{-4} = 2i$

b. $(2i)(3i) = -6$

c. $(2i)^2(-5i) = 20i$

d. $\sqrt{-25} = 5i$

CF-71. In many cases we get solutions that are written in the form $a + bi$ or $a - bi$. (for example $1 - 2i$). This number has a **real component**, the a part of the expression, and an **imaginary component**, the bi part. Numbers that can be expressed in the form of $a \pm bi$ are called **complex numbers**. Record this definition and the definition of i in your Tool Kit.

CF-72 should be discussed with the class. It develops the idea of the discriminant. Students should notice that they need to look at only one part of the quadratic formula to determine whether the roots will include an imaginary component.

CF-72. You are given the equation $x^2 + 2x + 5 = 0$.

a. Solve for x. $\dfrac{-2 \pm \sqrt{-16}}{2}$

b. Does this equation have real roots? What part of your quadratic formula determined whether the roots were real or imaginary? *The root part (discriminant)—more specifically, $b^2 - 4ac$*

c. Find a solution for x in the form $a \pm bi$. $-1 \pm 2i$

EXTENSION AND PRACTICE

CF-73. Simplify the following complex numbers (write in $a \pm bi$ form):

a. $-18 - \sqrt{-25}$ $-18 - 5i$

b. $\dfrac{2 \pm \sqrt{-16}}{2}$ $1 \pm 2i$

c. $5 + \sqrt{-6}$ $5 + i\sqrt{6}$

CF-74. To solve the system

$$y = x^2$$
$$y = 2x - 5$$

we need to solve the equation $x^2 = 2x - 5$. Use the quadratic formula to find the solutions of this equation, in the form $a \pm bi$. $1 + 2i,\ 1 - 2i$

CF-75. Explain why $i^3 = -i$. What does $i^4 = ?$ 1

CF-76. Solve for x:

$$16^{x+2} = 8^x$$

$x = -8$

CF-77. Is $(x - 5)^2$ equivalent to $(5 - x)^2$? Explain briefly. *Yes, both are $x^2 - 10x + 25$.*

CF-78.

a. Evaluate each of the following expressions:

i. $\left(\sqrt{7}\right)^2$ — *7* **ii.** $\left(\sqrt{18.3}\right)^2$ — *18.3* **iii.** $\left(\sqrt{d}\right)^2$ — *d*

b. Using what you observed above, evaluate $\left(\sqrt{-1}\right)^2$. *−1*

c. What does i^2 = ? *−1*

d. Is your answer to part c consistent with your answer to part b?

CF-79. Calculate each of the following expressions:

a. $\sqrt{-49}$ *7i* **b.** $\sqrt{-2}$ *$(\sqrt{2})i$*

c. $(4i)^2$ *−16* **d.** $(3i)^3$ *−27i*

CF-80. The function $h(x)$ is defined by the operations shown in the following table:

	Operations		
	1st	**2nd**	**3rd**
What h does to x:	add 2	$(\ )^3$	subtract 7
What h^{-1} does to x:	*add 7*	*cube root*	*subtract 2*

a. Copy and complete the table for $h^{-1}(x)$.

b. Write equations for $h(x)$ and $h^{-1}(x)$.

$h(x) = (x+2)^3 - 7$; $h^{-1}(x) = \left(\sqrt[3]{x+7}\right) - 2$

8.7 COMPLEX NUMBERS AS SOLUTIONS OF SYSTEMS OF EQUATIONS

In this section we continue to develop the idea of complex solutions to systems of equations, and we find out how complex conjugates make it easy for imaginary solutions to hide behind the real numbers in some equations.

CF-81. When a graph crosses the x-axis (in other words, when $y = 0$), the x-values of the intercepts are often referred to as the roots of the equation. We have seen that solutions to equations can be real or imaginary, so it must be possible for roots, and therefore intercepts, to be either real or imaginary.

a. Sketch the graph of $y = (x+3)^2 - 4$. What are the roots? *−1 and −5*

b. Sketch the graph of $y = (x+3)^2$. What are the roots? *Just one, at $x = -3$*

c. Sketch the graph of $y = (x+3)^2 + 4$. Find the roots by solving $(x+3)^2 + 4 = 0$. Where does the graph cross the x-axis? *$x = -3 \pm 2i$; It doesn't—hence the imaginary roots and intercepts.*

CF-82. Show by purely algebraic methods that the graphs of $y = \frac{1}{x}$ and $y = -x + 1$ do not intersect. What are the solutions for this system of two equations? Check the graphs on the graphing calculator. *$\frac{1}{x} = -x + 1$ means $1 = -x^2 + x$, so $x^2 - x + 1 = 0$. Since this equation has imaginary roots, $\frac{1 \pm i\sqrt{3}}{2}$, the graphs do not intersect.*

CF-83. Find the roots of each quadratic function by solving for x when $y = 0$. Do any of the graphs of these functions intersect the x-axis?

a. $y = (x + 5)^2 + 9$ *$-5 \pm 3i$*

b. $y = (x - 7)^2 + 4$ *$7 \pm 2i$*

c. $y = (x - 2)^2 + 5$ *$2 \pm i\sqrt{5}$*

CF-84. What do you notice about the complex solutions of the equations in CF-83? Describe any patterns you see. Discuss these with your group, and write down everything you can think of.

CF-85. Laurel and Patrick were working on a quadratic function $y = x^2 - 6x + 10$ in standard form. They had to put it into graphing form.

a. First they used the quadratic formula to find the roots of the parabola. What did they find? Are the roots real or imaginary? Explain how this affects the position of the graph relative to the x-axis. *The roots are imaginary, so the graph doesn't intersect the x-axis.*

b. Patrick thought they were stuck, but Laurel said, "No—we can still average these roots to get the graphing form." Will this work? Try it, and check your result by graphing both equations on the graphing calculator. *The roots are $3 \pm i$; $y = (x - 3)^2 + 1$*

EXTENSION AND PRACTICE

CF-86. Look for a pattern as you find the products in parts a through d. Then use your pattern to answer parts e and f.

a. $(2 - i)(2 + i)$ *5*

b. $(3 - 5i)(3 + 5i)$ *34*

c. $(4 - i)(4 + i)$ *17*

d. $(7 - 2i)(7 + 2i)$ *53*

e. Find a complex number to multiply with $3 + 2i$ to get a real number. *$3 - 2i$*

f. Find a complex number to multiply with $a + bi$ to get a real number. *$a - bi$*

CC-87. The complex numbers $2 - 3i$ and $2 + 3i$ are called **complex conjugates**. Write the complex conjugate of each number.

a. $4 + i$ *$4 - i$*

b. $2 + 7i$ *$2 - 7i$*

c. $3 - 5i$ *$3 + 5i$*

d. Put examples of complex conjugates in your Tool Kit. Explain what happens when you multiply them and when you add them.

CF-88.

a. Based on the following graphs, how many real roots does each polynomial function have?

i.

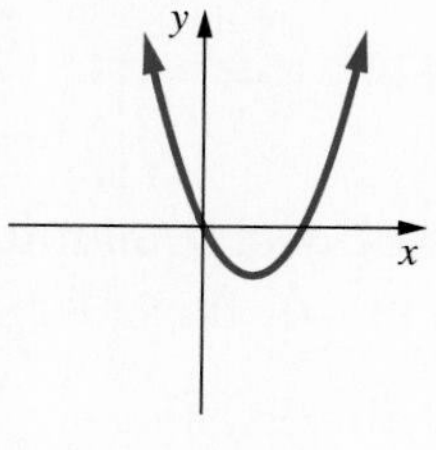

2

ii.

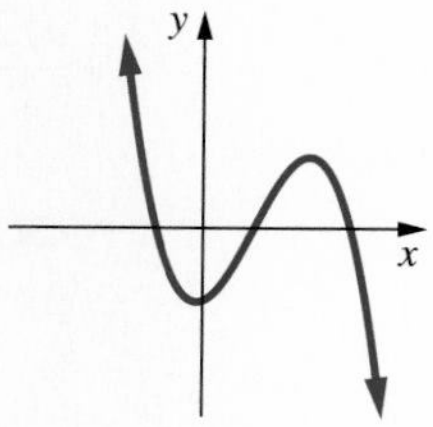

3

b. Something is added to each function in part a so that each graph is translated upward (resulting in the following graphs.) How many real roots does each of these polynomial functions now have?

i.

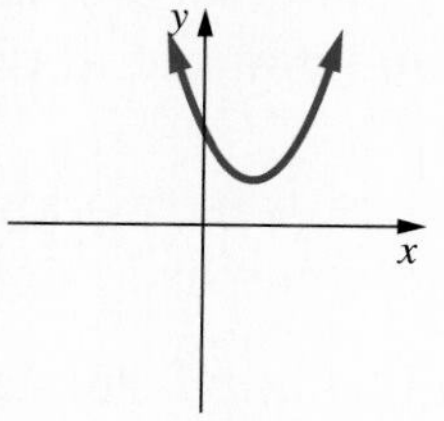

0

ii.

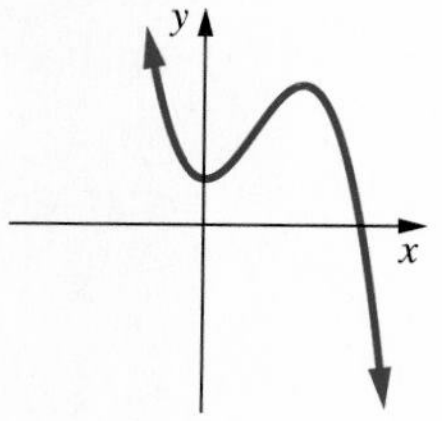

1

Actually, the polynomials in parts a and b have the same *total* number of roots. For example, in part b, polynomial i still has two roots, but now the roots are imaginary. Polynomial ii still has three roots: two are imaginary, and only one is real.

CF-89. Recall that a polynomial function with degree n crosses the x-axis *at most n* times. *Note*: It can cross less than n times, or even not cross at all. For instance, the graph of $y = (x + 1)^2$ intersects the x-axis once, while $y = x^2 + 1$ doesn't intersect at all. And $y = x^2 - 1$ intersects it twice:

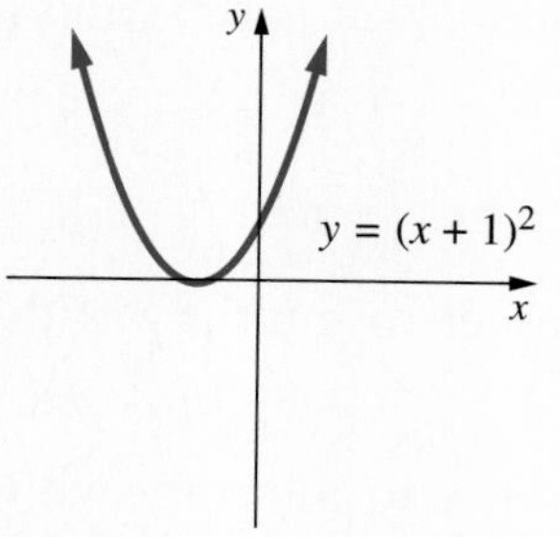

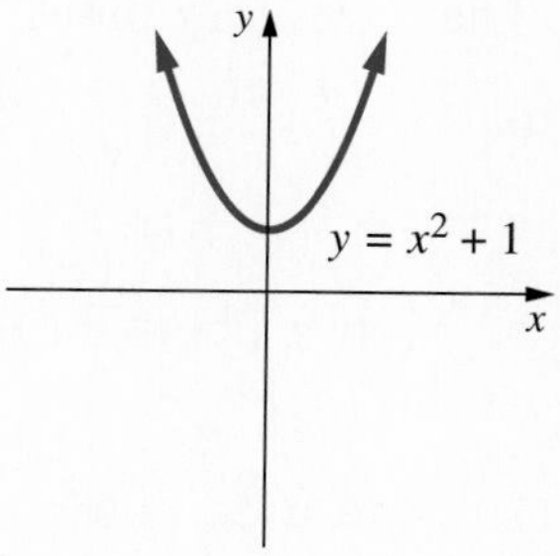

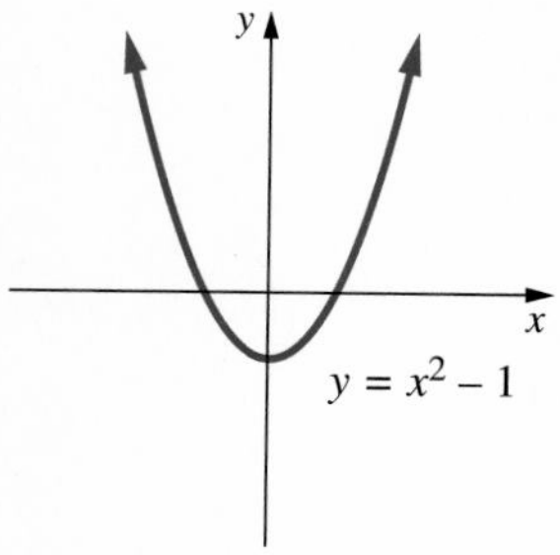

a. A polynomial of degree three might intersect the x-axis 0, 1, 2, or 3 times. Make sketches of all these possibilities.

b. Were all of the sketches in part a possible? Why couldn't one of the sketches be done?

CF-90. In our example with the parabolas in CF-89, $x^2 + 2x + 1 = 0$ has one real solution, and $x^2 - 1 = 0$ has two real solutions.

a. What about $x^2 + 1 = 0$? Solve for x. How does this relate to the graph of $y = x^2 + 1$? Explain. *$x = \pm\, i$, therefore it has no real roots and cannot cross the x-axis.*

b. Consider the factors of these three polynomials: $(x + 1)^2$, $x^2 - 1$, and $x^2 + 1$. What is the relationship between the factorization and the number and kind of roots? *The type of answer we are looking for: the first example, with one repeated linear factor, gives one real root; the second example, with two different linear factors, gives two real roots; and the third quadratic, which has no (real) factors, gives two imaginary roots.*

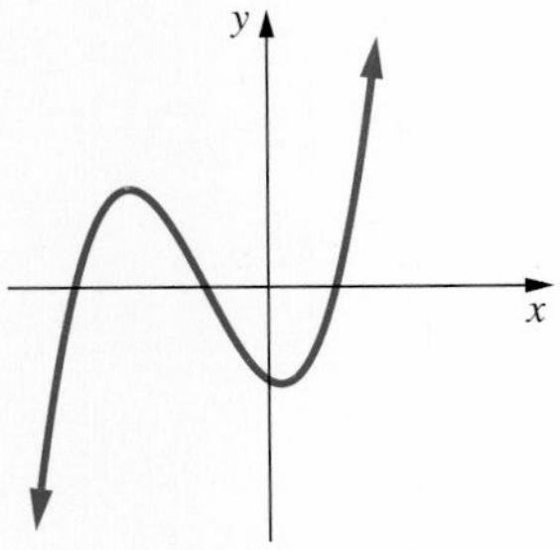

CF-91. For the graph of each polynomial function $f(x)$ shown below, tell how many linear and quadratic factors the factored form of its equation should have and how many real and complex (nonreal) solutions $f(x) = 0$ might have. Assume that each function is a polynomial of the lowest possible degree.

In the example to the left, $f(x) = 0$ will have three real linear factors, therefore three real solutions and no complex solutions.

a.

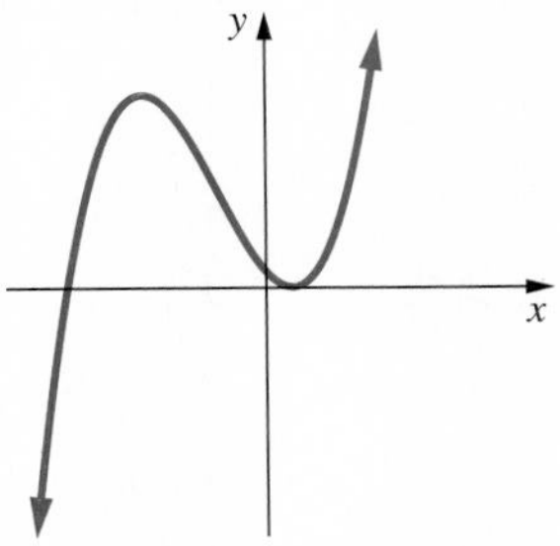

3 linear factors (1 repeated), 2 real roots (1 single, 1 double), and 0 imaginary roots

b.

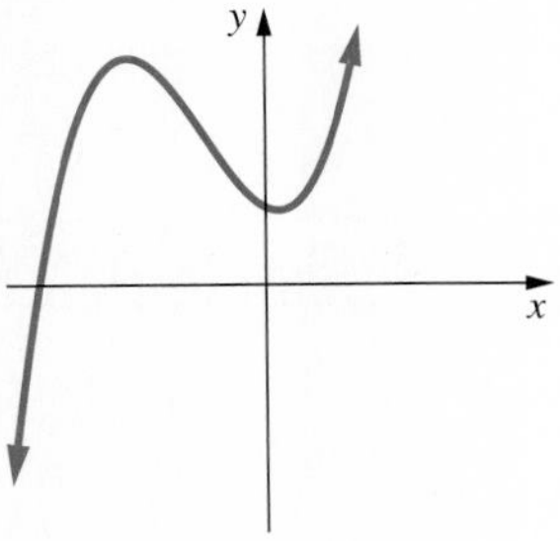

1 linear and 1 quadratic factor, 1 real and 2 imaginary roots

c.

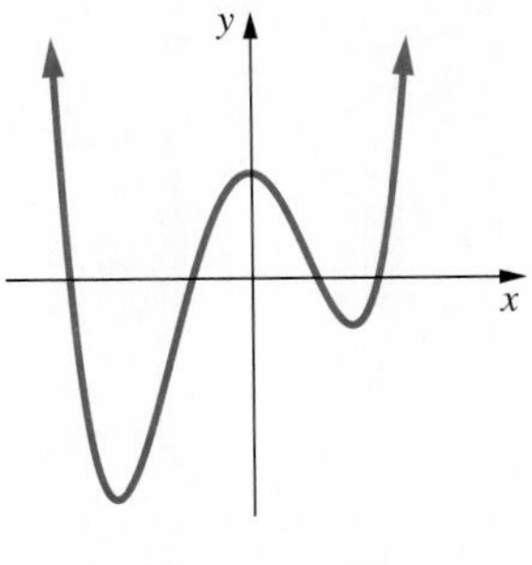

4 linear factors, so 4 real and 0 imaginary roots

d.

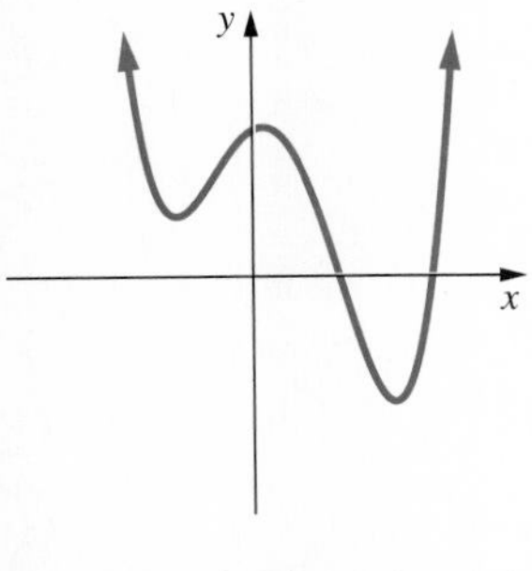

2 linear factors and 1 quadratic factor; 2 real and 2 imaginary roots

Students should understand that the degree of a polynomial function depends on the number of linear and quadratic factors and that the number of roots will be less than or equal to the degree because factors can be repeated.

CF-92. Make a sketch of a graph of polynomial function $P(x)$ so that $P(x) = 0$ could have the number and type of solutions indicated.

a. 5 real roots

b. 3 real and 2 imaginary roots

c. 4 imaginary roots

d. 4 imaginary and 2 real roots

e. For parts a through d, what is the smallest degree each function could be? *5, 5, 4, 6*

CF-93 could lead to a discussion of the concept of the discriminant

CF-93. You are given the equation $5x^2 + bx + 20 = 0$. For what values of b does this equation have real solutions? *$b \geq 20$ or $b \leq -20$*

CF-94. If $f(x) = x^2 + 7x - 9$, calculate the value of the following expressions:

a. $f(-3)$ *-21*

b. $f(i)$ *$-10 + 7i$*

c. $f(-3 + i)$ *$-22 + i$*

CF-95. Use algebra to solve the following system of equations:

$$y = x^2 + 5$$
$$y = 2x$$

Sketch the graphs to verify the fact that there are no real solutions. *$(1 + 2i, 2 + 4i)$, $(1 - 2i, 2 - 4i)$*

8.8 THE GAME TANK LAB

The Game Tank developed in CF-96 through CF-103 should be a portfolio problem. Remind students about the necessary work in creating a good portfolio item. CF-96 is a reading assignment only; there is nothing to answer yet. The Lab Report Format sheet you may have used in earlier labs can be used as a guide to students here.

CF-96. The Game Tank Lab (CF-96 through CF-103) The Mathamericaland Carnival Company wants to premiere a special game at this year's county fair. The game will consist of a tank filled with Ping-Pong balls. Most are ordinary white ones, but there are a limited numbers of orange, red, blue, and green ones. The blue ones have a prize value of $2.00, the reds $5.00, the orange ones $20.00, and the green Ping-Pong ball (there's only one!) is worth $1000.00. There are 1000 blue Ping-Pong balls, 500 red ones, and 100 orange ones. Fair-goers will pay $1.00 for the opportunity to crawl around in the tank for a set amount of time, blindfolded, trying to find one of the colored Ping-Pong balls.

The owner of the company thinks she will make the biggest profit if the tank has maximum volume. In order to cut down on the labor and materials costs of building this new game, she hires you to find out what shape to make the tank.

The tank will be rectangular, open at the top, and will be made by cutting squares out of each corner from a sheet of transparent aluminum that is 8.5 meters by 11 meters. (This is the only size available because the only other use for this material is whale aquariums.)

Since it is difficult to cut and bend transparent aluminum, we will work with paper. You will need to write a report of your findings to the Carnival Company. The report needs to include the following:

a. the data and conjectures found in the experiment making the paper tank

b. a well-drawn diagram of the tank with the dimensions clearly labeled using appropriate variables

c. a graph of the volume function that you found, including notes on a reasonable domain and range

d. an equation for your graph

e. an estimate of the number of Ping-Pong balls that will be needed and your adjusted recommendation based on this information

f. your final conclusions and observations

CF-97 through CF-103 will help guide you through the various aspects of the project.

CF-97. Use a standard 8.5-inch by11-inch sheet of paper (which is the same shape as the material for the tank, though at a slightly smaller scale). Draw a square at each corner. Measure the side of the square you drew (you may choose to use inches or centimeters, but be sure to use the same units consistently throughout the project). Cut out squares from each corner. Your paper should look like this:

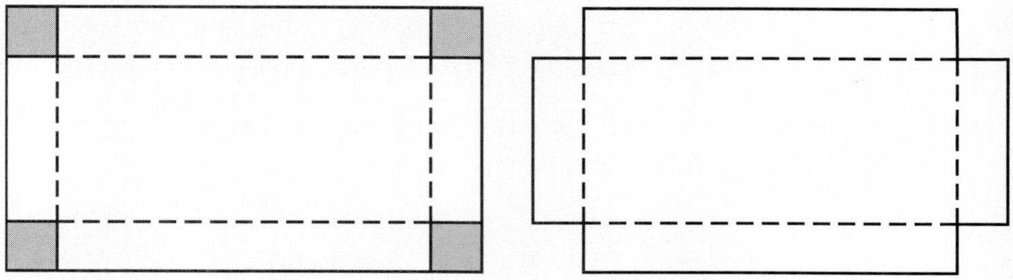

Now, fold it up (on the dotted lines) to create an open box. Then tape the cut edges together so that the box holds its shape. Measure the dimensions of the tank. Record the dimensions directly on the model tank.

CF-98. Make several different tanks. You only need to collect eight or so different examples. Having each member of the group do two tanks would be an appropriate division of labor. Make sure that you consider "extreme" tanks, those with the largest possible square cutout and those with the smallest possible square cutout. (Do not try to cut squares that would be physically impossible. You can't cut a square that is 1/100th of an inch on a side, but you can *imagine* cutting a square out of each corner that is zero inches on a side.) Make a table like the one shown below. Calculate the volume of each tank, record it in the table, and write it on the tank. Be sure to include the proper units. *If the students are using inches, the smallest cut might be a square with 1/2-in. sides; largest would have 4-in. sides.*

	Height	Width	Length	Volume
Tank # 1:				
Tank # 2:				

a. Discuss with the other members of your group (experimenting if you wish) what happens if the cutouts are not square. *The tank cannot be made; the sides will not meet evenly.*

b. Examine the data in the table for tanks with different sizes of cutout squares, and discuss it with your group. Make some conjectures about how to maximize the volume. *A very common conjecture is: the larger the cutout, the smaller the volume. Get them to write this conjecture in their notebooks.*

c. Label the height as x. Using x for the height, find expressions for the length and the width. *For inches: length* $= 11 - 2x$, *width* $= 8.5 - 2x$. *For centimeters: length* $= 27.9 - 2x$, *width* $= 21.6 - 2x$.

d. Find an equation for the volume of the tank. $V = x(11 - 2x)(8.5 - 2x)$ in.^{3} or $V = x(27.9 - 2x)(21.6 - 2x)$ cm^3.

e. Sketch the graph of your function by using the roots and determining the orientation.

CF-99. Graph the volume function on your calculator. Make sure you find appropriate values for the range on the graphing calculator.

a. Copy the graph onto graph paper with appropriate labels. What does x represent?

b. Find the maximum for the volume. And find the dimensions of the tank that will generate this volume. *For the model:* $V = 66.1$ in.^{3}, $h = 1.6$ in., $w = 5.3$ in., $l = 7.8$ in. *Or* $V = 1083$ cm^3, $h = 4.03$ cm, $l = 19.8$ cm, $w = 13.5$ cm.

CF-100 will assist the students in finding a reasonable domain for the volume function.

CF-100. Based on your table, your equation, and your graph in CF-98 and CF-99, find maximums and minimums for the length, width, and height of the tank. ***Length: max = 11, min = 3; width: max = 8.5, min = 0; height: max = 4.25, min = 0***

a. How are the dimensions of the tank related? In other words, what happens to the length as the height increases? ***The length and width decrease as the height increases.***

b. Make a drawing of the tank with the maximum volume. Label your drawing with its dimensions and its volume.

EXTENSION AND PRACTICE

In CF-101 we expect students to make a rough estimate based on using a cube with a 3.7-cm edge to represent each Ping-Pong ball. They might discuss the problems involved in packing spheres, but since that problem continues to be debated in some mathematical circles, discussion is enough. Note that while the new game tank in part b is to be similar to the original in CF-96, the space occupied by the Ping-Pong balls is not. The owners still want to keep the level of the balls 0.4 meter below the top.

CF-101. The Carnival Company decides to follow your recommendation on the dimensions of the tank and to fill the tank to a depth of 1.2 meters in order to avoid losing the Ping-Pong balls that might bounce out. At the retail price of six for $2.00 the balls would cost a fortune, but the carnival owners have found a real deal on army surplus Ping-Pong balls at $20.00 per case of 1000.

a. A standard Ping-Pong ball has a diameter of 3.7 cm (centimeters). Estimate the number of Ping-Pong balls the owners will need and the cost. ***about 980,000, for a cost of over $19,600***

b. Oh, oh! The company realizes that they were overly ambitious (or greedy) in going for the maximum dimensions. Now they want to keep the tank the same shape, but make it smaller. They want to know whether taking the height of the tank down to 1.2 meters, adjusting the length and width proportionally to the height, and keeping the level of the Ping-Pong balls 0.4 meters below the top of the tank will keep the cost of the balls under $8000. Work out the new dimensions for the tank, and present your conclusions as to whether they can fill the new tank as intended to a level of 0.8 meters for under $8000. ***The new tank will be 1.2m × 3.98 m × 5.85 m. About 367,000 balls will fill it to a depth of 0.8 m at a cost of $7355.***

CF-102. Based on your scaled-down tank, what is the probability of winning a prize? ***about 0.004***

CF-103. Prepare your lab report for the Game Tank Lab (CF-96 through CF-102). You may want to refer to the Lab/Investigation Resource Page used in the earlier chapters.

8.9 CHAPTER SUMMARY

Problem CF-105 is the last installment of Growth-over-Time No. 2. Now that students have completed the problem for a third time, they should look at their three attempts, compare them, and write a reflective essay describing their use of the new mathematical tools they have learned over the year. After comparing notes with their group, they also might include anything they would add if they were to do the problem a fourth time.

CF-104. Summary You can begin this problem in class with your group and complete the rest at home. Identify with your group five main topics from this chapter. For each topic, rewrite or create a problem that represents the main ideas in that topic. Carefully solve and explain the solution to each of those five problems so that someone else, someone who didn't know what to do, could use these problems and solutions as example problems to help them through the chapter. This should be done on a separate sheet of paper to be included with your portfolio.

CF-105. Portfolio: Growth-over-Time Problem No. 2—Final Round On a separate sheet of paper, explain everything you now know about

$$f(x) = 2^x - 3$$

CF-106. Growth-over-Time Reflection Look back at the three responses you wrote to this growth-over-time problem. Write a reflective self-evaluation of your growth based on those responses. Consider the following items while writing your answer.

a. What new concepts did you include the second time you did the problem? In what ways was your response better than your first attempt?

b. How was your final version different from the first two attempts? What new concepts did you include?

c. Did you omit anything in the final version that you used in one of the earlier problems? Why did you omit that item?

d. Draw some bars like the ones below, and shade them to represent the amount you knew when you did the problem at each stage:

First attempt: ☐

Second attempt: ☐

Third attempt: ☐

e. Is there anything you would add to your most recent version? If so, what?

f. Any additional comments about your mathematical growth this year?

EXTENSION AND PRACTICE

CF-107 suggests that students clean up their Tool Kits. There is a Resource Page that provides some guidance and it can be used at any time.

CF-107. Tool Kit Check Today might be a good day to clean up and reorganize your Tool Kit. Suggestions for "Building a Better Tool Kit" are included in the Resource Pages at the end of the text.

CF-108. A polynomial function has the equation $P(x) = x(x - 3)^2(2x + 1)$. What are the x-intercepts of its graph? *(0, 0), (3, 0), and (–0.5, 0)*

CF-109. Sketch a graph of a degree 4 polynomial that has no real roots.

CF-110. A polynomial function has the equation $y = ax\,(x - 3)\,(x + 1)^2$ and goes through the point (2, 12). Use the coordinates of the point to figure out what a must equal, and write the specific equation. *$a = -\frac{2}{3}$, $y = (-\frac{2}{3})x(x - 3)(x + 1)^2$*

CF-111. Which of the following equations have real roots and which have imaginary roots?

a. $y = x^2 - 6$ *R*

b. $y = x^2 + 6$ *I*

c. $y = x^2 - 2x + 10$ *I*

d. $y = x^2 - 2x - 10$ *R*

e. $y = (x - 3)^2 - 4$ *R*

f. $y = (x - 3)^2 + 4$ *I*

CF-112. Use algebra to solve the following system of equations:

$$y = x^2$$
$$y = 2x - 5$$

$(1 \pm 2i, -3 \pm 4i)$

a. Sketch the graphs.

b. How does your graphical solution relate to your algebraic solution? *Since the graphs do not intersect, the system has no real number solution.*

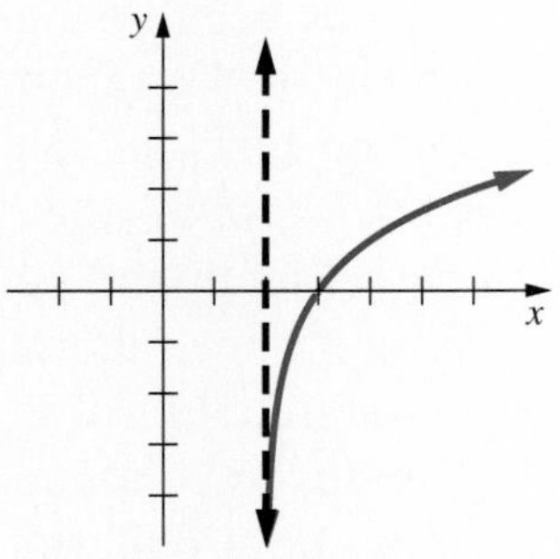

CF-113. Consider the graph shown here.

a. What is the parent for this function? *log x*

b. What are the coordinates of the locator point? *(2, 0) is the x-intercept of the asymptote. This locator point will not change if there is a stretch factor. (3, 0), which is the shifted x-intercept for the function, will also do.*

c. Write an equation for this graph. *$y = \log_2(x - 2)$ is one possibility.*

CF-114. Sketch the graph of each polynomial function. Be sure to label the x- and y-intercepts.

a. $y = x(2x + 5)(2x - 7)$
x: (–5/2, 0), (0, 0), (7/2, 0); y: (0, 0)

b. $y = (15 - 2x)^2(x + 3)$
x: (–3, 0), (15/2, 0) double root; y: (0, 675)

8.10 COURSE SUMMARY

*Now might be a good time to show the **Polynomials** video by Project Mathematics (see the introduction to this chapter). Although it deals with some issues not covered by this chapter, it also provides a nice summary.*

In this section we ask the students to reflect on the entire course. Have your class debate, either in groups or as a class, which ideas they consider to be most important. The discussion is important here.

CF-115. Course Summary Identify with your group at least six, but no more than ten, big ideas from Chapters 1 through 8. After describing each big idea, rewrite or create one or two problems that illustrate the main point(s). Just as you did in the Chapter Summary, solve and explain the solution to each of those problems thoroughly enough so that you could convince someone who had taught this course that you really understood those ideas. This problem should be done separately from your other assigned problems and could be included with your portfolio.

EXTENSION AND PRACTICE

The remainder of the problems are review, to be assigned as needed.

CF-116. Solve for x and y in each of the following systems. If you have a graphing calculator it would be a good idea to check the graphs of these systems so you will know what to look for when solving algebraically. Remember that in order to use the calculator to graph either $xy = 16$ or $xy = 20$, you have to write the equation in y-form.

a. $2x + y = 12$
$xy = 16$
(2, 8), (4, 4)

b. $y = -2x + 12$
$xy = 20$
$(3 + i, 6 - 2i), (3 - i, 6 + 2i)$

c. What is the difference between the graphs of the two systems?

CF-117. Farmer Ted is having trouble with rabbits. He has been losing more of his alfalfa crop each month because of the growth of the rabbit population. Ted has been advised to trap the rabbits, but he feels he can simply increase his production to compensate for the alfalfa they destroyed. Currently Ted produces 600 tons of alfalfa each month and he plans to increase his monthly production by 5 tons. Last month the rabbits destroyed 3 tons of the crop. The rabbit population around Ted's farm is increasing by 15 percent each month. If Ted allows the rabbits to continue destroying his crop, at what point will they eat everything he produces? Write a system of equations to represent his dilemma, and estimate how long it will take. *$y = 600 + 5x$, $y = 3(1.15^x)$; in 40 months*

CF-118. A degree 5 polynomial has exactly two x-intercepts. What does this tell you about the roots of the function. Try sketching several examples. *The function could have three real roots (one x-intercept must be a double root) and two imaginary roots. Or it could have one real triple root, one real double root, and no imaginary roots. Or it could have one quadruple real root, another single real root, and no imaginary roots.*

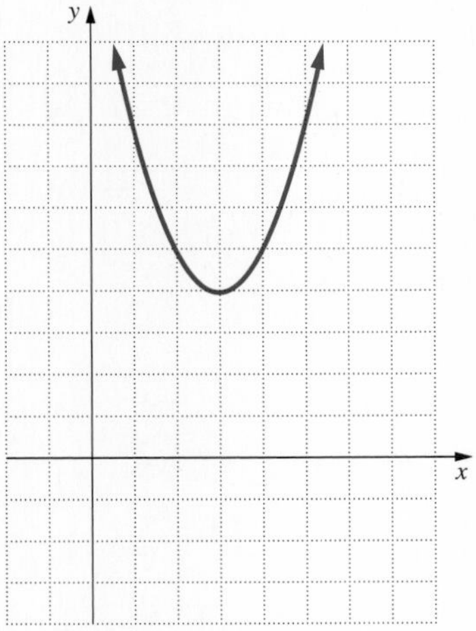

CF-119. Consider the graph of the function shown here.

a. What can you tell about the roots? *The roots are imaginary.*

b. If the graph has a vertex at (3, 4) and a y-intercept at (0, 13), what is the equation of the function? $y = (x-3)^2 + 4 = x^2 - 6x + 13$

c. Find the roots of the equation you found in part b. $x = 3 \pm 2i$

CF-120. If a system of equations has only imaginary number solutions, what do we know about the graphs of the equations? *The graphs do not intersect in the real plane.*

CF-121. If $f(x) = \frac{1}{2}\sqrt{3(x-1)} + 5$, then $f^{-1}(x) = ?$ $\frac{(2(x-5))^2}{3} + 1$

CF-122. Consider the geometric sequence $i^0, i^1, i^2, i^3, i^4, i^5, \ldots, i^{15}$.

a. You know that $i^0 = 1$, and $i^1 = i$, and $i^2 = -1$. Calculate the result for each term of this sequence up to i^{15}, and organize your answers so a pattern is obvious. *repeat 1, i, −1, −i, 1 . . .*

b. Use the pattern you found in part a to calculate i^{396}, i^{397}, i^{398}, and i^{399}. *1, i, −1, −i*

CF-123. Use the pattern you found in CF-122, part a, to help you to evaluate the following expressions:

a. $i^{592} = ?$ *1* **b.** $i^{797} = ?$ *i* **c.** $i^{10{,}648{,}202} = ?$ *−1*

CF-124. Describe how you would evaluate i^n where n could be any integer.

CF-125. The management of the County Fair Theater was worried about breaking even on their movie "Elvis Returns from Mars." To break even they had to take in \$5000 on the matinee. They were selling adult tickets for \$8.50 and children's for \$5.00. They knew they had sold a total of 685 tickets. How many of those would have to be adult tickets for them to meet their goal? *450*

CF-126. Multiply:

a. $2(x+3)^2$
$2x^2 + 12x + 18$

b. $(3+2x)(4-x)$
$12 + 5x - 2x^2$

CF-127. Show that each statement is true.

a. $(i-3)^2 = 8 - 6i$ **b.** $(2i-1)(3i+1) = -7 - i$ **c.** $(3-2i)(2i+3) = 13$

CF-128. Verify that $x = -2 + 5i$ is a root of the quadratic equation $x^2 + 4x + 29 = 0$.

CF-129. Carmel picks integers from 1 to 12 as possible values for c in the equation $x^2 + 4x + c = 0$. What is the probability that his equation will have imaginary roots? *8/12*

CF-130. A parabola has intercepts at (2, 0), (–4, 0) and (0, 4). What is the equation of the parabola? There are several ways to solve this problem. Can you show more than one? *$y = -0.5(x-2)(x+4)$ or $y = -0.5x^2 - x + 4$*

CF-131. Solve each equation:

a. $x^2 - 4x + 3 = 0$ *1, 3* **b.** $x^2 - 8x + 25 = 0$ *$4 \pm 3i$*

c. $x^2 - 4x + 20 = 0$ *$2 \pm 4i$* **d.** $x^2 + 100 = 0$ *$\pm 10i$*

CF-132. Write an explanation, including an example, that tells how to find out what f^{-1} does to x if you know what f does to x.

CF-133. Find the point where the graph of $y = 2^x + 5$ intersects the line $y = 50$. Find your answer to three decimal places. *≈(5.492, 50)*

CF-134.

a. Solve the following equation for x:

$$\frac{x+3}{x-1} - \frac{x}{x+1} = \frac{8}{x^2-1}$$

b. In part a the result of solving the equation is $x = 1$, but what happens when you substitute 1 for x? What does this mean in relation to solutions for this equation?

CF-135. Consider $x^3 - 3x^2 + 3x - 2 = 0$. Since the number of different real solutions is the same as the number of points of intersection of the graph of $y = x^3 - 3x^2 + 3x - 2$ with the x-axis, how many real solutions could this equation have? *One, two, or three—this equation actually has just one, and students should realize that a cubic has to cross at least once.*

a. Check to verify that $x^3 - 3x^2 + 3x - 2 = (x-2)(x^2 - x + 1)$. How many real solutions does $x^3 - 3x^2 + 3x - 2 = 0$ have? How many imaginary? What are all the solutions? *one real, two nonreal; $x = 2$, $(1 \pm i\sqrt{3})/2$*

b. How many roots does $x^3 - 3x^2 + 3x - 2 = 0$ have? How many real and how many imaginary? How many times does the graph of $y = x^3 - 3x^2 + 3x - 2$ intersect the x-axis? Check it out.

APPENDIX A

SOLVING EQUATIONS

Solving an equation is like solving a puzzle. If someone says to you, "I'm thinking of a number, and 5 times the number is 15," could you figure out the number? Probably so. But the real question is, do you know *how* you figured out the number?

Most likely, you said to yourself, "Five times what number is 15?" and you figured out that the answer is 3. You may not be aware, however, of how you got 3 as an answer. Think about this for a minute. Did you guess a number and check? Did you do some operation (+, −, ×, ÷)? Try to decipher the following:

- Four times my number is 42. What is my number?
- Fourteen is 6 less than my number. What is my number?
- Three times my number plus 8 equals 17. What is my number?

Each of these puzzles can be solved with an equation. Becoming aware of how you solve these puzzles will help you solve equations. Let's look at these puzzles and see how they can be represented by equations.

"Four times my number is 42. What is my number?" To solve this puzzle, we need to figure out what the number is. All we know is that four times the number is 42:

$$4 \; ? \; = 42$$

Of course, in mathematics we represent unknown quantities (in this case, the number the person is thinking of) with **variables**. So we write the above puzzle as:

$$4x = 42$$

To solve this puzzle we do the same thing you did when you solved it in your head. We **undo** the puzzle by dividing both sides of the equation by 4:

$$4x = 42$$

Think: "Divide $4x$ by 4 and divide 42 by 4."

$$x = 10.5$$

When solving equations, think of solving these puzzles by undoing what the thinker did. We have to work backward, doing each step in reverse, to get the answer. Let's look at another problem.

"Fourteen is 6 less than my number. What is my number?" Again we can represent this with an equation where x is the number we are trying to figure out:

$$14 = x - 6$$

Be sure this equation makes sense to you: "six less than my number" means subtract six from the unknown number. How can we undo this equation? We undo "subtract six" by adding six:

$$14 = x - 6$$

Think "add 6 to 14 and to $x - 6$."

$$20 = x$$

Always perform the same operations to both sides of the equation. An equation is like a scale in balance. We need to keep the scale in balance while trying to solve for x.

"Three times my number plus 8 equals 17. What is my number?" This can be represented by the equation

$$3x + 8 = 17$$

We undo this equation by performing two different steps. First, we undo the "+8." Think "subtract 8 from $3x + 8$ and from 17":

$$3x = 9$$

We still want to know what x is, so there is more to do. We undo "three times" by dividing both sides by 3: Think "divide $3x$ by 3 and divide 9 by 3" to get

$$x = 3$$

The following examples represent a variety of equations. Following the list of equations a complete solution is given for each. Try solving the equations on your own before reading through the solutions. If you do not get the same answer as the solution shows, carefully compare your solution to the one given. Learn from your mistakes! Try to decipher what went wrong and why you made the mistake. If, after having worked through these examples, you still can't solve equations, then you need to see your instructor. Perhaps you should enroll in a course that could better prepare you for this material.

We chose the following examples of equations because they are often the ones that pose problems for students. Even though these may seem different from our introductory examples, the undoing idea is the key to solving them. The idea of balance—doing the same operation to both sides of the equation—is the other key idea. Keep these two ideas in mind as you do the problems.

A-1. $5x - 3 = 38$ A-2. $4(x - 6) = 14$ A-3. $-5(x - 2) = 25$

A-4. $\frac{x}{7} + 2 = 18$ A-5. $4 + \frac{2x}{6} = -3$ A-6. $-3\left(5 - \frac{x}{4}\right) = 8$

SOLUTIONS

A-1. $5x - 3 = 38$ Think "undo subtracting 3," so add 3.

$5x = 41$ Think "undo multiplying by 5," so divide by 5

$x = 8.2$

A-2. $4(x - 6) = 14$

We will look at two different ways to start this problem. Here is one way.

$$4(x-6) = 14$$
$$x - 6 = 3.5$$
$$x = 9.5$$

Divide both sides by 4
Then add 6 to both sides

Another way of solving this problem is to use the **distributive property**. The distributive property says:

$$a(b + c) = ab + ac$$

$4(x-6) = 14$	Use the distributive property:
$4x - 24 = 14$	Now add 24 to both sides:
$4x = 38$	And divide both sides by 4:
$x = 9.5$	

A-3. $-5(x - 2) = 25$

As in A-2, there is more than one way to solve this equation for x. We show one solution here. See if you can produce a different solution on your own. Be sure you get the same answer for x.

$-5(x-2) = 25$	Think "distributive property":
$-5x + 10 = 25$	Think "subtract 10":
$-5x = 15$	Think "divide by -5":
$x = -3$	

A-4. $\frac{x}{7}+2=18$ Subtract 2:

$\frac{x}{7}=16$

$\frac{x}{7}$ means "x divided by 7." So to undo this we will multiply every term on both sides by 7:

$x = 112$

A-5. $4+\frac{2x}{6}=-3$ Subtract 4:

$\frac{2x}{6}=-7$ Multiply by 6:

$2x=-42$ Divide by 2:

$x=-21$

A-6. $-3\left(5-\frac{x}{4}\right)=8$ Use the distributive property.

$-15+\frac{3x}{4}=8$ Then add 15:

$\frac{3x}{4}=23$ Now multiply by 4:
And divide by 3:

$3x=92$

$x\approx 30.67$

The symbol ≈ means **approximately equal to**.

FRACTION BUSTERS

Often people do not like working with fractions, either because they are not used to them or because they are afraid of making mistakes when using them. Well, we have the perfect solution, called **fraction busters**. Here is the way they work. Suppose we are given an equation with a fraction in it, such as:

$$\frac{2}{3}x+6=10$$

To rewrite this equation without any fractions, we ask ourselves, "How can we eliminate the three?" Although the fraction is $\frac{2}{3}$, the part we would like to eliminate is the denominator. Fraction busting uses the multiplication property of equality to rearrange the equation. Start by multiplying *every term* on *both* sides of the equation by 3:

$3\left(\frac{2}{3}x\right)+3(6)=3(10)$ Now write each term in simplest form to get:

$2x+18=30$

Notice that we now have an equation with no fractions. We can solve this equation as usual.

Solve the following equations using the fraction buster method.

A-7. $\frac{3}{4}x+2=\frac{29}{4}$ **A-8.** $\frac{4}{5}x-4=6$ **A-9.** $3x+\frac{2}{7}=5$

SOLUTIONS

A-7.

$$\frac{3}{4}x+2=\frac{29}{4}$$
$$4\left(\frac{3}{4}x+2\right)=4\left(\frac{29}{4}\right)$$
$$4\left(\frac{3}{4}x\right)+4(2)=29$$
$$3x+8=29$$
$$3x=21$$
$$x=7$$

A-8.

$$\frac{4}{5}x-4=6$$
$$5\left(\frac{4}{5}x-4\right)=5(6)$$
$$5\left(\frac{4}{5}x\right)-5(4)=30$$
$$4x-20=30$$
$$4x=50$$
$$x=12.5$$

A-9.

$$3x+\frac{2}{7}=5$$
$$7\left(3x+\frac{2}{7}\right)=7(5)$$
$$7(3x)+7\left(\frac{2}{7}\right)=35$$
$$21x+2=35$$
$$21x=33$$
$$x=\frac{33}{21}$$
$$x=\frac{11}{7}$$

MORE FRACTION BUSTERS

Suppose you wanted to solve this equation, which has two different denominators:

$$\frac{3}{4}x+\frac{1}{6}=3.$$

We can still use the idea of fraction busters to solve this equation. We want to multiply every term in the equation by some number that *all* the denominators will divide evenly. In this particular equation, we need something divisible by both 4 and 6. Certainly their product, 24, will do, but can you think of another *smaller* number that will work? The smallest possibility is called the **least common multiple**. For 4 and 6, the LCM is 12.

$$\frac{3}{4}x + \frac{1}{6} = 3$$

Multiply by 12:

$$12\left(\frac{3}{4}x + \frac{1}{6}\right) = 12(3)$$

Distribute the 12:

$$12\left(\frac{3}{4}x\right) + 12\left(\frac{1}{6}\right) = 36$$

Divide 12 by 4, and divide 12 by 6:

$$3(3x) + 2(1) = 36$$

Multiply what is left.

$$9x + 2 = 36$$

If you had multiplied by 24 you would have the equation $18x + 4 = 72$, which is also easy to solve. The important thing is to find *some* number that will eliminate the fractions. If it is the smallest number, you will save some time at the end, but do not worry too much about it.

In either case, the process results in **no more fractions**!

Solve the following equations using the fraction busters method:

A-10. $\frac{2}{3}x = \frac{1}{2} + 4$

A-11. $\frac{5}{6} - 2x = \frac{x}{3}$

SOLUTIONS

A-10.

$$\frac{2}{3}x = \frac{1}{2} + 4$$

$$6\left(\frac{2}{3}x\right) = 6\left(\frac{1}{2} + 4\right)$$

$$2(2x) = 3(1) + 6(4)$$

$$4x = 3 + 24$$

$$4x = 27$$

$$x = \frac{27}{4}$$

A-11.

$$\frac{5}{6} - 2x = \frac{x}{3}$$

$$6\left(\frac{5}{6} - 2x\right) = 6\left(\frac{x}{3}\right)$$

$$1(5) - 6(2x) = 2(x)$$

$$5 - 12x = 2x$$

$$5 = 14x$$

$$x = \frac{5}{14}$$

A SHORTCUT

Some equations that involve fractions can be simplified in one step. When two fractions (or ratios) are equal, the resulting equation is called a proportion. For example,

$$\frac{x}{3} = \frac{7}{8}$$

is called a proportion.

To solve this algebraically using fraction busting, we begin by multiplying by 24:

$$24\left(\frac{x}{3}\right) = \left(\frac{7}{8}\right)24$$
$$8x = 21$$
$$x = \frac{21}{8} = 2.625$$

Some people use a shortcut to go directly to the equation $8x = 21$. They call it **cross-multiplying**. They say that the result from multiplying both sides by 24 is the same as multiplying the numerator on each side by the demoninator from the other side. Why does this work?

$$\frac{x}{3} = \frac{7}{8}$$
$$x(8) = 3(7)$$
$$8x = 21$$

Caution! If you choose to save some work by simply cross-multiplying, *be sure* you have a proportion to start with. The answer to the following equation is $x \approx 8.36$: Show what would happen if you simply cross-multiplied in this problem:

$$\frac{2}{3}x - 5 = \frac{4}{7}$$

Then show what you have to do to solve the problem correctly.

Use cross-multiplication to solve the following proportions:

A-12. $\frac{3x}{5} = \frac{6}{11}$

A-13. $\frac{12}{x} = \frac{37}{18}$

A-14. $\frac{2x+1}{4} = \frac{3}{8}$

A-15. $\frac{-2}{4-x} = \frac{2x}{5}$

SOLUTIONS

A-12.

$$\frac{3x}{5} = \frac{6}{11}$$
$$3x(11) = 5(6)$$
$$33x = 30$$
$$x = \frac{30}{33} = \frac{10}{11}$$

A-13.

$$\frac{12}{x} = \frac{37}{18}$$
$$18(12) = 37x$$
$$216 = 37x$$
$$x = \frac{216}{37} \approx 5.84$$

A-14.

$$\frac{2x+1}{4} = \frac{3}{8}$$
$$8(2x+1) = 4(3)$$
$$16x + 8 = 12$$
$$16x = 4$$
$$x = \frac{1}{4}$$

A-15.

$$\frac{-2}{4-x} = \frac{2x}{5}$$
$$-5(2) = (4-x)(2x)$$
$$-10 = 8x - 2x^2$$
$$2x^2 - 8x - 10 = 0$$

Note: This equation is a quadratic equation. In Chapter 1, PS-47, we show how to solve these. Some quadratic equations can be solved by factoring and some must be solved using the quadratic formula. Refer to PS-47 if you need help solving this type of equation.

$$x = \frac{8 \pm \sqrt{(-8)^2 - 4(2)(-10)}}{2(2)}$$
$$x = \frac{8 \pm \sqrt{64 + 80}}{4}$$
$$x = \frac{8 \pm \sqrt{144}}{4}$$
$$x = \frac{8 \pm 12}{4}$$
$$x = 5 \text{ or } x = -1$$

COMPLETING THE SQUARE

Completing the square is an algebraic manipulation that provides another way to write expressions in a graphing form. It also is used to derive the quadratic formula. We have not provided that derivation; however, the following problems will guide you through this algebraic method for rewriting quadratic expressions.

In Chapter 5 you averaged x-intercepts to change a quadratic function from the standard form $y = ax^2 + bx + c$ to the graphing form $y = a(x - h)^2 + k$. You probably noticed from Chapter 5 that it is easier to identify the vertex of a parabola if the quadratic expression is written in graphing form. But it takes some work to get from one form to the other.

Completing the square is another way to change from standard form to graphing form. But what does it mean to complete the square?

Taken literally it means filling in the missing corner of a square. Some elementary algebra texts represent expressions such as $x^2 + 8x + 10$ with algebra tiles where

x^2 represents the area of a square tile with dimensions x by x.
x represents the area of a rectangle with dimensions x by 1, and
1 represents the area of a square with dimensions 1 by 1.

The expression $x^2 + 8x + 10$ is represented below:

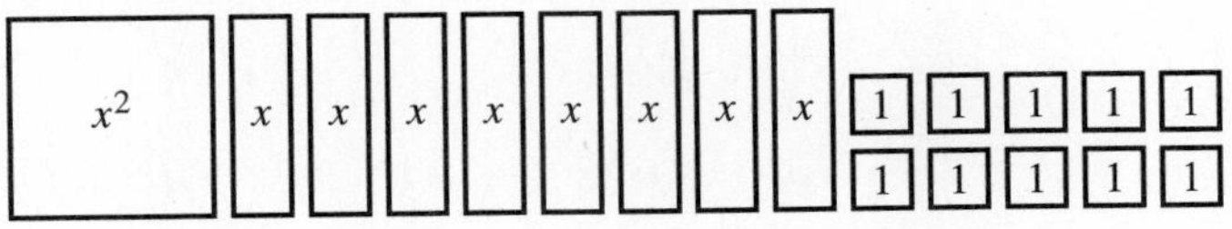

If we look at just the $x^2 + 8x$ part, we can create a square that needs completing.

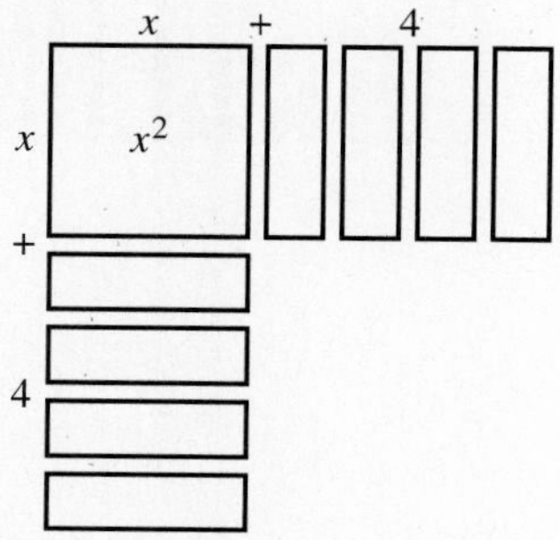

To complete this square we need 16 small squares But we already have 10. So we need to add 6 more small squares:

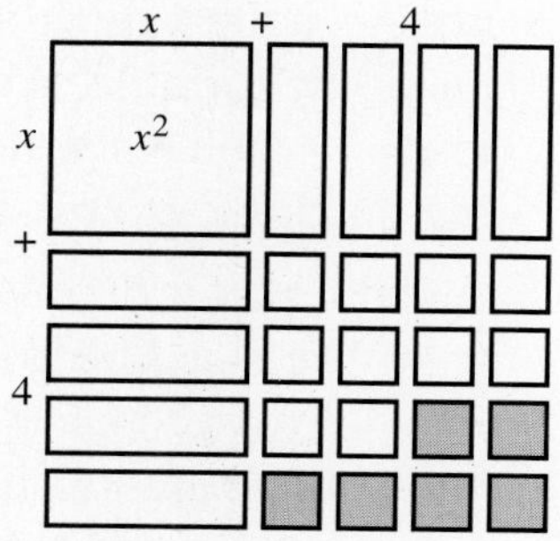

The shaded squares in the above figure are the extra squares we would need to complete the square.

This picture now shows

$$x^2 + 8x + 16 = (x + 4)(x + 4)$$
$$= (x + 4)^2$$

But that's 6 more small squares than we started with! The original expression is $x^2 + 8x + 10$, not $x^2 + 8x + 16$.

What do we need to do to get the expression back to what we started with? Subtract 6, of course, but notice the clever way in which we do it:

Start with: $x^2 + 8x + 10$

Add 6 to get: $x^2 + 8x + 10 + \mathbf{6}$

Subtract 6 to get: $x^2 + 8x + 10 + \mathbf{6} - \mathbf{6}$, which is equivalent to $x^2 + 8x + 10$.

Now we have an expression that is equivalent to the one we started with. Here's the clever part. We use parentheses to show grouping:

$(x^2 + 8x + 10 + \mathbf{6}) - \mathbf{6}$

$(x^2 + 8x + 16) - 6$

$(x + 4)^2 - 6$

Since $x^2 + 8x + 16$ is the same as $(x + 4)^2$, our result is $(x + 4)^2 - 6$.

So our original expression $x^2 + 8x + 10$ is equivalent to

$$(x^2 + 8x + 16) - 6$$
$$= (x + 4)^2 - 6$$

B-1. Consider the function $y = (x + 4)^2 - 6$.

a. Where is the vertex of this parabola?

b. Sketch the graph.

B-2. How could you complete the square of the quadratic expression and find the vertex for

$$f(x) = x^2 + 5x + 2?$$

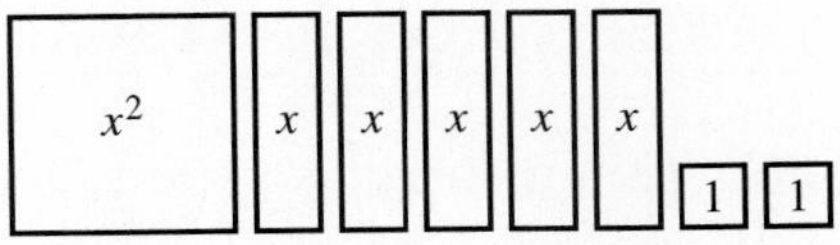

What about the five x bars? Use force! Split one in half. See Figure A.

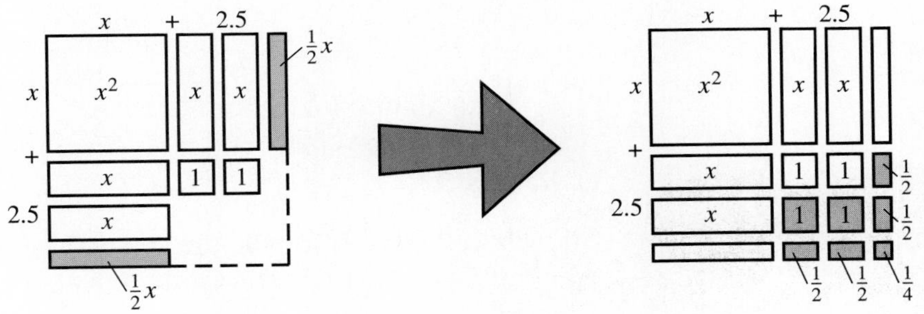

Figure A

Figure B

a. How many 1-by-1 squares do we already have in the bottom corner of Figure A?

b. How many 1-by-1 squares, including parts of squares, do we need to complete the bottom corner (that is, fill in the corner to make a square)? See Figure B.

c. So how many 1-by-1 squares do we need to add (and then later subtract)? Examine Figure B if you're not sure.

d. Write the original expression on your paper with the correct amount added and then subtracted:

$$x^2 + 5x + \_\_\_ - \_\_\_$$

e. Now put the parentheses where you want them to group the part that makes the complete square.

f. Last step: write the square part in factored form.

g. Name the vertex and sketch the graph.

B-3. Try some. For each quadratic function complete the square to get the graphing form. Then give the vertex of each parabola, and sketch the graph.

a. $f(x) = x^2 + 6x + 7$

b. $f(x) = x^2 - 10x$

B-4. Explain how to find the number that has to be added to and subtracted from any expression of the form $x^2 + bx + c$ to change it into graphing form. Use drawings and examples.

B-5. Now that you have seen two different methods for changing a quadratic expression from standard form to graphing form—by averaging x-intercepts or by completing the square—which method do you prefer? What is the advantage of one method over the other?

EXTENSION AND PRACTICE

B-6. Complete the square to get an equivalent expression for each of the following:

a. $x^2 + 2x$

b. $x^2 - 4x + 2$

c. $x^2 - 10x + 21$

d. $x^2 + 7x + 2$

B-7. Ryan thinks that completing the square is only good for changing quadratic *functions* into graphing form, but cannot be used for quadratic *equations*.

Show him how to complete the square to rewrite $x^2 + 6x - 10 = 0$. Then see whether you can use that form to solve for x.

CIRCLES

To get the equation for a circle we need to represent the relationship between a point on the circle, (x, y), and its distance from the center of the circle. In other words, an equation describes all the points on the circle itself. The distance between any two points, found by using the Pythagorean theorem, offers an easy way to define and sketch a circle.

Let's say you had a circle with a radius of 5 and a center at the origin (0, 0):

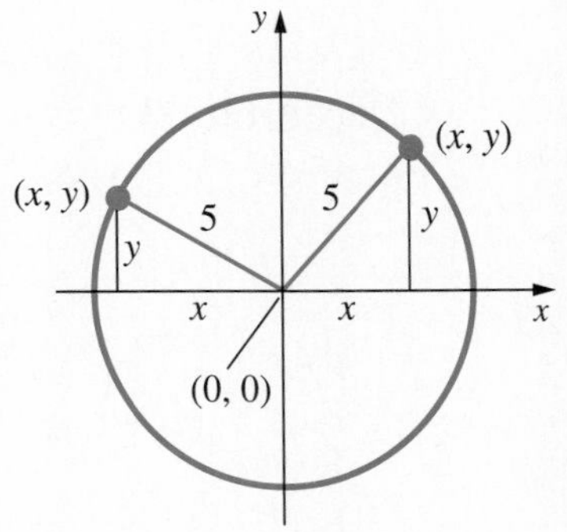

Then any point (x, y) on that circle is a fixed distance 5 units from the origin. By using the Pythagorean theorem, we can get an equation for this circle:

$$x^2 + y^2 = 25$$

This equation is in what is called the **standard form for the equation of a circle with its center at the origin**.

C-1. Think about a circle located at the origin with a radius of 5.

a. Draw this graph carefully on graph paper. How many points on the graph do you know to be accurate? What are they?

b. Find coordinates of other points that will satisfy the equation $x^2 + y^2 = 25$. To get started, consider (−4, 3). Look for a total of 12 points with integer coordinates, including those in part a. *(4, 3), (3, 4), (−3, 4), (−4, 3), (−4, −3), (−3, −4), (3, −4), (4, −3)*

C-2. Write the standard form of an equation for each circle described or shown.

a. center at (0, 0) and radius of 3. **b.** center at (0, 0) and radius of 7

c.

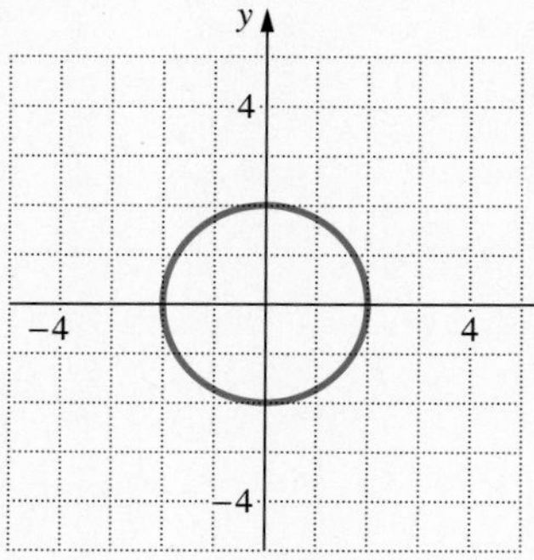

d.

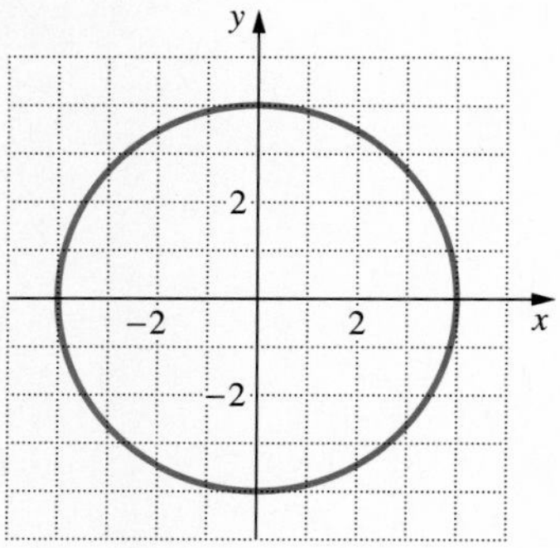

C-3. What do you think the graph of $(x-2)^2 + (y-3)^2 = 25$ would look like? Think about moving the whole circle the way you moved other graphs. (What would be the locator point for a circle?) Sketch what you think should be the graph.

C-4. Try to graph $(x-2)^2 + (y-3)^2 = 25$ on your graphing calculator.

a. There is an immediate problem. What is it?

b. Rearranging the equation to get $y =$ __ requires several steps. Be sure to write each of these steps on your own paper. Start with the original equation:

$$(x-2)^2 + (y-3)^2 = 25$$

Begin isolating y:

$$(y-3)^2 = 25 - (x-2)^2$$

Take the square root of both sides:

$$y - 3 = \sqrt{25 - (x-2)^2}$$

Remember, we just want y:

$$y = \sqrt{25 - (x-2)^2} + 3$$

c. Try graphing the result from part b.

d. Did you use another set of parentheses? Remember, there is no bar to do the grouping on the calculator. You will need parentheses around the whole expression $25 - (x-2)^2$.

$$y = \sqrt{\left(25 - (x-2)^2\right)} + 3$$

e. Why did you get only half a circle? How could you get the other half? To answer this, think about why your instructor has fits when you solve $x^2 = 7$ and get just $\sqrt{7}$ for an answer. What's the other answer? Use this idea to get the equation for other half of the circle.

C-5. Tamar thinks that the equation $(x-4)^2+(y-3)^2=25$ is equivalent to the equation $(x-4)+(y-3)=5$ because you just take the square root of both sides of the first equation. Are the two equations equivalent? Explain why or why not.

C-6.

a. Use the idea of a locator point and the radius, *without* using the calculator, to sketch the graph of $(x+5)^2+(y-2)^2=49$.

b. Now rewrite the equation and use the calculator. Be sure you graph both the "top" and "bottom" of the circle.

c. Which is easier—by calculator or by hand? Does your group agree?

C-7. Write a general equation for a circle. A standard approach is to use (h, k) for the center and r for the radius.

C-8. Write an equation for each circle, and sketch its graph:

a. center $(9, -3)$; radius 4 **b.** center $(-5, 0)$; radius $\sqrt{23}$

C-9. Sketch the graph of $(x-2)^2+y^2=20$.

C-10. Find the center and radius of each circle:

a. $(x+9)^2+(y-4)^2=50$ **b.** $x^2+(y+5)^2=16$

c. $(y-7)^2=25-(x-3)^2$ **d.** $y+(x-3)^2=1$ (oops!)

C-11. Write the equation for each circle graphed below:

a.

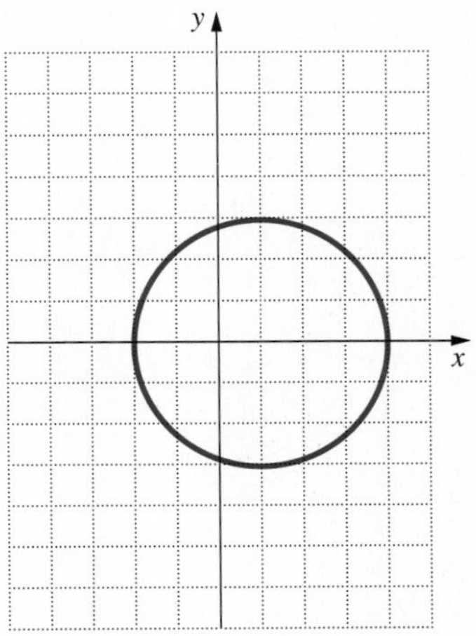

b.

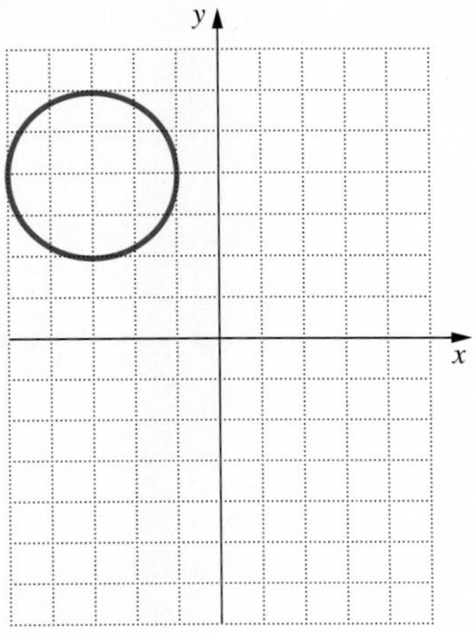

C-12. Choosing your own values for k, sketch on the same set of axes four different examples of graphs that fit the equation: $(x-3)^2+(y-5)^2=k$. Then describe the graphs in words.

C-13. Sketch three different graphs that fit the equation $(x-d)^2+(y-5)^2=16$ by choosing different values of d. Describe the graphs in words.

C-14. Choosing different values for k and j, sketch three different examples of graphs that fit the equation: $(x-k)^2+(y-j)^2=9$. Then describe the graphs in words.

The remaining problems will round out the topic of circles by looking at their equations and graphs from different viewpoints. Drawing a sketch for each of these as you begin will be a big help.

C-15. Write the equation of a circle with a center at $(-3, 5)$ that is tangent to the y–axis. Sketching a picture will help.

C-16. A circle is tangent to the lines $y=6$, $y=-2$, and the y-axis. What is the equation of the circle? Draw a picture!

C-17. Write the equation for the circle whose center is $(-4, 7)$ and which is tangent to the y-axis. Sketching a graph will help.

C-18. Graph the inequality $x^2+y^2\leq 25$; then describe its graph in words.

C-19. Sketch a graph of the system

$$\begin{aligned} x^2+y^2 &\leq 25 \\ x-2y &> 5. \end{aligned}$$

3 X 3 TECHNOLOGY

Graphing calculators are very efficient at solving systems of equations. The suggestions that follow are intended to be general guidelines for any graphing calculator. You will need to refer to the manual for your model of graphing calculator to learn how to enter a system of equations in matrix form.

Consider this system:

$$\begin{aligned} x - 2y + 3z &= 10 \\ 2x - y + z &= 10 \\ x - y + 2z &= 14 \end{aligned}$$

Enter the coefficients of x, y, and z for each equation. 1,–2, 3; 2, 1, 1; and 1, 1, 2 as matrix **A** in your graphing calculator. Matrix **A** represents a 3×3 matrix (a matrix with three rows and three columns).

It should look like this:

$$\begin{bmatrix} 1 & -2 & 3 \\ 2 & 1 & 1 \\ 1 & 1 & 2 \end{bmatrix}$$

Next enter

$$\begin{bmatrix} 10 \\ 10 \\ 14 \end{bmatrix}$$

as matrix **B** in your graphing calculator. Matrix **B** represents a 3×1 matrix (three *rows* by one *column*). (Notice that the number of rows is always given first.)

Using some algebra you could solve **A·X** = **B** where **X** is a 3 × 1 matrix for $\begin{bmatrix} x \\ y \\ z \end{bmatrix}$.

$$\mathbf{A} \cdot \mathbf{X} = \mathbf{B}$$

$$\frac{1}{\mathbf{A}} \cdot \mathbf{A} \cdot \mathbf{X} = \frac{1}{\mathbf{A}} \cdot \mathbf{B}$$

$$x = \mathbf{A}^{-2}\mathbf{B}$$

Remember, we use the negative exponent to represent

$$\frac{1}{\mathbf{A}} = \mathbf{A}^{-1}.$$

On the calculator you can just enter the result $\mathbf{A}^{-1}\mathbf{B}$ (this is read "the inverse of matrix **A** times matrix **B**"). Press "Enter" to ask the calculator to carry out the multiplication. The output should be a 3 × 1 matrix:

$$\begin{bmatrix} 1 \\ 3 \\ 5 \end{bmatrix}.$$

This matrix represents the solution to the system:

$$\begin{bmatrix} x \\ y \\ z \end{bmatrix} = \begin{bmatrix} 1 \\ 3 \\ 5 \end{bmatrix} \quad \text{or} \quad x = 1,\ y = 3, \text{ and } z = 5$$

The solution can be verified by substituting these values into the original system.

Now, solve the following system of equations using your graphing calculator:

$$5x - y + 2z = 6$$

$$3x - 6y - 9z = -48$$

$$x - 2y + z = 12$$

Solution: $x = -2.3,\ y = -3.7,\ z = 7$

Mathematics Tool Kit

Mathematics Tool Kit

PS-39. Digger Dog

Note: Shed and yard approximately to scale.

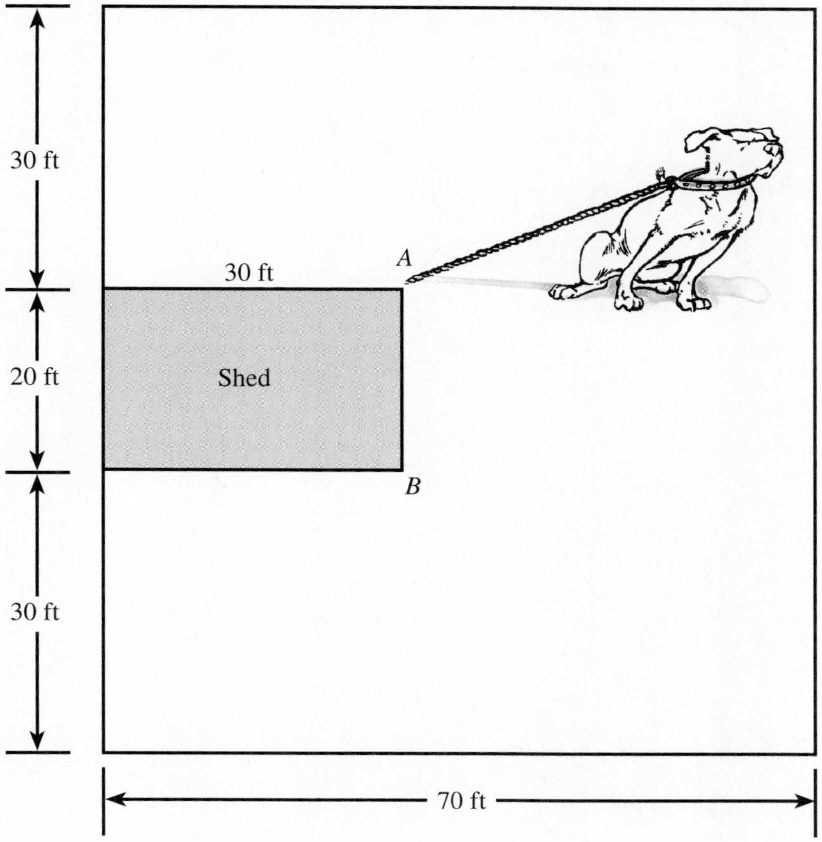

Lab/Investigation Write-Up

1. Description and Purpose of the Investigation

- Briefly explain the investigation—describe what you were required to do. Do not go into too much detail. People reading it should understand the major requirements of the investigation or lab even if they don't necessarily have enough information to recreate it on their own.
- What was the purpose of the investigation?

2. Data Compilation

- Include all your data—tables, graphs, drawings, and so forth—in a clear, organized format with appropriate labels.

3. Data Analysis

- Include your mathematical work, and write a thorough description of what you learned from your data and from this investigation overall. Give specifics! Address the questions from the investigation that ask you to draw conclusions. Write your answers in complete sentences and in paragraph form.
- Describe the mathematical concepts and ideas you used or discovered during this investigation.

4. Questions and Comments

- What questions do you have now? These can be questions you don't feel you can answer yet or extension questions that the lab/investigation raised. Come up with at least one good question. You may include constructive comments about this activity in this section.

Note: Carefully read the grade sheet/checklist (if provided) for more specific instructions about how components of this activity are weighted.

Lab/Investigation Write-Up Grade Sheet

	Value	First-Draft Score	Finished Copy Score
Description and Purpose • Concise • Clearly stated • Purpose stated			
Data Compilation • Tables • Graphs • Organized • Items labeled			
Data Analysis • Mathematical development • Description of results			

EF-21. Domains and Ranges

Function | Domain | Range

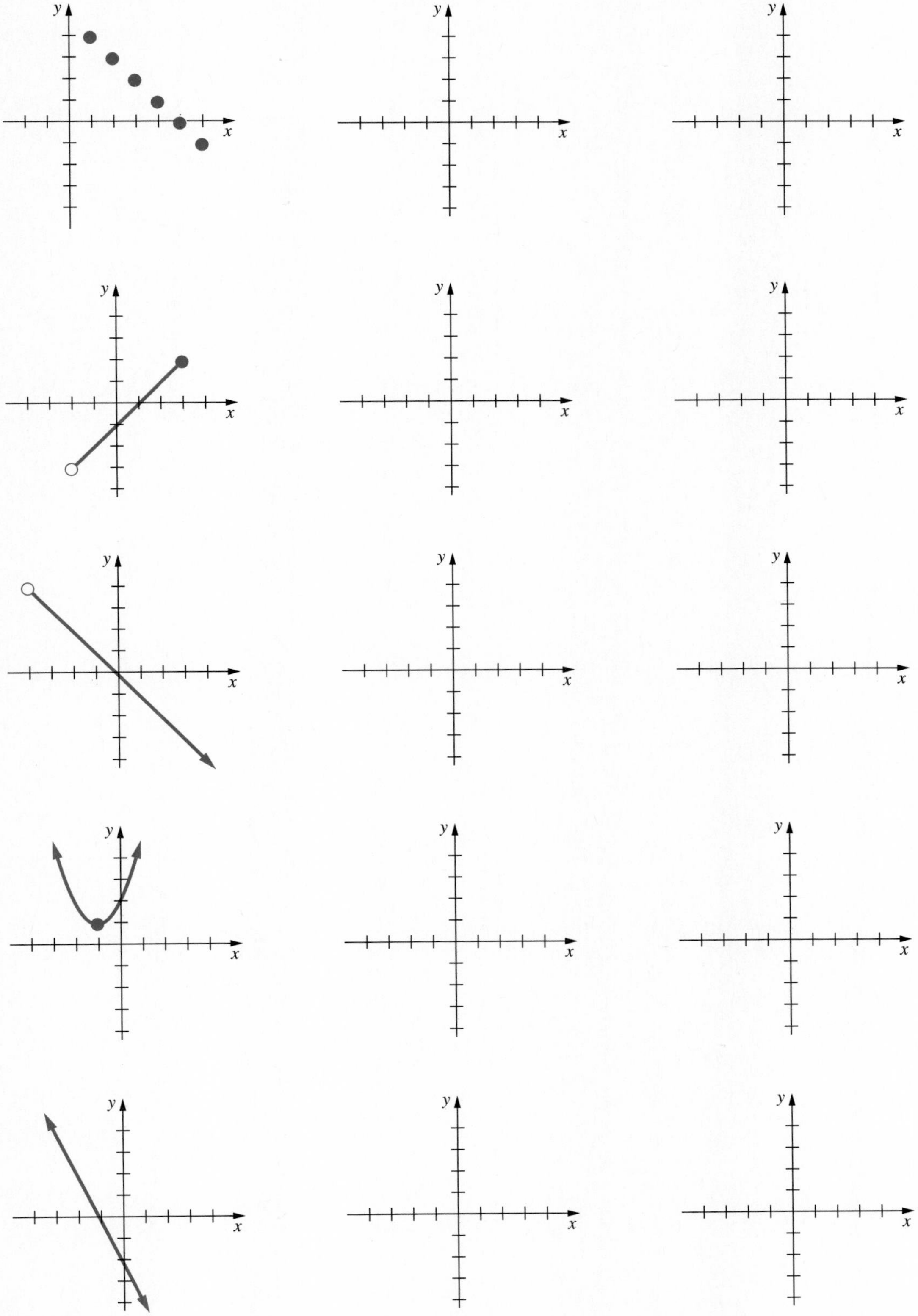

Calculator Tool Kit

Entering Variables

Setting the Range

Trace

Zoom

Clear Screen

Two or More Graphs

Shifting the Window

Calculator Tool Kit

Contrast

Inequalities

Powers and Roots

ADDITIONAL GRAPHING CALCULATOR PRACTICE

Easier Problems

EFX-1. Use the graphing calculator to find the intercepts of $y = x^2 + x - 7$ accurate to two decimal places.

EFX-2. Find the coordinates of the vertex of $y = x^2 + x - 7$.

EFX-3. Find the coordinates of the intersection of $y = 2x - 5$ and $y = x^2 - 5x$ accurate tø two decimal places.

EFX-4. Find the coordinates of the intersection of $y = -x + 3$ and $y = 0.5x^2 + 4x$ accurate to two decimal places.

Harder Problems

EFX-5. Use the graphing calculator to find the intercepts of $y = x^2 + x - 17$ accurate to two decimal places.

EFX-6. Find reasonable values to set your viewing window for the graph $y = x^2 - 16x + 44$ so that the intercepts and the vertex can be seen. Find the coordinates of the vertex.

EFX-7. Find the coordinates of the intersection of $y = 5x - 8$ and $y = \frac{1}{3}x^2 - 4x - 3$ accurate to two decimal places.

EFX-8. Find the intercepts of $y = -0.1x^2 + 5x + 100$.

Answers

EFX-1: (0 –7), (2.19, 0), (–3.19, 0)

EFX-2: (–0.5, –7.25)

EFX-3: (0.81, –3.39), (6.19, 7.39)

EFX-4: (0.57, 2.43), (-10.57, 13.57)

EFX-5: (0, –17), (7.96, 0), (–14.96, 0)

EFX-6: $-5 \le x \le 20$, $-40 \le y \le 70$ *is a resonable answer. V(8,–20)*

EFX-7: (0.57, –5.16), (26.43, 124.16)

EFX-8: (0, 100), –15.31, 0), (65.31, 0)

HANDY GRAPHING GRIDS

Use these grids under a sheet of regular binder paper to save using a whole sheet of graph paper when only one graph is to be drawn

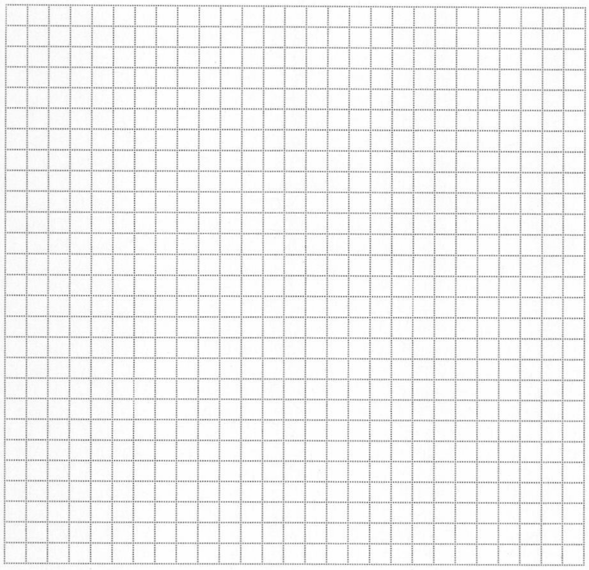

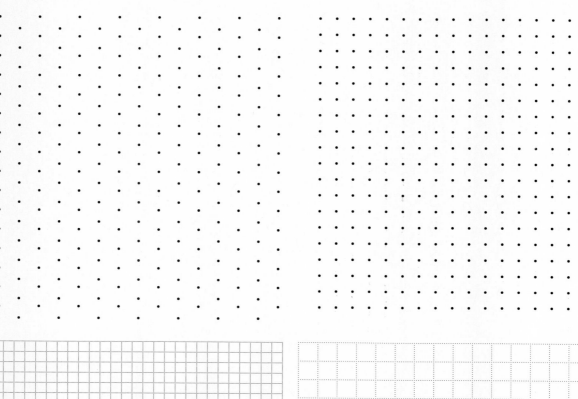

BB-16. Graphs of Sequences

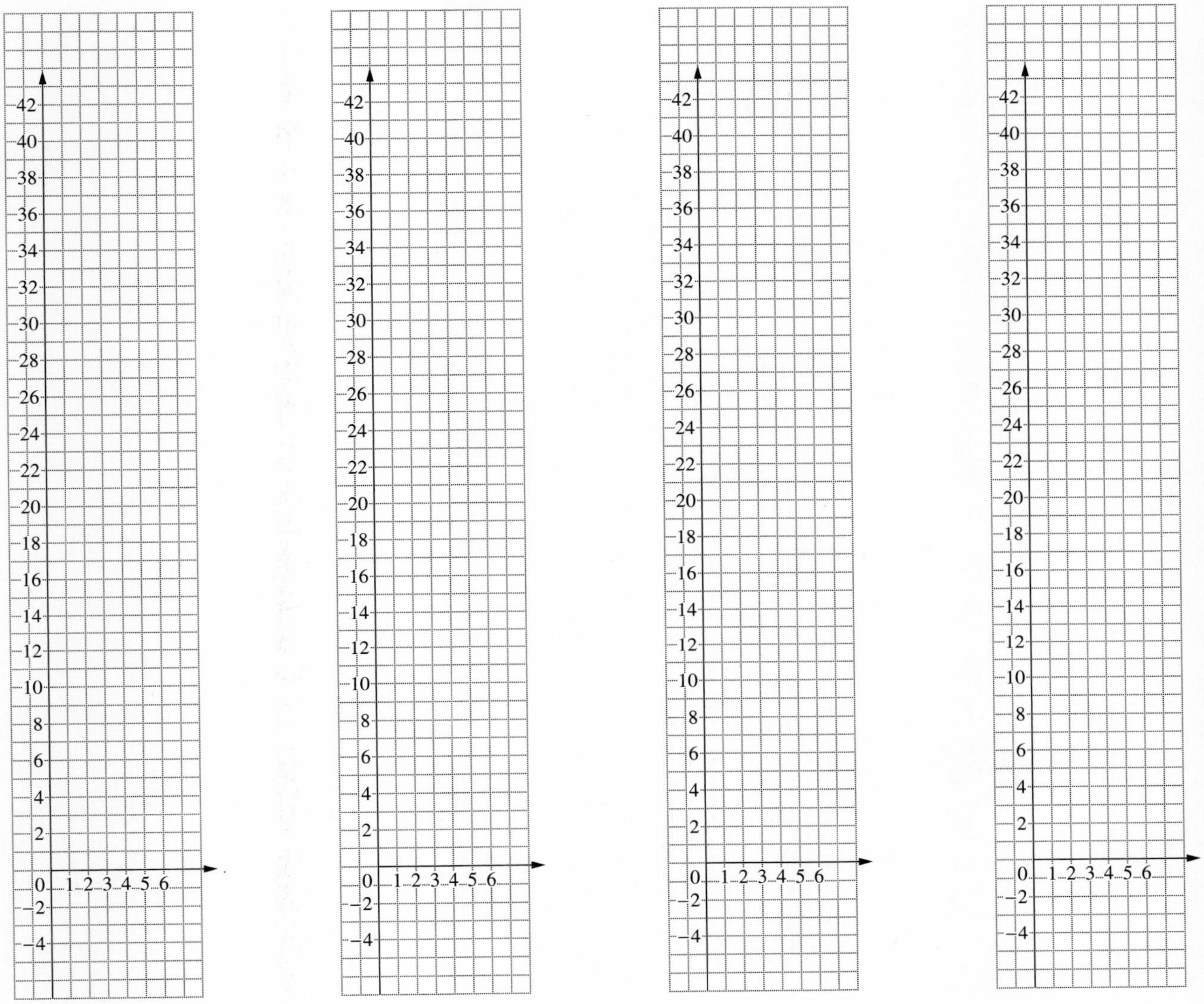

PG-57. Moving Other Functions (Page 1 of 2)

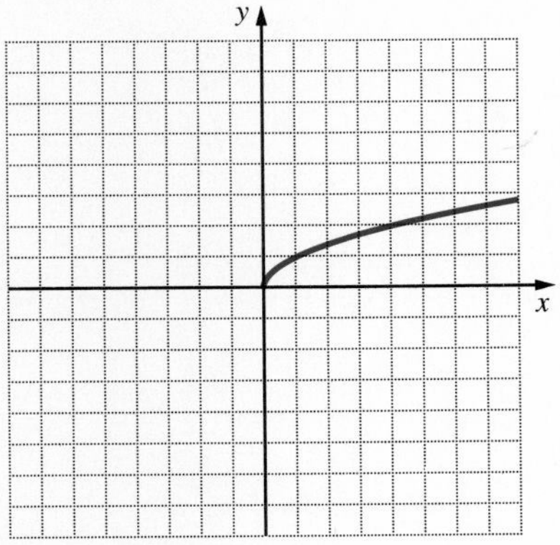

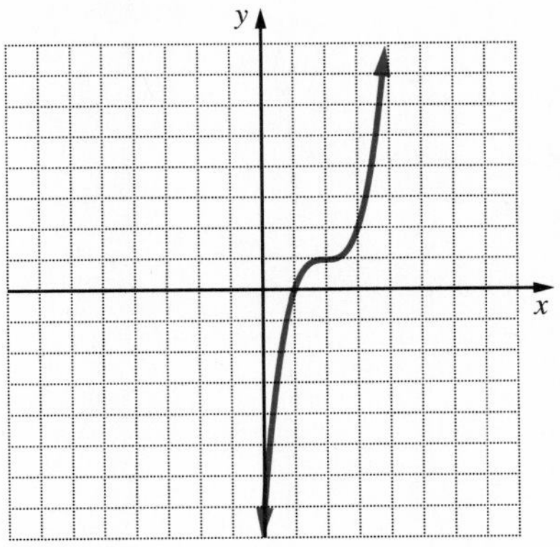

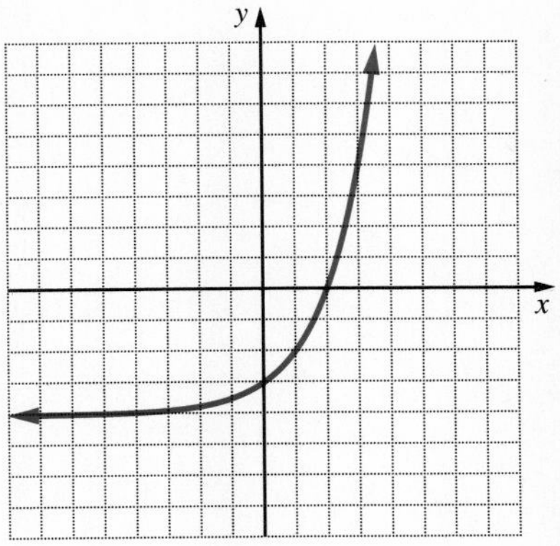

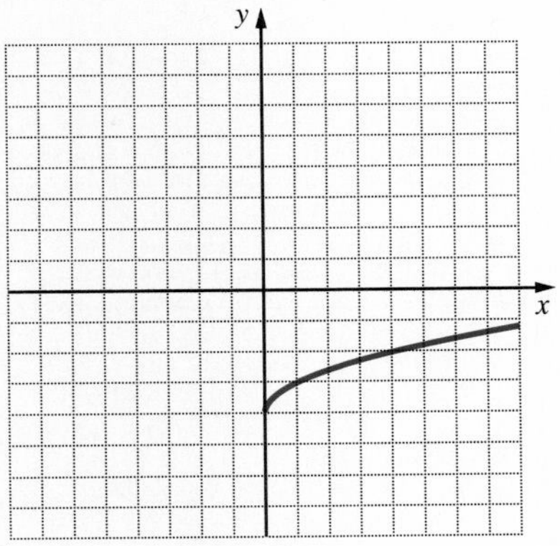

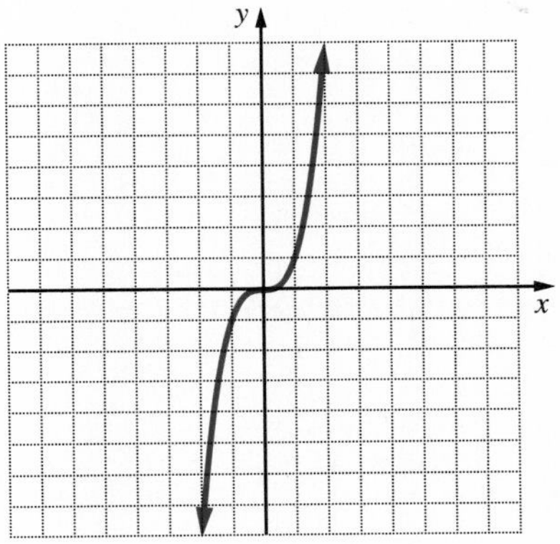

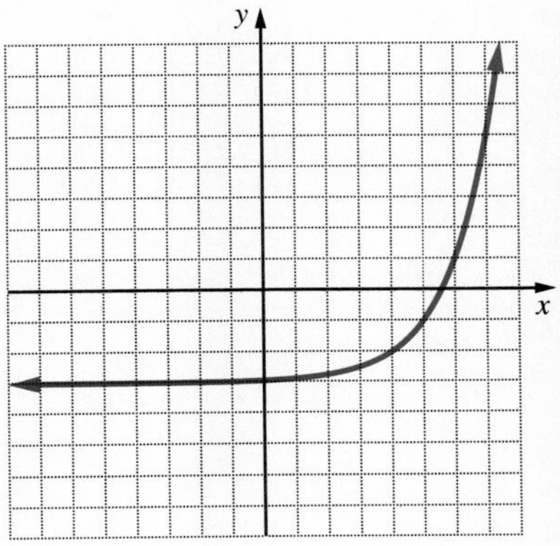

PG-57. Moving Other Functions (Page 2 of 2)

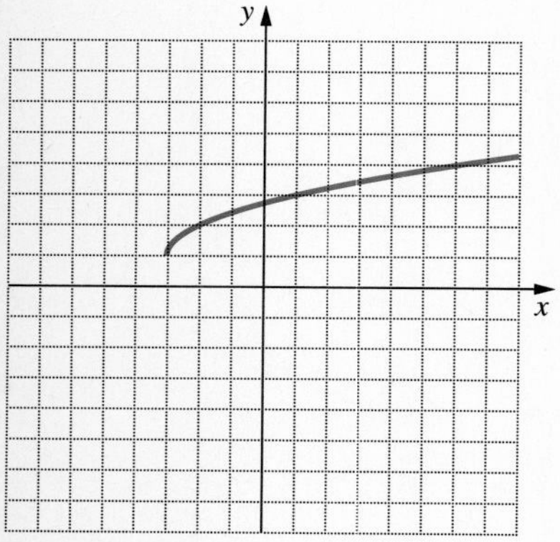

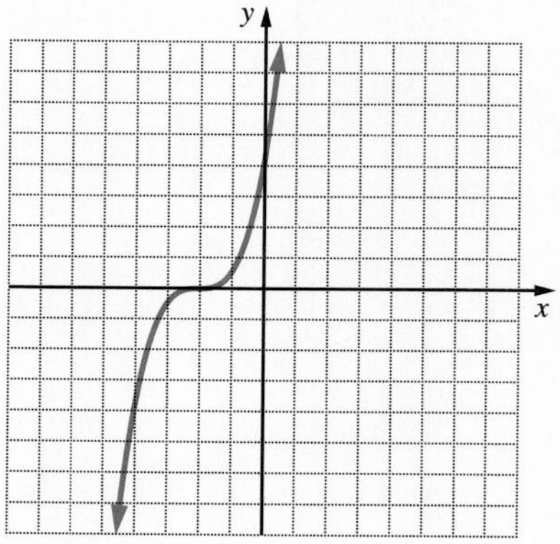

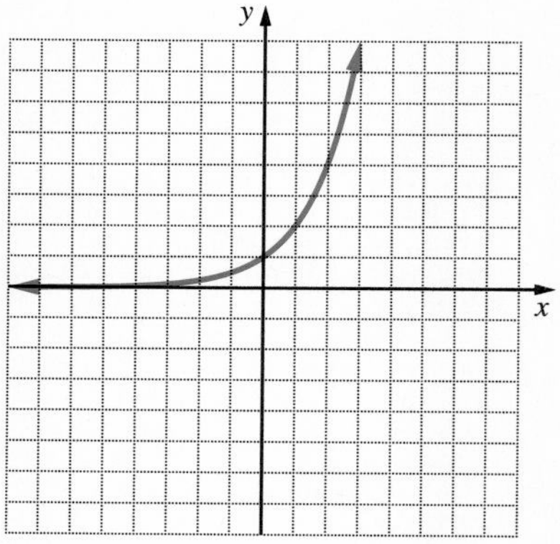

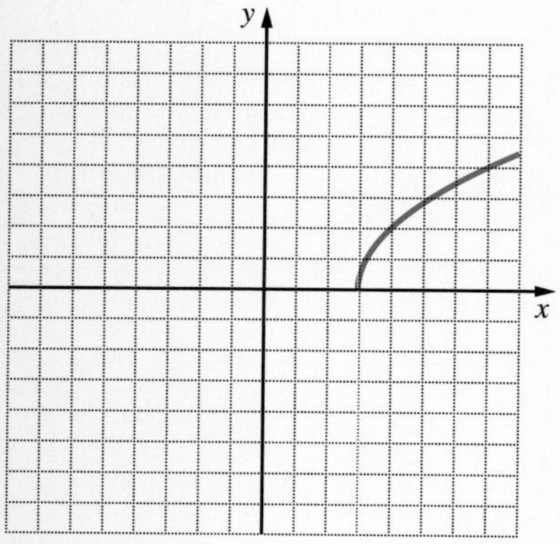

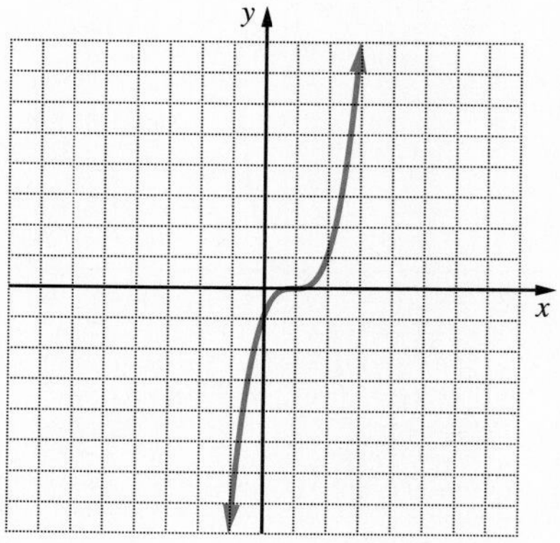

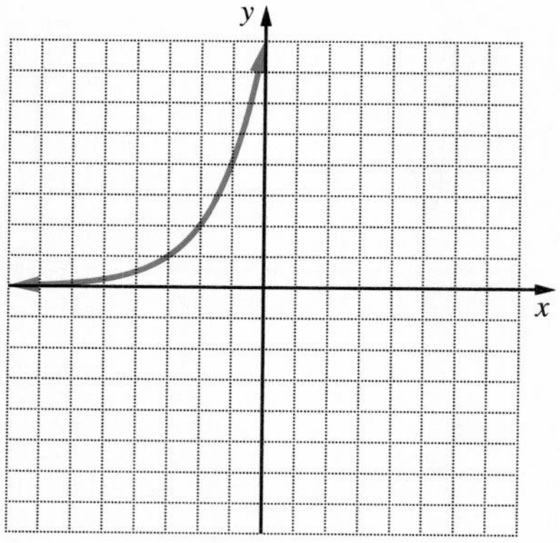

PARENT GRAPH TOOL KIT

Parent graph:

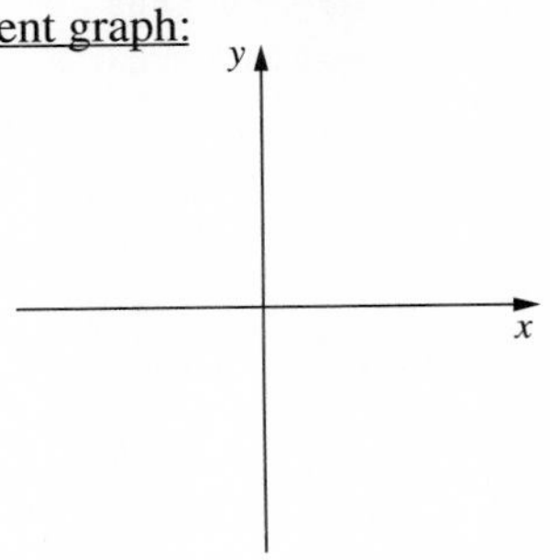

Parent equation:

Locator:

General equation:

Special properties:

Example 1

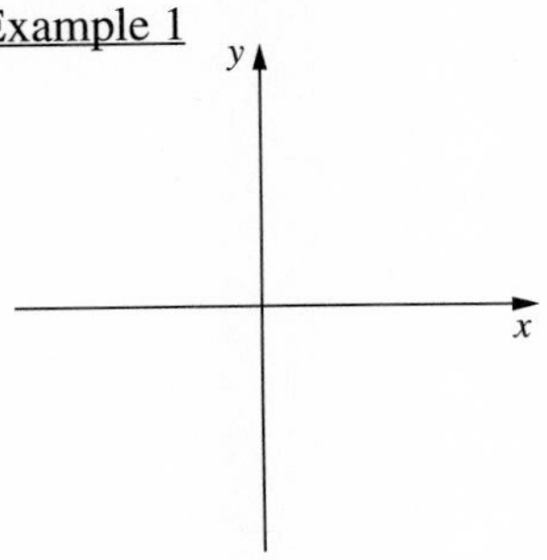

Equation:

Example 2

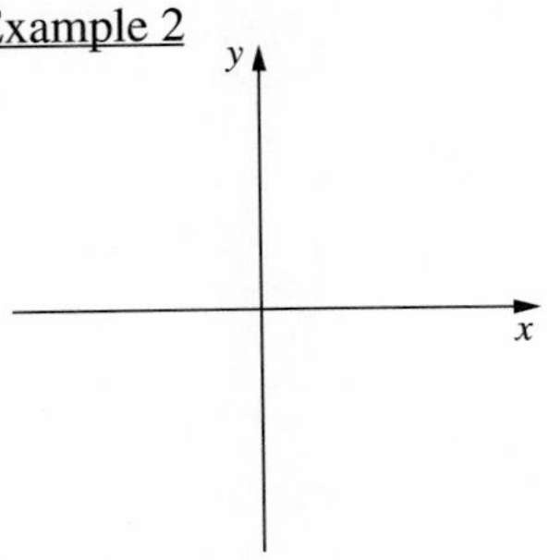

Equation:

Parent graph:

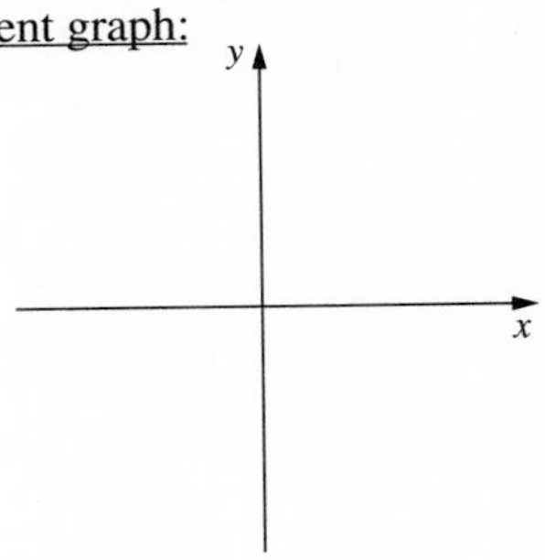

Parent equation:

Locator:

General equation:

Special properties:

Example 1

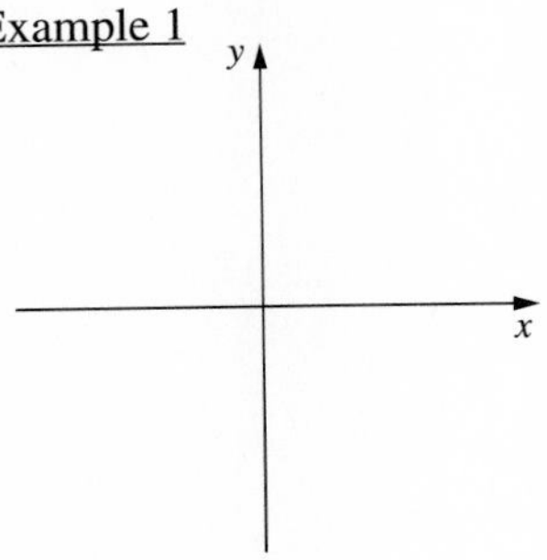

Equation:

Example 2

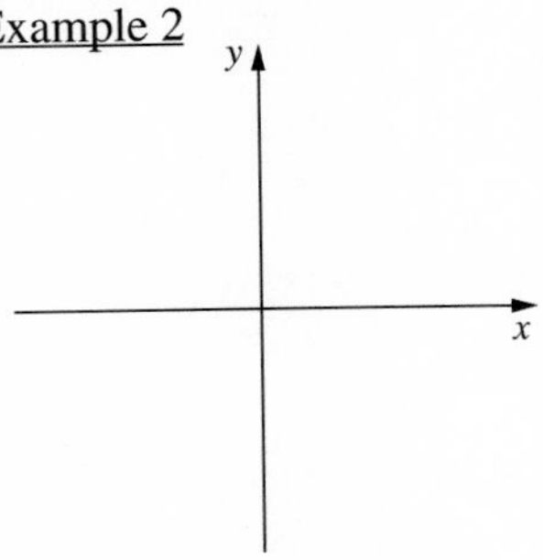

Equation:

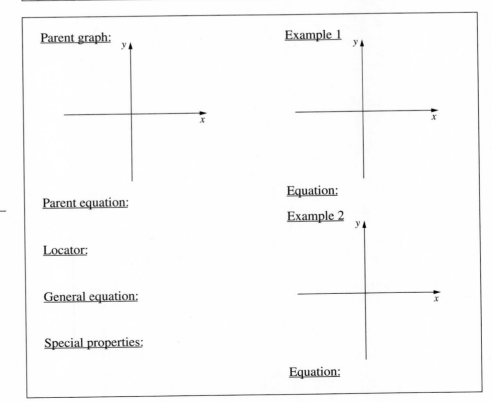

LS-1. Number Lines

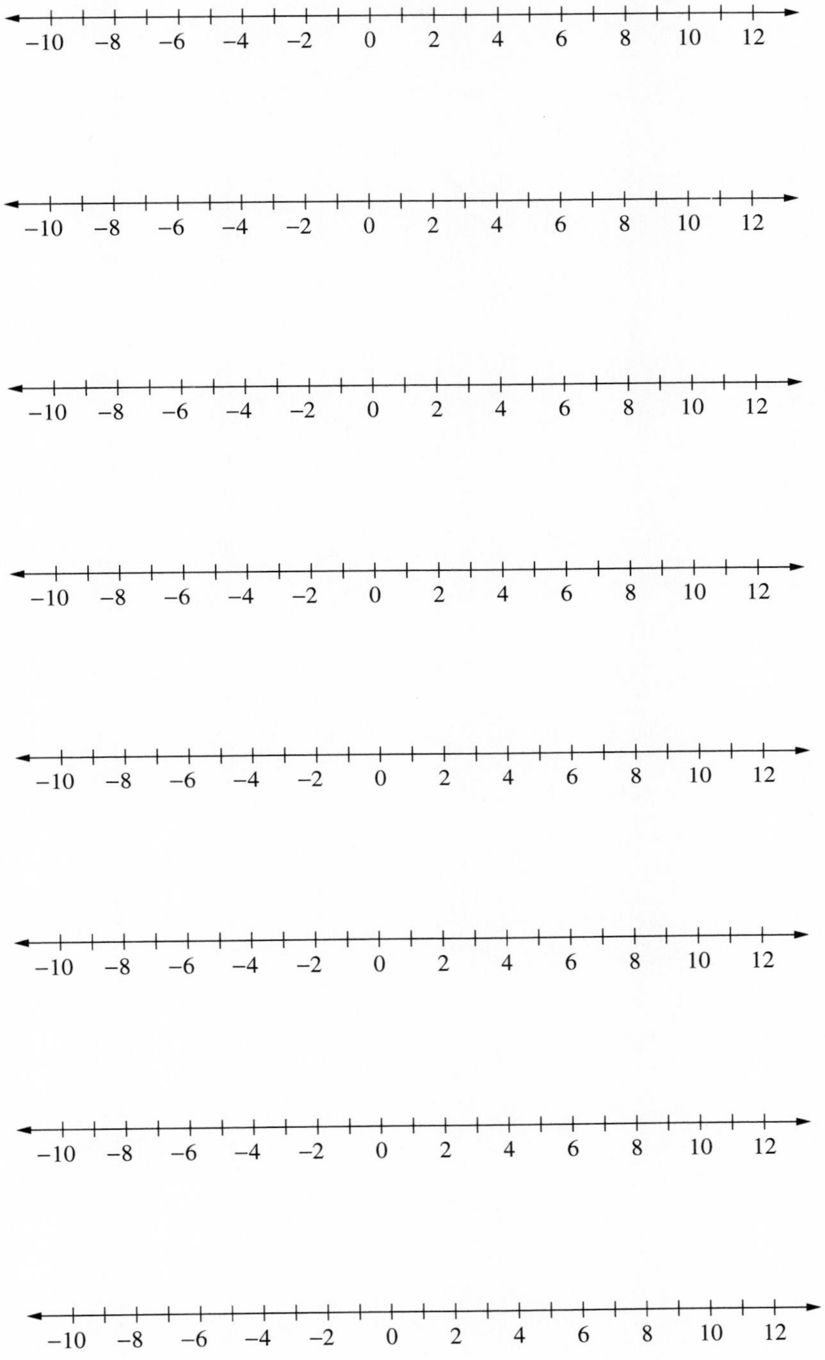

CF-2 AND CF-4. Polynomial Functions Lab

For each problem, complete the following:

a. Graph the function. Draw what you see in the window

b. Use the zoom box feature to get a good view of how the graph curves. Sketch this view.

c. Use the trace feature to find the x-intercepts of the function. List the roots of the corresponding equation. p(x) = 0.

d. Shade on the number line where the values (outputs) of the function are positive.

e. Find the degree and describe the graph.

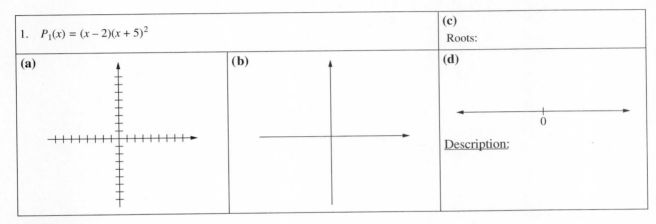

1. $P_1(x) = (x - 2)(x + 5)^2$

(c) Roots:

(a)

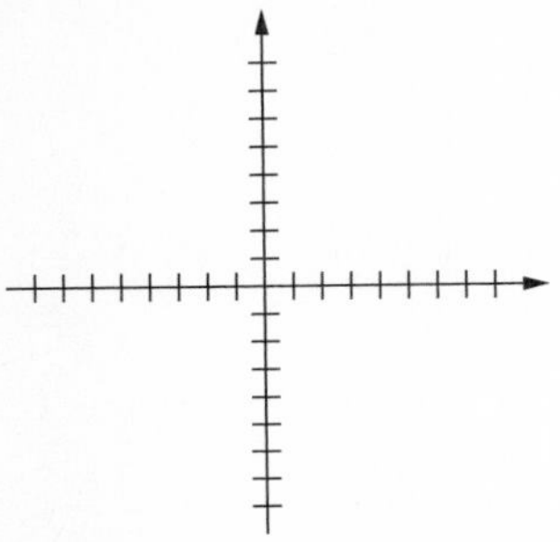

(b)

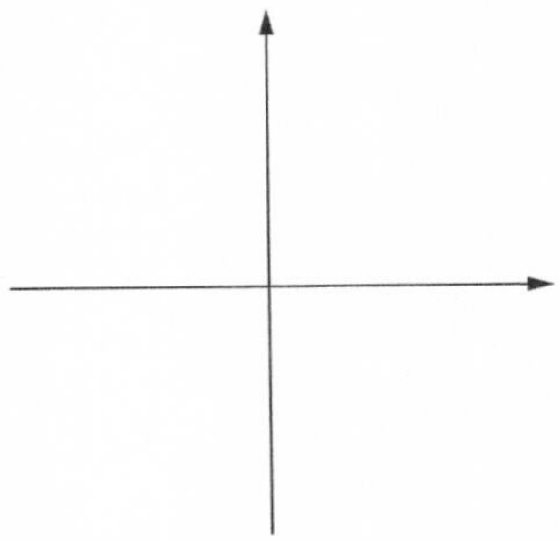

(d)

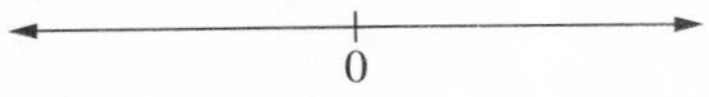

Description:

Notes

2. $P_2(x) = (x - 1)(x + 1)(x - 3)$

(c) Roots:

(a)

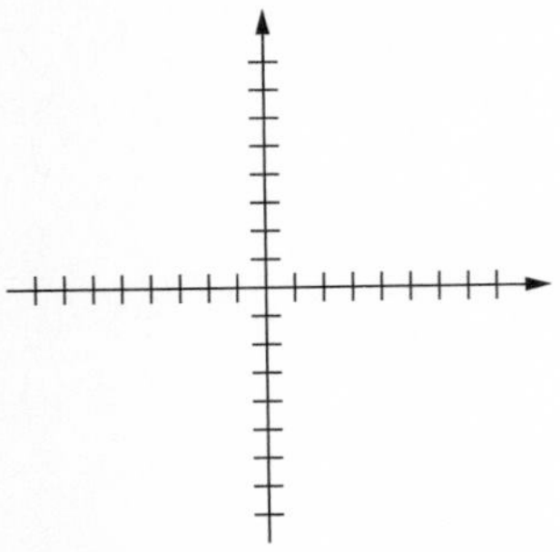

(b)

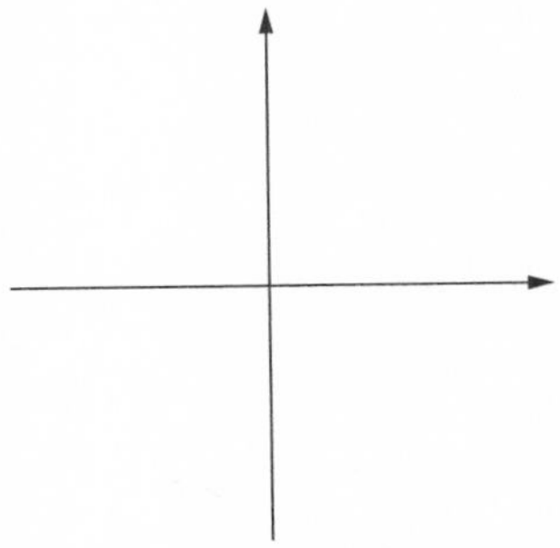

(d)

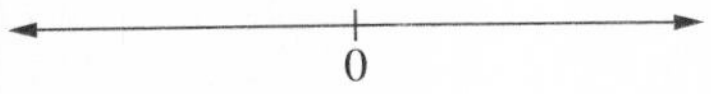

Description:

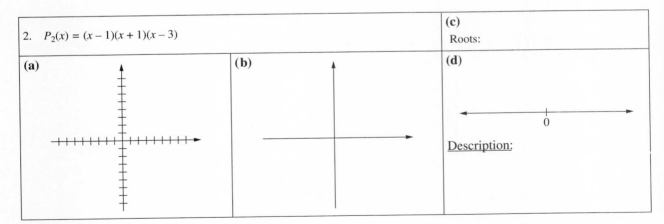

Notes

3. $P_3(x) = 0.2x(x + 1)(x - 3)(x + 4)$		**(c)** Roots:
(a) 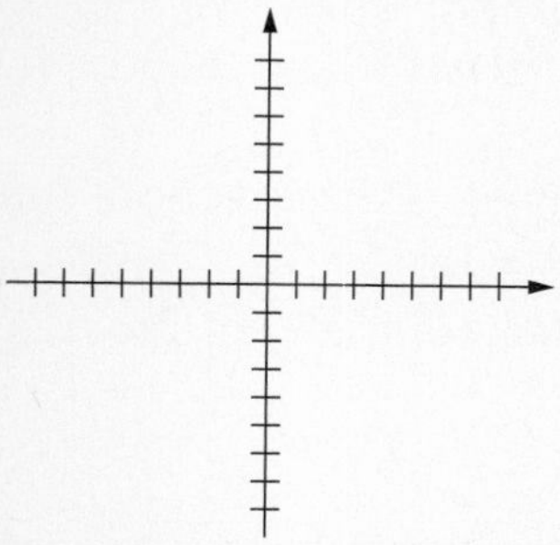	**(b)** 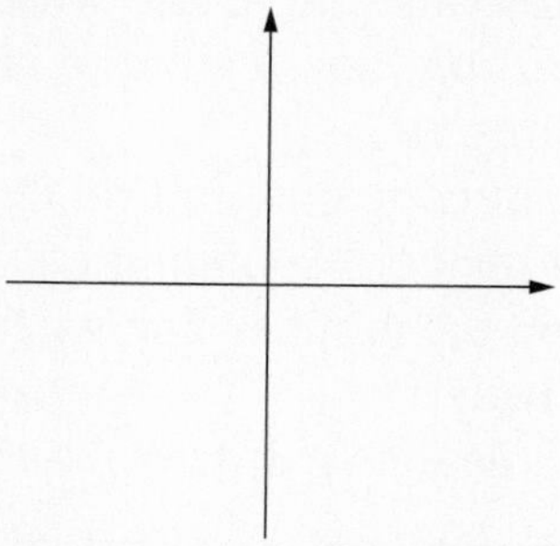	**(d)** 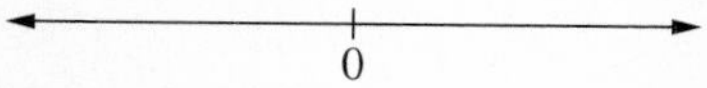 Description:

Notes

4. $P_4(x) = (x + 3)^2(x + 1)(x - 1)$		**(c)** Roots:
(a) 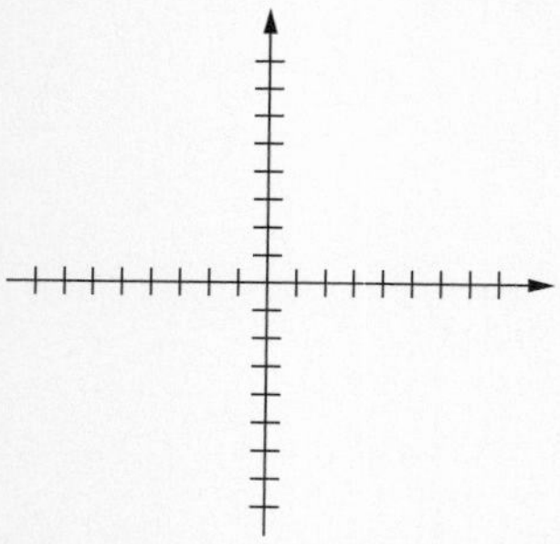	**(b)** 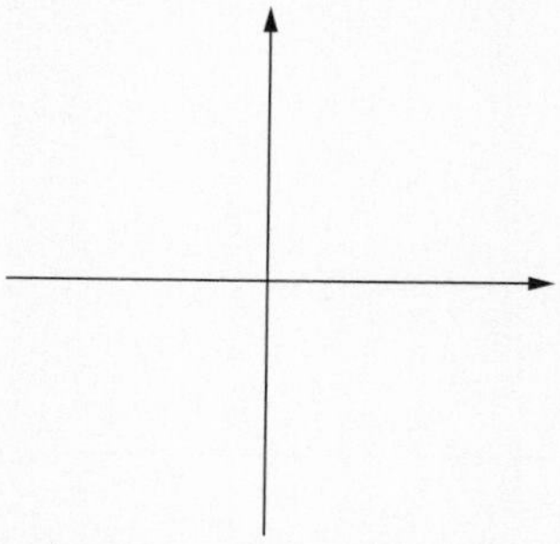	**(d)** 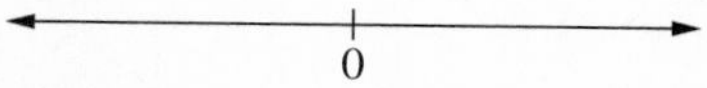 Description:

Notes

5. $P_5(x) = -0.1x(x+4)^3$		**(c)** Roots:
(a) 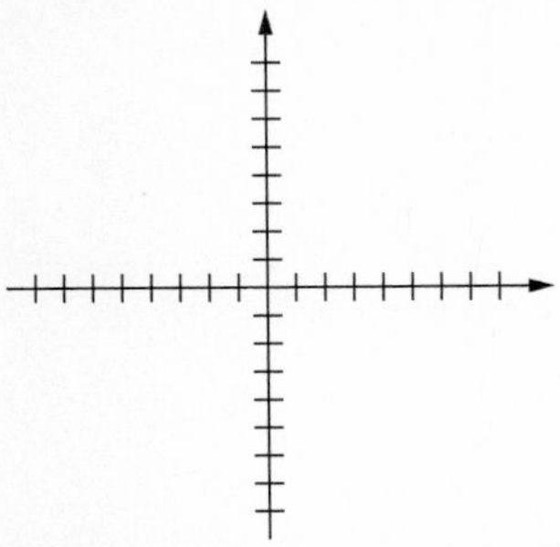	**(b)** 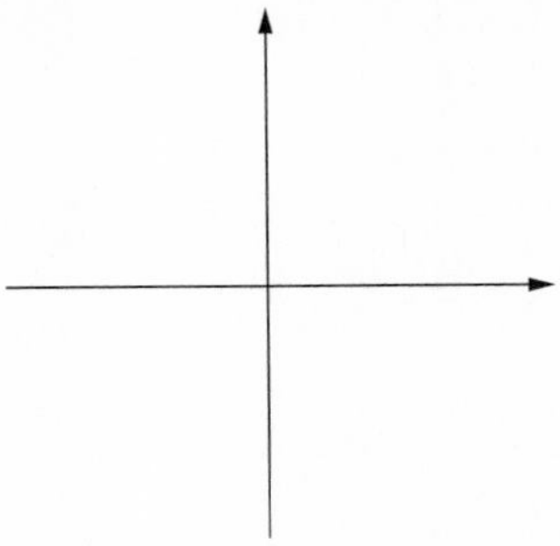	**(d)** 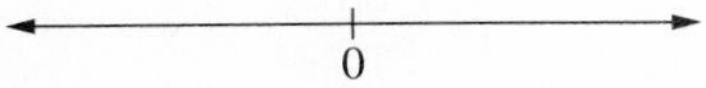 Description:

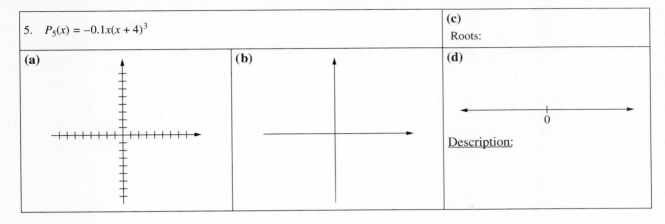

Notes

6. $P_6(x) = x^4 - x^2$		**(c)** Roots:
(a) 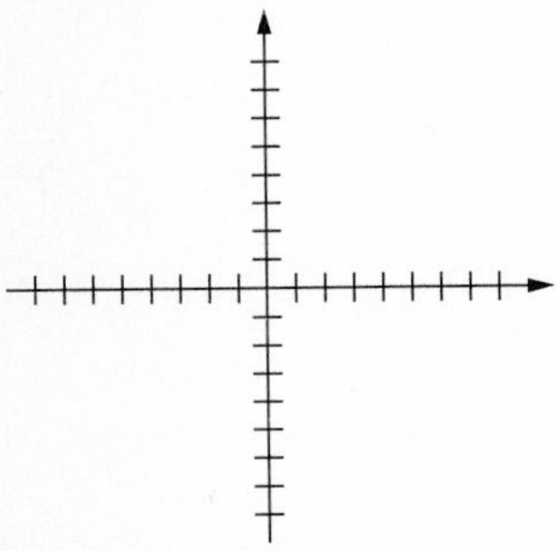	**(b)** 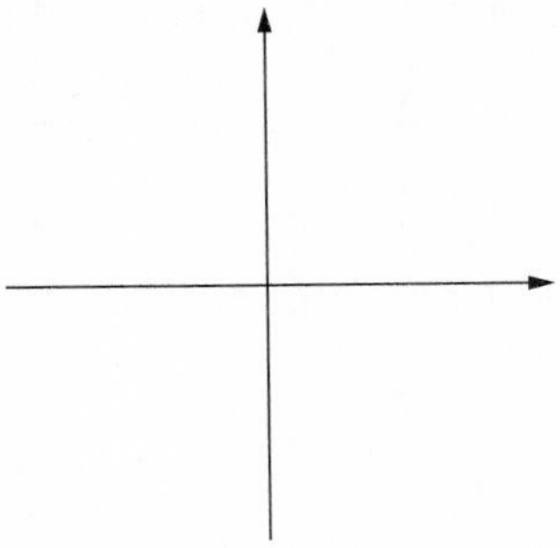	**(d)** 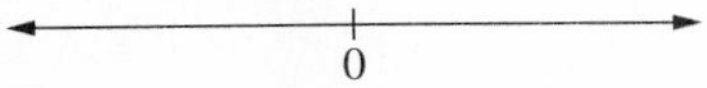 Description:

Notes

CF-107. Building a Better Tool Kit

Part 1: Parent Graph Tool Kit

This section of your Took Kit should include the names, general equations, and graphs of all the parent graphs we have studied:

cubic, exponential, (hyperbola), linear, logarithmic, quadratic, square root

Be sure to include descriptions of any important characteristics, such as domain, range locator point, asymptotes, symmetry, general equations, along with an example or two of graphs this graph is a parent graph for.

Part 2: Vocabulary/Examples Tool Kit

Use the following as a checklist, and include an example or definition and any other useful information. Use this opportunity to reorganize your Tool Kit by categories that make sense to you or possibly alphabetically. Try to include only the things you think you'll need. Do not include what you already know and are unlikely to forget.

area formulas
arithmetic sequences/ linear functions
averaging intercepts
circles
completing the square
complex conjugates
continuous;discrete/ other
dependent and independent variables
domain and range
elimination method
equation of a line

exponent laws
- dividing like bases
- fractional exponents
- multiplying like bases
- powers of powers
- reciprocals (negative exponents)
- zero power

exponential growth
- initial value/multiplier

factoring
fraction busters
geometric sequences/ exponential functions

intercepts (x- and y-)
inverse functions (undoing)
imaginary number
line of symmetry
logarithms
matrices (Appendix D only)
polynomial functions
- degree
- roots

probability
pythagorean theorem
quadratic formula

quadratic functions, equations, expressions
- graphing form
- standard form

sample space
similarity (portions)
sketch vs. graph
slope
special products
- difference of squares
- perfect squares

substitution method
translations (of functions)
volume formulas

Part 3: Calculator Tool Kit

Include in this section the calculator operations that you don't use everyday, the ones you are likely to forget if you don't use them for a couple of weeks.

TI-82 Linear Regression:

Finding the Equation of the Line of Best Fit

Note: [ ] = key or menu selection.

▶ ▶ = use arrow keys to choose menu.

In numbered menus either press the number indicated or use arrow keys and press [ENTER].

1. Before you begin:

Clear the regular screen

Clear the **Y** = screen.

Set [MODE].

Float 0 1 2 3 . . .

Select [2] and press [ENTER].

2. To adjust the viewing screen:

Set [WINDOW].

Set **XMin** and **XMax** to fit your domain.

Set **YMin** and **YMax** to fit your range.

3. To clear data:

First clear previously stored data.

Set [STAT] [statistics key].

Select [4: ClrLst].

Type [2nd] [1] [comma] [2nd] [2].

The screen should now read:

ClrLst L1, L2

Press [ENTER].

Now you will see the regular screen with **Done.**

4. To enter data:

Set [STAT].

Select [1: EDIT].

This will show you the empty data columns. Enter all the independent variable values in **L1**.

Press [ENTER] after each one. Then enter the dependent values in **L2**.

Example: (1, 1) (3, 2) (5, 3) (7, 5) (9, 8)

5. To see the scatterplot:

Press [2nd] [Y=] for **STAT PLOT** menu.

Select [1: Plot 1] and press [ENTER].

Select [On] and press [ENTER].

Select [1st type of graph], press [ENTER].

Select [square marks] and press [ENTER].

Press [GRAPH].

6. To calculate the equation:

Press [STAT] [▶] [CALC].

Select [5: LinReg] and press [ENTER].

a is the slope and *b* is the *y*-intercept.

The closer the *r* value is to ±1, the better the fit of the line will be.

7. To see the equation:

Press [Y=], then [VARS].

Select [5: Statistics].

Press [▶] [▶] [EQ].

Select [7: RegEq].

Press [GRAPH].

8. Now that you are done:

Set [MODE].

Select [FLOAT], press [ENTER].

Press [2nd] [Y=] for **STAT PLOT** menu.

Select [4: PlotsOff].

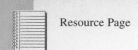

TI-83 Linear Regression:

Finding the Equation of the Line of Best Fit

Note: [ ] = key or menu selection.

▶ ▶ = use arrow keys to choose menu.

In numbered menus either press the number indicated or use arrow keys and press [ENTER].

1. Before you begin:

Clear the regular screen.
Clear the **Y** = screen.
Set [MODE].

Float 0 1 2 3 . . .

Select [2] and press [ENTER].

2. To adjust the viewing screen:

Set [WINDOW].

Set **XMin** and **XMax** to fit your domain.
Set **YMin** and **YMax** to fit your range.

3. To clear data:

First clear previously stored data.

Set [STAT] [statistics key].

Select [4: ClrLst].

Type [2nd] [1] [comma] [2nd] [2]

The screen should now read:

ClrLst L1, L2

Press [ENTER].

Now you will see the regular screen with **Done.**

4. To enter data:

Set [STAT].

Select [1: EDIT].

This will show you the empty data columns. Enter all the independent variable values in **L1**.

Press [ENTER] after each one. Then enter the dependent values in **L2**.

Example: (1, 1) (3, 2) (5, 3) (7, 5) (9, 8)

5. To see the scatterplot:

Press [2nd] [Y=] for **STAT PLOT** menu.

Select [1: Plot 1] and press [ENTER].

Select [On] and press [ENTER].

Select [1st type of graph] and press [ENTER].

Select [square marks] and press [ENTER].

Press [GRAPH].

6. To calculate the equation:

Press [STAT] [▶] [CALC].

Select [4: LinReg], press [ENTER].

a is the slope, *b* is the *y*-intercept.

7. To see the equation:

Press [Y=] then [VARS].

Select [5: Statistics].

Press [▶] [▶] [EQ].

Select [1: RegEq], press [ENTER].

Press [GRAPH].

8. Now that you are done:

Press [MODE].

Select [FLOAT], then press [ENTER].

Press [2nd] [Y=] for **STAT PLOT** menu.

Select [4: PlotsOff].

CASIO CFX-9850G Linear Regression:

Finding the Equation of the Line of Best Fit

Note: [ ] = key or menu selection.

In numbered menus either press the number indicated or use arrow keys and press [EXE].

1. Before you begin:

Clear the **Y** = screen.

2. To clear data:

Set [MENU].

Select [4].

Use the arrow keys to move the cursor to the name of the list you want to delete.

Press [F4] [F1].

The list should be clear.

Clear List 1 and 2.

3. To enter data:

First clear previously stored data.

Use the arrow keys to move the cursor to **1** in List 1 and enter the independent variables in List 1.

Press [EXE] after each one.

Then enter the dependent values in List 2.

Example: (1, 1) (3, 2) (5, 3) (7, 5) (9, 8)

4. To see the scatterplot:

Press [MENU] [2] [F1] [F1].

5. To calculate the equation:

Press [F1].

a is the slope, b is the y-intercept.

The closer the r value is to ±1, the better the fit of the line will be.

6. To see the equation:

Press [F5] [EXE] [F5].

7. To graph the equation:

Press [EXE] [F6].

If you need a better viewing window for the graph:

Press [SHIFT] [F2] and then either [F3] (to zoom in) or [F4] (to zoom out).

TI-82 Quadratic Regression:

Finding the Equation of the Curve of Best Fit

Note: [ ] = key or menu selection.

►► = use arrow keys to choose menu.

In numbered menus either press the number indicated or use arrow keys and press [ENTER].

1. Before you begin:

Clear the regular screen
Clear the **Y** = screen.
Set [MODE].

Float 0 1 2 3 . . .

Select [2] and press [ENTER].

2. To adjust the viewing screen:

Set [WINDOW].

Set **XMin** and **XMax** to fit your domain.
Set **YMin** and **YMax** to fit your range.

3. To clear data:

First clear previously stored data.

Press [STAT] [statistics key].

Select [4: ClrLst].

Type [2nd] [1] [comma] [2nd] [2].

The screen should now read:

ClrLst L1, L2

Press [ENTER].

Now you will see the regular screen with **Done.**

4. To enter data:

Set [STAT]

Select [1:EDIT]

This will show you the empty data columns. Enter all the independent variable values in **L1**. Press [ENTER] after each one. Then enter the dependent values in **L2**.

Example: Enter the three points from LS-79 in Chapter 6: (2, 3) (−1, 6) (0, 3)

5. To see the scatterplot:

Press [2nd] [Y=] for **STAT PLOT** menu.

Select [1: Plot 1], and press [ENTER].

Select [On], then press [ENTER].

Select [1st type of graph], press [ENTER].

Select [square marks] and press [ENTER].

Press [GRAPH].

6. To calculate the equation:

Press [STAT] [►] [CALC].

Select [6: QuadReg].

Press [ENTER].

a is the coefficient of x^2, b is the coefficient of x, and c is the constant term.

7. To see the equation:

Press [Y=], then [VARS].

Select [5: Statistics].

Press [►] [►] [EQ].

Select [7: RegEq].

Press [GRAPH].

8. Now that you are done:

Press [MODE].

Select [FLOAT], then press [ENTER].

Press [2nd] [Y=] for **STAT PLOT** menu.

Select [4: PlotsOff].

Press [ENTER].

TI-83 Quadratic Regression:

Finding the Equation of the Curve of Best Fit

Note: [ ] = key or menu selection.

▶ ▶ = use arrow keys to choose menu.

In numbered menus either press the number indicated or use arrow keys and press [ENTER].

1. Before you begin:

Clear the regular screen
Clear the **Y** = screen.
Set [MODE].

Float 0 1 2 3 . . .

Select [2] and press [ENTER].

2. To adjust the viewing screen:

Set [WINDOW].

Set **XMin** and **XMax** to fit your domain.
Set **YMin** and **YMax** to fit your range.

3. To clear data:

First clear previously stored data.

Set [STAT] [statistics key].

Select [4: ClrLst].

Type [2nd] [1] [comma] [2nd] [2].

The screen should now read:

ClrLst L1, L2

Press [ENTER].

Now you will see the regular screen with **Done.**

4. To enter data:

Set [STAT].

Select [1: EDIT].

This will show you the empty data columns. Enter all the independent variable values in **L1**. Press [ENTER] after each one. Then enter the dependent values in **L2**.

Example: Enter the three points from LS-79 in Chapter 6: (2, 3) (−1, 6) (0, 3)

5. To see the scatterplot:

Press [2nd] [Y=] for **STAT PLOT** menu.

Select [1: Plot 1] and press [ENTER].

Select [On], then press [ENTER].

Select [1st type of graph], press [ENTER].

Select [square marks], press [ENTER].

Press [GRAPH].

6. To calculate the equation:

Press [STAT] [▶] [CALC].

Select [5: QuadReg].

Press [ENTER].

a is the coefficient of x^2, b is the coefficient of x, and c is the constant term.

7. To see the equation:

Press [Y =] then [VARS].

Select [5: Statistics].

Press [▶] [▶] [EQ].

Select [1: RegEq], press [ENTER].

Press [GRAPH].

8. Now that you are done:

Press [MODE].

Select [FLOAT], then press [ENTER].

Press [2nd] [Y=] for **STAT PLOT** menu.

Select [4: PlotsOff].

Press [ENTER].

CASIO CFX-9850G Quadratic Regression:

Finding the Equation of the Curve of Best Fit

Note: [ ] = key or menu selection.

In numbered menus either press the number indicated or use arrow keys and press [EXE].

1. Before you begin:

Clear the **Y** = screen.

2. To clear data:

Set [MENU].

Select [4].

Use the arrow keys to move the cursor to the name of the list you want to delete.

Press [F4] [F1].

The list should be clear.

Clear List 1 and 2.

3. To enter data:

First clear previously stored data.
Use the arrow keys to move the cursor to 1 in List 1, and enter the independent variables in List 1.

Press [EXE] after each one.

Then enter the dependent values in List 2.
Example: Enter the three points from LS-79 in Chapter 6: (2, 3) (−1, 6) (0, 3).

4. To see the scatterplot:

Press [MENU] [2] [F1] [F1].

5. To calculate the equation:

Press [F3].

a is the coefficient of x^2, b is the the coefficient of x, and c is the constant term.

6. To see the equation:

Press [F5].

7. To graph the equation:

Press [EXE] [F6].

If you need a better viewing window for the graph, Press [SHIFT] [F2] and then either [F3] (to zoom in) or [F4] (to zoom out).

If you are graphing the parabola from problem LS-79, press [F4].

SOME WAYS TO GET STARTED AND SOME ANSWERS

CHAPTER 1

PS-6. If the monkey has 15 bananas and he has 17 more coconuts than bananas, how many coconuts does he have? Does this equal a total of 53 or is it too small? What should you guess now?

PS-7. If Carlos ate 20 turnips, then Judy ate 8 and David ate 4. Does this give 83 turnips total? Now what should you do?

PS-8. See Appendix A for a note on the distributive property, $x = -37$.

PS-9. Appendix A will help you here as well.
a. $x = -17$
b. $x = 8.4$
c. $x = 11$

PS-10. Here's a similar situation. It should help. The probability of flipping a fair coin and getting heads is $\frac{1}{2}$ because there is only one head out of two possible sides (outcomes). Also, in part c, notice that we are only interested in numbers *less* than 5.

PS-11. A square with side lengths of 6 has an area of $6 \cdot 6 = 36$ square units. That's too small. What now?

PS-19. On a 50-mile trip, traveling 25 miles per hour, it will take $\frac{50 \text{ miles}}{25 \text{ mph}} = 2$ hours.

PS-20. The error occurs from line 2 to line 3. What is it?

PS-21. See Appendix A.
a. The answer is less than 1.
b. The answer is not a whole number.
c. There are two answers here.

PS-22. If Howie has one \$50 bill, how many \$10 bills does he have? How much money is that? Make a table.

PS-23. Look back at a graph and write the coordinates of the y-intercept. That should help.

PS-24. Try writing an equation with two ratios to solve the problem, $\frac{11}{3} = \frac{x}{30}$.

PS-25. You can write equations of two ratios to help you solve these also. For instance:
a. $\frac{42}{100} = \frac{112}{x}$ and $x \approx 266.67$
b. $x = 37.5\%$
c. $x = 27\%$
d. $x = 135$

PS-31. Substitute the x-value and calculate the y-value:
a. $y = 3(2) + 15 = 6 + 15 = 21$. When $x = 0$, $y = 15$ so the y-intercept is $(0, 15)$. For equation (a) when $y = 0$, $x = -5$ and the x-intercept is $(-5, 0)$.

PS-32. Shouldn't you always do things to *both* sides of an equation? The two solutions are -1 and 5.

PS-33. If the first side is 10 centimeters, that makes the second side 20 centimeters. How long does that make the third side? We find the perimeter by adding up all the sides. Will these three sides give us a perimeter of 76 cm?

PS-34. See Appendix A.
a. $x = 15.4$
b. $x = 1$
c. $x = -13$

PS-35.

x	0	1	2	3
y	7	6	5	4

PS-36. If a cube has an edge length of 3, the volume is $3 \cdot 3 \cdot 3 = 27$ cubic units. What is the volume of a cube with edge length of 5?

PS-37. Part a has been shown in detail to get you started.

a. $\frac{1}{10} = \frac{y}{8}$, $10y = 8$, $y = 0.8$

b. $y = 1.6$

c. $y = 2.4$

PS-38. How many kings are in the deck of 52 cards?

PS-43. Here are some more input-output combinations: $5 \to 25$, $-10 \to 100$, $4 \to 16$.

PS-44. Here's how you can figure out

a. $3^2 + 2(3) + 1 = 9 + 6 + 1 = 16$.

b. 9

c. 478.384384

PS-45. Guess and check; start small. Or you could solve the equation $x^2 + 2x + 1 = 1$. To solve it you can factor the left side of $x^2 + 2x = 0$ to get $x(x + 2) = 0$, and there are two solutions.

PS-46. Go ahead to PS-47 for some help in seeing the error.

PS-48. **a.** $x = \frac{5 \pm \sqrt{13}}{2}$, decimal form: 4.30, 0.70

b. $x = \frac{-3 \pm \sqrt{21}}{2}$

c. $x = \frac{7 \pm \sqrt{193}}{6}$

d. $x = -2, \frac{3}{2}$

PS-49. The two consecutive sides on a rectangle are marked *L* and *W* in the figure below.

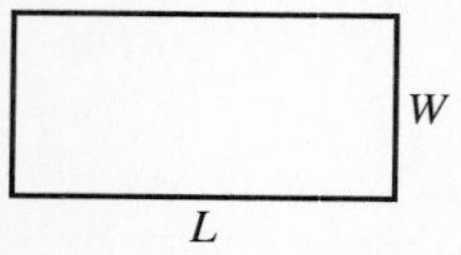

PS-51. How many x's are in the alphabet? With that one removed, how many letters are left? Of those, how many are y's?

CHAPTER 2

EF-8. **a.** $\frac{5(y-1)}{3}$

b. $x = \left(-\frac{2}{3}\right)y + 3$

c. $x = \pm\sqrt{y}$

EF-9. The volume of water is the independent variable, and the height is dependent. The graph of *C* should be steepest and of *B* the least steep.

EF-10. Use the problem-solving strategy of working backwards. Start with the outputs that are given and see if you can figure out the inputs. What would you do to 10 to get 100? to 4 to get 16?

EF-12. **a.** \$24.95

b. \$259.95

c. Make a table for several numbers of hours of use and the cost. Then use x as the number of hours, or think, "base cost \$9.95 plus \$2.50 per hour."

EF-13. **a.** $g(x) = 3x$

EF-15. 7 feet and 369.6 feet

EF-16. $\frac{5 \pm \sqrt{57}}{-8}$ or the decimal approximations -1.57, 0.32

EF-18. There is only one Queen of Hearts, and there are how many cards?

EF-27. To see the relationship for the equation you could make a table using the following points from the graph: (0, 5), (1, 4.5), (2, 4), (3, 3.5), (4, 3), ... Then try x.

EF-28. Check the step just before the last one.

EF-29. Try using a number for x. Then try some other numbers.

EF-30. To figure out an equation, use a guess and check table. You might start with a guess of 1000 for the total number of registered voters.

EF-31. **a.** 7

b. 55

c. -5

EF-32. **a.** Domain: all real numbers

b. Range: $-3, -2, 1, 3$

c. Domain: $x \geq -2$, Range: all real numbers

e. Domain: 2, Range: all real numbers

EF-33. $x = 10.5$

EF-40. Check for appropriate scaling, complete graphs, and axis labels on each graph.

EF-41. Try making a table for 1, 2, 3, and 4 pounds. That could lead you to an equation for x pounds.

EF-42. How many days are there in August? How many is Ashley interested in?

EF-43. **a.** $-11/8$

b. 8

c. 0.30, 0.42

d. -3.83, 1.83

EF-44. **a.** -60

b. 3

c. 31

EF-45. 1/12

EF-47. If Kendall's results are 9, 4, 1, 0, 1, 4, 9, then each of Amy's will be just 5 less.

EF-48. This is a right triangle, so you can use the Pythagorean theorem to get 14.42.

EF-55. Something very strange happened on line 3. What did this student do?

EF-56. **a.** 2, 5
b. –7, 6
c. –5, 0
d. –1, 3/2

EF-61. **a.** Domain: $x \geq -2$ and $x < 1$. Range: $y > -1$ and $y \leq 3$.
b. Domain: 2 and –2 and all the numbers in between. Range: –2 and –1 and all the numbers in between.

EF-63. Try multiplying the input values by 5 and checking the outputs.

EF-64. Make a list to see how many arrangements there are for the three digits 6, 7, and 9.

EF-71. **a.** (10, 48)
b. (–1, –5)

EF-72. **a.** (0, 6)
c. (0, 0)
e. (0, 25)

EF-73. **a.** (–2, 0)
c. (0, 0)
e. (5, 0)

EF-74. **a.** –3
c. 12
e. 0

EF-75. For example, graph no. 3 could represent situation C.

EF-76. Draw a picture of the wall, the ground and the ladder leaning against Sean and Ayla's house. That should form a right triangle. Label the lengths of the legs and use the Pythagorean theorem.

EF-77. It helps to draw a box marking the edges of the domain and range. Then you can put the function inside. There are many possibilities.

EF-82. **a.** $y = 2x : (0,0)$

$y = \frac{1}{2}x + 6$: $(0,6)(12,0)$

c. $0 \leq x \leq 12,\ 0 \leq y \leq 4.8$
d. Remember, a triangle is half of a rectangle, so its area is one-half of its height times its base, 28.8 square units.
e. 90°
f. They are reciprocals and opposites.

EF-84. (3, 0) and (7, 0) are the x-intercepts

EF-85. **a.** $x \approx 781.36$
b. $x = -4/3$ or 2
c. $x = 1$ or 1/5

EF-86. **b.** 23.78 cm

EF-87. **a.** $D = \{-2, -1, 2\}$, $R = \{-1, 0, 1\}$
The following answers describing the domains and ranges are written in a kind of "shorthand" called **interval notation**. See if you can decipher it in order to check your answers.
b. $D = (-1, 1)$, $R = (-1, 2)$
c. $D = [-2, 2]$, $R = [-2, 1]$
d. $D = (-2, 2]$, $R = [-2, 2)$

EF-88. 18.11 cups per day

EF-89. $y = 18.11x$

CHAPTER 3

BB-4. **a.** square root
b. One—if you continue to keep press the "√" button, after 30 to 40 times you will get 1.

BB-5. **c.** 1, 2, 4, 8, 16, 32, 64, . . .

BB-6. **a.** $5^2 = 25$
b. 3^{51}
c. $(3 \cdot 4^4)/7$
d. 6^{104}

BB-7. **a.** $y = -2x + 7$
b. $y = -\frac{3}{2}x + 6$

BB-8. **a.** Table 1, add 1; Table 2, add 3; Table 3, add 4.
c. 27, 75, 98
d. $t(n) = n + 2$; $s(n) = 3n$; $p(n) = 4n - 2$

BB-9. You will need to use graph paper to draw the triangle and then use the Pythagorean theorem to calculate the length of each side to determine whether the triangle is equilateral, isoceles, or scalene. Side $AB = \sqrt{5^2 + 2^2} = \sqrt{29} \approx 5.4$. You'll need to calculate BC and AC.

BB-10. **a.** $(x - 2)/(x + 2)$
b. $(x - 3)/(2x + 1)$

BB-11. (0, 0), (–6, 0)

BB-12.

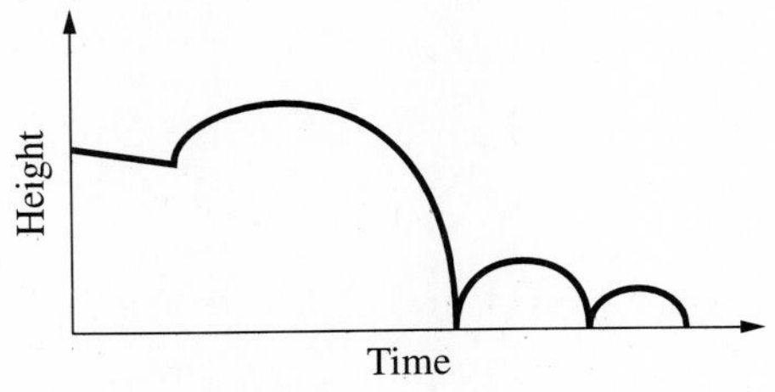

BB-13. $x = \dfrac{-5 \pm \sqrt{137}}{-8}$

BB-14. **a.** 1/4
b. How many sections fall outside the center triangle? How many sections are there total?

BB-15. **a.** Multiply both sides by common denominator.
b. Use the distributive property.
c. Simplify.
d. Subtract 8 from both sides.
e. Divide both sides by 3.

BB-21. **a.** (−1, −2)
b. (3, 1)

BB-22. Consider the issue of accuracy when writing your response.

BB-23. soup, \$0.79; tuna, \$1.39

BB-24. Calculate the slope of the segment joining each pair of points. $m_{AB} = -1/5$; $m_{BC} = -1/3$; $m_{AC} = -1/4$. What do you conclude?

BB-26. **a.** $y = -3x + 7$
b. $y = -x - \frac{2}{5}$

BB-27. **a.** x: 0, 1, 2; y: −2, 0, 1
b. $-1 \leq x \leq 1; -1 \leq y \leq 2$
c. $-2 < x \leq 2; -2 \leq y < 2$

BB-28. **a.** 3
b. $1/b$
c. $1/b$

BB-29. 4—isolate h in the volume equation.

BB-30. **a.** $x = -4$
b. $x = 56/3$
c. (1, 3)
d. Use quadratic formula, if you cannot factor easily.

BB-31. Three gallons of guava to 5 of mango would be the right ratio but would only total 8 gallons of punch, so start with the ratio of guava to punch. Then you can write an equation to figure out how much guava for 15 gallons of punch.

BB-32. **a.** 1/8
b. 3/8 for exactly 2 tails

BB-42. **a.** 25
c. It is linear, and in the form $y = mx + b$
e. For every unit of change horizontally there is a certain vertical change. What term from your Tool Kit represents this verticle change for this kind of sequence?

BB-43. $m = 4, b = -5$

BB-44. $y = 4x - 5$

BB-45. $y = -3x + 9$

BB-46. **b.** $t(46) = 93$

BB-47. **b.** The lacrosse ball rebound is 0.625 – 0.681, and the handball rebound is 0.62 – 0.65.

BB-48. $y = 3x + 10$
a. 64
b. 24
c. Slope—that a 3:1 ratio was used to adjust the points.

BB-49. 1/7

BB-51. **b.** First: (0, 8); second: (0, 15), these both represent the original height of the candles.
c. Intersect at (5.6, 2.4)

BB-58. **a.** yes, 90th term
b. no
c. yes, 152nd term
d. no
e. no, $-64 = n$ is not in the domain

BB-60. 10
a. 248
b. $t(n) = 14n + 10$

BB-62. **a.** Geometric; multiply by 1/2.
c. No, the sequence approaches zero. Half of a positive number is still positive.

BB-64. **b.** The soup cost \$0.28 and the bread cost \$1.08.

BB-65. **a.** 5
b. $4\pi/7$

BB-66. $(0, -17)$, $(-2 \pm \sqrt{21}, 0)$, or $(2.58, 0)$, $(-6.58, 0)$

BB-67. **a.** $y = -1$
b. $x = 0$

BB-68. **a.** $1.03y$
b. $0.8z$
c. $1.002x$

BB-69. **a.** 5, 7.5, 10, 12.5, 15, …

BB-71. All domains: real numbers. Ranges:
a. $y \geq -1$;
b. $y \leq 1$

BB-72. **a.** Distribute.
b. Add $10x$.
c. Subtract 28.
d. Divide by 2.

BB-73. **a.** arithmetic; 3 $t(n) = 3n + 1$
b. neither
c. geometric, $r = 2$
e. arithmetic; 1 $t(n) = x + n$

BB-81. **a.** 5.3 to 5.8 feet
b. 0.0175 to 0.0431 ft.
c. $10(0.53^n)$ to $10(0.58^n)$ ft.

BB-82. **a.** $x = 7$, or x could also equal −7 because $(-7)(-7) = 49$.
c. $x = 3.11$
d. As in part a, this one also has two solutions.

BB-83. **a.** 1, 8, 15, 22, 29
b. −5, −5, −5, −5, −5
c. 1/16, 1/8, 1/4, 1/2, 1
d. 1, −2, 4, −8, 16
e. whole numbers
f. the result or sequence values
g. What do all the domains have in common? all the ranges?

BB-84. **a.** −3, 1, 5, 9; arithmetic; yes, the terms can be generated knowing the common difference and the initial value.
b. Yes, it is term 81; let $4n - 3 = 321$ and solve for n.

BB-85. Create your own initial value, say \$10 or \$100, or any amount you choose.

BB-86. **a.** $0.05x$
b. $1.05x$
c. Multiply by 1.07.
d. 1.03, 1.0825, 1.0208

BB-87. **b.** 110
c. If n is the number of dots in the base, how many dots are in the height? What do you need to do to get the number of dots in the figure?

BB-88. 17 marbles to start, add 13 per year
a. 13
b. 94 years
c. 17
d. $t(n) = 17 + 13n$
e. in 59 years

BB-89. $t(1) = 6$; $t(3) = 24$; set up a table to answer the rest of the problem.

BB-90. Look at the changes in the outputs for each table, and match them with the graphical trends.

BB-95. **a.** 3252
b. $1250(1.27^n)$
c. Show how to guess and check to get 8 weeks

BB-96. **a.** 285
b. 56

BB-97. **a.** $0.20x$
b. $0.8x$
c. Multiply by 0.85

BB-98. **a** .0.97
b. 0.75
c. 0.925

BB-100. Set up ratios.

BB-101. **a.** 0, −50; 75, 125
b. 25, 12.5, $50\sqrt{2}$, $100\sqrt{2}$
c. $t(n) = 100 - 50n$,
$f(x) = 50 + 25x$;
$t(n) = 100(0.5^n)$,
$f(x) = 50(\sqrt{2})^x$

BB-102. **a.** 120
b. Go back to BB-87. How are these triangles related to those rectangles?
c. 20,100

BB-105. In 14 days they will earn close to the same amount: Plan A: $c(n) = 11.50n$; Plan B: $c(n) = 0.01(2^n)$.

BB-106. \$1,396,569.60
a. \$575,918.47
b. $s(n) = 673{,}500(1.20^{n-1})$;
$s(n) = 575{,}918(1.15^{n-1})$
c. \$5,011,917; \$3,883,058

BB-107. **i.** $f(n) = \sqrt{(n+1)}$
ii. $f(n) = \dfrac{\sqrt{n^2}}{2}$

CHAPTER 4

FX-9. **a.** $x = 3$
b. $x = 5$
c. $x = 2$
d. $x = 3$

FX-10. **a.** Using the pattern, 1/2 of 2 is 1 so $2^0 = 1$.

FX-11. **a.** zeroth
b. just one, 2^0

FX-12. **a.** 1, 2
c. 3, 8

FX-14. $\dfrac{3 \pm \sqrt{65}}{4}$, or 2.77 and −1.27

FX-15. **a.** 2^6
b. 2^9
c. 5^{2x}
d. 2^{4x+4}. Try changing 16 into 2 raised to a power.
e. $(2/3)^4$. Use the same approach as in part d.
f. 3^8

FX-16. **a.** (4, −1)
b. (−1, −2)

FX-17. **a.** Try the quadratic formula, or factoring.
x: (6, 0), (−12, 0); y: (0, −72)
b. x: (4/5, 0); y: (0, 4)

FX-20. **b.** 1/2

FX-24. the first bet 14 times and the second bet 33 times

FX-25. **a.** 1.05
b. 0.96
c. Remember, there are 12 monthly increases in 1 year; $1.03^{12} = 1.426$
d. $0.98^{12} = 0.785$

FX-26. **b.** x^5, x^3
c. x^{A-B}

FX-27. second step; distributive property:
$3 + 2x - 10 = 5x - 10$; $x = 1$

FX-28. **a.** (0, 1)
c. (0, 1)

FX-29. **b.** They do not.
c. undefined; error

FX-30. Review what you did in FX-9 and FX-16; $x = 6$.

FX-31. **a.** $x = 3$
b. $x = 1$
c. $x = 80$
d. $x = 210$

FX-32. Try isolating y first; then put the equation into $y = mx + b$ form.

FX-33. Replace $f(x)$ with 5, and use the methods of Appendix A; 11/5 or 2.2.

FX-34. (3, 2)

FX-35. **a.** $(x+3)(x+4)$
b. $x^2 + 7x + 12$

FX-42. 2, 1/2

FX-43. **a.** 1.06
b. 24
c. 368 years
d. $V(t) = 24(1.06^t)$
e. Use equation from part d; 4.93×10^{10}

FX-44. **a.** $(5)^{-3}$
d. 2^{5x-5}

FX-45. **a.** −3
b. 3
c. −2
d. −3

FX-46. **b.** Answers vary; $y = -x + 5$ is one possible response.

FX-47. **a.** 0.96; 126,000; 5 years
b. 1.10; 0.35; 24 years in 1994
c. 0.85; 11,000; 6 years

FX-48. (2, −4)

FX-49. **a.** arithmetic
b. $7 - \frac{2}{3}n$
c. −3
d. 42nd term

FX-50. no x-intercept; y: (0, −5)

FX-51. **a.** 3
c. −5

FX-52. How many whole numbers are between 10 and 20? How many are divisible by 3?

FX-62. **a.** $x = 0.7$
b. $x = 2/3$
c. $x = 0.8$

FX-64. **c.** yes
d. yes
e. yes
f. yes

FX-65. All are the same

FX-66. **a.** 1.0123 for 1 year
b. 0.97 for 1 month
c. 0.9712 = 0.694, multiplier for 1 year

FX-67. **a.** $x = 3$
b. $x = 4$
c. $x = -3/2$

FX-69. $x = \frac{-7}{2}$

FX-70. **a.** $x = 0$
c. $x = 0$

FX-71. **a.** Answers vary; $y = 0.024x + 0.50$ is reasonable.

FX-72. **a.** 5/10 = 1/2
b. 3/10
c. 2/10 = 1/5

FX-78. $y = 1.04^x$. To double, use $2 = 1.04^x$, then guess and check.

FX-79. **a.** $x = 7$
b. $x = 103.82$
c. $x = 9$
d. $x = 1.5$
e. $x = \pm 1.75$
f. $x = 3/2$

FX-80. 8, 2^3, $(16^3)^{1/4}$, $(16^{1/4})^3$, and several $\sqrt{\ }$ forms

FX-82. **a.** 120
b. 22,204

FX-83. **a.** $x(x+8)$
b. $6x(x+8)$
c. $2(x+8)(x-1)$
d. $2x(x+8)(x-8)$

FX-84. **a.** 0.3125, 0.15625, 0.07813
b. geometric
c. Discrete—it is a sequence.

FX-85. **a.** $x = 23$
b. Graph each side of the equation, and find the point of intersection of the two expressions. The x-coordinate of the point of intersection represents the solution.

FX-86. The answer depends on the context: on a number line, a point; in a plane, a line; or in space, a plane.

FX-88. **a.** $x = \frac{-3 \pm \sqrt{23}}{2}$, or 0.898, −3.898
b. $x = 0.610, -24.610$

FX-89. **a.** 1/3
b. choice a.

FX-95. **a.** $x = 2$
b. $x = 13/3$
c. $x = 8/3$

FX-96. **a.** $2(x+2)(x+2)$
b. $6(x+3)(x-4)$

FX-97. **a.** 3
c. 4
e. 1/4
g. 1/3
h. 4
i. a

FX-98. **a.** 3/2
b. 3
c. 6
d. 2
e. never; (0, 3)

FX-100. **a.** $x = 2$
b. $x = 3$
c. $x = -3$

FX-101. **a.** $y = \frac{2x-7}{3}$
b. $y = \left(\frac{-1}{2}\right)^x - 2$

FX-102. **a.** $x = -15$
b. $x = \pm 5\sqrt{2}$

FX-103. **a.** $(-8, 2)$
b. $(5/3, -1)$

FX-104. (6, something)

FX-105. $D = 3$, $F = 15$, $E = -1/3$

FX-106. $31^2 = 961$

FX-113. **a.** 5
b. 1/625
c. 1/5
d. 625
e. not a real number

FX-114. **a.** 2, 6, 18, 54

FX-115. **a.** no solution
b. $x \approx 3.17$

FX-116. **a.** $x = 3$, $y = 2$
b. $x = 2$, $y = 1$

FX-117. **a.** $\left(\frac{1}{5}\right)^2 = \frac{1}{25}$
b. $\left(\frac{1}{4}\right)^3 = \frac{1}{64}$
c. $\sqrt{9} = 3$
d. $\left(\sqrt[3]{64}\right)^2$

FX-118. **a.** 5600
b. 5627.54
c. 5634.13

FX-119. Use $500 = 3000(m^3)$ to find m.

FX-120. $(-60, -26)$

FX-122. x: $(-1, 0)$, $(-13, 0)$, $(0, 13)$; $x = -7$

FX-123. **a.** 3%, for a multiplier of 1.03
b. 4.45578 grams

FX-124. **a.** Drop 5% per month; multiplier = 0.95.
b. \$26,600

FX-125. **a.** neither arithmetic nor geometric, but quadratic.
b. discrete
c. 5, 10, 17

FX-126. 24/7 and 18/7

FX-127. **a.** 4
c. 243

CHAPTER 5

PG-5. **b.** shifted to the right 2 units

PG-6. **a.** 4, 1, or 0.25, $t(n) = 256(0.25)^n$
b. They get smaller.
c. They get closer to the x-axis.

PG-7. **a.** Use $y = mx + b$.
b. $y = 2$
c. If you know the answer to part b, you can figure out the answer to part c.
d. $y = \frac{2}{3}x - \frac{8}{3}$

PG-8. **a.** a cylinder
b. $45\pi \approx 141.37$

PG-9. $f(-10) = 101$ also, and $f(12) = 145$, which is 1 more than 144.

PG-10. **a.** $(2x - 3y)(2x + 3y)$
b. First factor out the largest common factor, then factor again.
c. $(x^2 + 9y^2)(x - 3y)(x + 3y)$
d. $2x^3(4 + x^4)$

PG-11. **c.** a circle

PG-12. **a.**

n	$t(n)$
0	0
1	3
2	6
3	9

n	$s(n)$
0	1
1	2
2	4
3	8

b. Yes. The points lie on a line (since the sequence is arithmetic), so they have a constant slope. Thus the ratio of the sides of triangles is the slope.
c. No—the points aren't collinear, so the slope and therefore the ratios of the sides vary.

PG-13. Try substitution and solve the equations.

PG-14. **a.** Did you use a multiplier of 1.04?
b. $y = 150(1.04^x)$
c. ≈\$384.50

PG-18. **b.** $(-5, 0)$, $(-1, 0)$, $(0, 5)$

PG-19. **a.** Try to rewrite 4 with a base of 2, and 8 with a base of 2.
b. -2

PG-20. **a.** 0.625 hours or about 37.5 minutes
b. 0.77 hrs or about 46 min
c. Find how much time she saved first.

PG-21. **a.** popcorn: \$3.75; soft drink: \$2.75

PG-22. **a.** $6x^4 + 8x^5y$
b. Did you raise x^3y^2 to the fourth power first?
c. $\frac{x+3}{2}$

PG-23. $x = \dfrac{-by^3 + c + 7}{a}$

PG-24. **a.** 8
b. ≈75.44
c. Try to solve $32 = 2(x + 3)^2$.
d. −3

PG-26. **a.** Did you use $a^2 + b^2 = c^2$?
b. $\sqrt{27} = 3\sqrt{3} \approx 5.20$

PG-27. It doesn't matter.

PG-34. **a.** $y = 0, 6$
b. Try to solve one side of the equation for zero.
c. $t = 0, 7$
d. $x = 0, -9$
e. They all have zero as a solution.

PG-35. **a.** $(7, -16)$; $y = (x - 7)^2 - 16$
b. $(2, -16)$; $y = (x - 2)^2 - 16$
c. Try factoring or using the quadratic formula.
d. $(2, -1)$; $y = x^2 - 4x + 3$

PG-37. **a.** $y = (1/3)x - 4$
b. Multiply each term in the equation by y.
c. To undo the square root, you'll need to square both sides of the equation. $y = x^2 + 4$
d. Remember $(x + 3)^2 = (x + 3)(x + 3)$
e. (a) x: $(12, 0)$; y: $(0, -4)$;
(b) x:$(1/6, 0)$; y: $(0, -1/5)$

PG-38. 10.5 and 7.5 pounds

PG-39. Notice that x is a common factor. If you factor out x, what is the other factor? Three solutions: $\dfrac{-23 \pm \sqrt{561}}{8}$ and 0.

PG-40. **a.** 7.656 gigatons
b. $C(x) = 7(1.01^{x+9})$ or $7.656(1.01^x)$

PG-41. **a.** Did you draw a diagram first? 15-foot radius for the largest pool.
b. $528.76

PG-48. **a.** Let $x = 0$ and solve for y; then let $y = 0$ and …
b. x: $(2, 0)$; y: none

PG-49. **a.** $g(1/2) = -4.75$
b. $g(h + 1) = h^2 + 2h - 4$

PG-50. **a.** Note that 20 is the output and solve for x.
b. $x = \pm\sqrt{11}$

PG-51. $y = \dfrac{8}{25}(x - 5)^2 + 8$, if you set up the axes so you are standing at (0,0).

PG-52. **a.** $A(n)$ is arithmetic, but $a(n)$ is something else—it is quadratic and its rule involves squaring.
b. $A(n)$ is a line, $a(n)$ is a parabola.

PG-54. **a.** Did you divide by 2 and then take the square root?
b. To deal with the cube root you will need to raise the expressions on both sides of the equation to the third power: $x = (y + 7)^3 - 5$ or the longer (and harder) version, which is the result of doing the multiplication $(x + 7)(x + 7)(x + 7)$. $x = y^3 + 21y^2 + 147y + 338$

PG-55. **a.** 8 is the solution that fits the diagram.
b. ≈30.8

PG-56. **b.** As distance from your corner increases, what happens to the sound?

PG-63. $y \approx 2(x - 5)^2 + 2$ and $y \approx \frac{-1}{2}(x - 5)^2 + 2$ The stretch numbers are estimates.

PG-64. **a.** Did you use $y = a(x - h)^2 + 3$?
b. $y = (x - 2)^3 + 3$
c. $y = -2(x + 6)^2$

PG-66. x: $(-10, 0)$, $(8, 0)$; y: $(0, -80)$; V: $(-1, -81)$

PG-67. **a.** $h(3) = 1/5$
b. $h(-3) = -1$
c. $h(a - 2) = \dfrac{1}{a}$

PG-68. **a.** $x^2 - 1$
b. $2x^3 + 4x^2 + 2x$
c. $x^3 - 2x^2 - x + 2$

PG-69. Try letting $y = 0$ to figure out x, and $x = 0$ to figure out y.

PG-70. $x \approx 21.14$ so the other width is 528.59. There is another solution to the equation, but it is negative.

PG-71. **a.** Did you use $y = k(m^x)$?
b. $342.59
c. Domain: $x \geq 0$. Range: $y \geq 100$.

PG-80. For example: $y = (x - 4)^2$, $y = 5(x - 4)^2$, $y = -3(x - 4)^2$

PG-81. The second graph shifts the first 5 units to the left and 7 units up, and also stretches it by a factor of 4.

PG-82. The second graph is a reflection of the first over the x-axis.

PG-83. **a.** $y = 2x^2 - 4x + 6$
b. none

PG-84. Did you try substituting some values for x and y? Or what happens when you multiply $(x + y)(x + y)$?

PG-85. **a.** 4.116×10^{12}
b. $y = 1.665(10^{12})(1.0317)^t$

PG-86. **a.** x: $(-3, 0)$; y: $(0, 27)$
b. x: none; y: $(0, 2)$

PG-95. $y = 0.01(x - 100)^2$ or $y = 0.01x^2$ depending on where you decided to place the axes.

PG-96. Try graphing $P(n)$. What is the vertex?

PG-97. Explain why Brooke is right.

PG-98. Move it up 6 units or redraw the axes 6 units lower.

PG-99. Try all the different ways to factor 18.

PG-100. **a.** $y = x$

b. $\left(\frac{1}{2}, \frac{1}{3}\right)$

PG-101. One idea: What could you buy for about \$2000 that would be worth about 91% of its original value 1 year later?

PG-102. **a.** Did you raise each factor inside the parentheses to the third power?

b. $25x^2 + 5x + 0.25$

c. $20s^2 - 245$

d. $-30{,}375$

PG-103. **a.** x: $(-1, 0)$; y: $(0, 2)$; V: $(-1, 0)$

b. x: $(0, 0)$, $(2, 0)$; y: $(0, 0)$, V: $(1,1)$

PG-104. $-\frac{3}{4}(x-2)^2 + 3$

PG-105. **a.** Try redrawing the figure as two separate triangles. $y = 10.125$

b. $x \approx 6.18$

PG-106. **a.** Try $A = s^2$.

b. $15\sqrt{2} \approx 21.21$ cm

PG-113. **a.** $y = \frac{1}{x}$

b. $y = \frac{1}{x+3} - 2$

PG-114. The second graph shifts the first 5 units left and 7 units up, and also stretches it by a factor of 4.

PG-115. yes, stretch factor 4

PG-116. **a.** only two—at $(-2, 0)$ and a double root at $(3, 0)$

b. i. add a positive number

ii. add a negative number

iii. impossible

PG-119. **b.** $y = 3x + 2$

c. 2, 5, 8, 11

d. One is continuous and one is discrete.

PG-120. **a.** $y = x^3$

b. $4\pi/3$

c. cm^3

e. D: $r \geq 0$. R: $V(r) \geq 0$.

PG-121. **a.** x: $(-1/2, 0)$ $(-1, 0)$; y: $(0, 1)$

b. Try averaging the x-intercepts.

c. $(-0.75, -0.125)$

PG-122. Move it up 0.125 units: $y = 2x^2 + 3x + 1.125$

PG-123. It's not a parabola anymore.

PG-124. **a.** ≈ 16.755 m^3

b. no, remember the r, r^2, r^3 relationship. V: ≈ 83.776 m^3

c. $V = (4/3)\pi r^3 + 4\pi r^2$

PG-125. When $x \geq 0$ represents possible volumes.

PG-134. **a.** yes

b. $(1, -1)$

c. D: all real numbers, R: $y \geq -1$

PG-135. **b.** 28

c. Solve $9 = x^3 + 1$.

d. 1

e. $x = -1$

f. $x = -3$

g. no solution

PG-136. For $f(x)$ x-intercept $(-1,0)$, y-intercept $(0,1)$, locator $(0,1)$, line of symmetry $x = 0$. For $g(x)$: x-intercept $(-1, 0)$, y-intercept $(0, 1)$, locator point $(-1, 0)$, line of symmetry $x = -1$.

PG-137. $y = \frac{-5}{9}(x-3)^2 + 6$

PG-140. **a.** $(0, -6)$

b. $(-6, 0)$ and $(1, 0)$

c. x: $(0, 0)$, $(-5, 0)$, y: $(0, 0)$. The graph of $p(x)$ is 6 units lower than $q(x)$.

CHAPTER 6

LS-12. **a.** $\approx$4 hours, cost $\approx$ \$73–74

b. Cadillac = \$92

c. Cadillac = 9 hrs and 10 min

LS-13. **b.** $(-2, 3)$

LS-14. **b.** $y = 0$

c. $x = 0$

LS-15. **a.** Did you substitute $8 - 3y$ for x in the top equation? What happened?

c. Parallel lines do not intersect.

d. parallel planes

LS-16. Have you tried to solve $18x - 30 = -22x + 50$?

LS-17. **a.** Did you let $x = 0$ to find the y-intercept, and then let $y = 0$ to find the x-intercept?

b. x-intercept: $(-6, 0)$; y-intercept: $(0,4)$

LS-18. **a.** Did you factor out the common factor w?

b. 0, 2/5

c. Did you solve the quadratic equation for zero first?

LS-19. **a.** Approximately 1750

b. 4.14 pounds

LS-20. They don't cross.

LS-21. **a.** Does time depend on area or does area depend on time?

b. D: 0 to the distance from the pebble's "plip" to the edge of the pond. R: 0 to πr^2, where r is the shortest distance to the edge of the pond.

LS-26. Did you solve the top equation for x first?

LS-27. **a.** $C = 800 + 60m$

b. $C = 1200 + 40m$

c. To obtain your solution did you use the method given in LS-16?

d. 5 years

LS-28. 26.5%
a. $x(1.05^{10})$
b. $0.2x$
c. $0.2x(1.08^{10})$
d. $0.2\left(\frac{1.08}{1.05}\right)^{10} = \frac{p}{100}$. Now use your calculator to solve for p.

LS-29. **c.** Did you make a table to help you see the pattern? $4N - 3$, arithmetic

LS-30. **a.** Did you draw a diagram of the cube?
b. 2/9
c. 4/9
d. 8/27
e. Use your diagram to help you visualize this answer.

LS-31. The second one is shifted to the right 4 units and up 2 units.

LS-32. $x \approx 36.78$

LS-33. Rewrite as $\frac{7}{2} \cdot (4)^{-4}$ or $\frac{7}{2 \cdot 4^4}$

LS-34. Did you compare their results and then use $y = k(m^x)$ in your explanation?

LS-35. **a.** The skater slows, then speeds up again.
b. $y = (x - 4)^2 + 1$
c. If you chose a situation involving time and speed, it should involve slowing down and speeding up.

LS-39. Did you multiply every term in the bottom equation by 2? What happened when you combined equations?

LS-40. **a.** Show that if you input c, the output will be c.

LS-41. Did you use $y = k(m^t)$? What variable are your solving for?

LS-42. **a.** Remember $\sqrt[2]{x} = x^{1/2}$.
b. $9^{1/3}$
c. $17^{x/8}$

LS-43. Clearly 1 is not a solution. Is 2 a solution? What about 1.5?

LS-44. $1/11 \approx 0.0909$

LS-45. **b.** line parallel to the y-axis at $x = 4$
c. plane parallel to yz-plane at $x = 4$
d. The first is just a point, the second a line, and the third a plane.

LS-46. You can set this problem up as two equal ratios.

LS-47. **a.** $x = \frac{5 \pm \sqrt{157}}{6}$
b. Start by multiplying every term in the equation by $2(x - 1)$.
c. $x = \frac{-14}{3}$
d. The substitution method is useful here.

LS-48. **a.** The first is a line.
The next is the shading below the line $y = x$.
The next is the same shading, but now the line is also included.
b. The first is a parabola
The next is the same parabola shaded above, and the parabola is dashed.
In the last, the parabola is also included.

LS-50. 27 square units

LS-51. First rewrite 16 as 2^4 and 1/8 as 2^{-3}.

LS-52. Try some numbers such as $a = 16$ and $b = 9$, or square both expressions and compare. Be careful when you square the second one:
$$\left(\sqrt{a} + \sqrt{b}\right)^2 = \left(\sqrt{a} + \sqrt{b}\right)\left(\sqrt{a} + \sqrt{b}\right).$$

LS-53. **a.** You can draw a box that marks the boundaries of the domain and range.
b. Your "box" for this one will have only two sides.

LS-54. To get an approximate answer don't worry about the height of your eye level—just use the Pythagorean theorem as if you were looking up from the ground.

LS-55. $(-2, 3, -5)$

LS-56. $a = 18.5$, $b = 5.5$

LS-57. $\frac{50\pi}{4} \div 400 \approx 9.8\%$

LS-58. The x-axis is $y = 0$; the y-axis is $x = 0$. This may be something you need to memorize in addition to keeping it in your Tool Kit.

LS-59. Substitute $x = -4$ and $x = 2$ into $y = x^2$ to find the y-coordinates of these points. Then use the two points to find the equation of the line.

LS-60. **a.** bond investment: $0.07(0.3x)$ or $0.021x$
b. stock investment: $0.09(0.7x)$ or $0.063x$
c. solve: $0.021x + 0.063x = 5000$

LS-65. **a.** $y = -2(x + 4)^2 + 2$
b. $y = \sqrt{(x-2)}$
c. $y = -x^3 + 3$

LS-66. **a.** $3x = b$, so $x =$?
b. $\frac{b}{5a}$
c. Remember: $x + ax = x(1 + a)$

LS-67. All three planes don't intersect in a single point or line; pairs of planes may cross to form two or three parallel lines, or all the planes are parallel and there are no intersections at all.

LS-68. **a.** Be sure to check some negative numbers. Input equals output only if $x \geq 0$.

LS-69. **b.** 6 square units

LS-70. **a.** $2a^2 - 4$

b. Did you substitute $3a$ for x in $f(x)$?
c. $2a^2 + 4ab + 2b^2 - 4$
d. $2x^2 + 28x + 94$
e. $50x^2 + 60x + 14$
f. Substitute $5x + 3$ for x in $g(x)$. Result is $10x^2 - 17$.

LS-71. **a.** Did you use an initial value of \$6.75 and a multiplier of 1.15?
b. \$2.21

LS-73. **a.** 1/5
b. 1/3
c. Set up a proportion.
d. $\frac{x+60}{x+300} = \frac{2}{5}$

LS-74. Remember: you can find the x-intercepts by solving $x^2 + 4x - 2 = 0$. Then average the solutions and substitute the result to find the vertex.

LS-82. $y \le -x + 4;\ y > \frac{1}{3}x$

LS-83. Substitute 7 for z in two of the original equations. Be sure you use the one Arturo didn't use. Check your solution by substituting your solution into the original equations.

LS-84. Multiply each term by x. Then you will have to reorganize and factor as in LS-66 c.

LS-86. **a.** 1/4
b. $1 - \frac{x}{12}$

LS-87. The exponents for each prime factor must be equal. So $2^x = 2^3$ and $x = 3$. Now solve for y and z.

LS-88. **a.** −7
b. 94
c. 94
d. 124

LS-89. Remember: the arrows mean segments CP and AB are parallel, so angle A = angle C. Also, vertical angles are equal. These two facts should allow you to write two equations that you can then solve for x and y.

LS-91 (6, 0) and $\left(\frac{1}{2}, \frac{21}{4}\right)$ are the solutions. But you do not have to solve this, just explain how you would go about solving it.

LS-92. 1/4

LS-107. $\left(-1, \frac{1}{2}, 2\right)$

LS-108. **a.** $y = x^2 - 2x + 3$
b. (1, 2)

LS-110. **b.** 1, 3, 6, 10; no; no
c. $0.5n^2 + 0.5n$

LS-111. Try writing a system of equations, and then solve it. red = 10 cm, blue = 14 cm

LS-112. The solution of the original system is (25, −3); the solutions of the new systems are (5, −3) and (−5, −3).

LS-113. **a.** −15, 0, 0
b. (0, −15) is the y-intercept. Since the other two numbers gave an output of 0, the points (−1.5, 0) and (5,0) are the x-intercepts.

LS-114. **a.** In a nonleap year, 1/59; in a leap year, 1/60.
b. In a nonleap year, 1/59; in a leap year, 1/60.
c. Sammy, who will be happy with any day except Super Bowl Sunday

CHAPTER 7

CC-7. **a.** 9
b. 4
c. $x \approx 1.89$

CC-8. Yes. You could use two function machines, each with the same function, one to "do," the other to "undo."

CC-9. The bases are equal.

CC-10. One parabola opens up the other opens down. Show where their solution sets intersect.

CC-11. **a.** First add 3, then …
b. Yes, $y = x$.

CC-12. Remember: there are two solutions.

CC-13. $x \approx 0.53$

CC-14. Eliminate y first. (−3, 0, 5)

CC-15. If she adds nothing else to the account and it just sits there making interest, she will have \$440.13 on her 18th birthday.

CC-16. Solve: $x + 4 = 3x - 1$

CC-24. Trejo is correct.

CC-25. **a.** $L(x) = x^2 - 1$; $R(x) = 3(x + 2)$
b. 30
c. Begin by substituting numbers in the equations.

CC-26. 36

CC-27. 2

CC-28. Yes, order matters; not necessarily equal

CC-29. Does the graph reflect across $y = x$?

CC-30. Did you use $y = k(m^x)$? Then you'll need to guess and check.

CC-31. Remember to draw and label a diagram, then use $a^2 + b^2 = c^2$.

CC-32. 1/5; Remember LS-86 about the stick?

CC-36. **a.** $e(x) = (x - 1)^2 - 5$
b. One machine undoes the other, so $e(f(-4)) = -4$.
c. They would be reflections of each other across the line $y = x$.

CC-37. Begin by subtracting 7.

CC-38. No solution: the planes don't intersect in a single point—parallel lines are formed.

CC-39. The vertices $(-5, -6)$ are for one inequality and $(-4, -1)$ for the other. Does the parabola bounding each graph have a solid or dashed line?

CC-40. **a.** Temperature depends on time.
b. *D*: 0 to ∞, *R*: room temperature to starting temperature, presumably boiling or close to it.
c. Theoretically, yes; actually, no?

CC-41. **a.** $x \approx \pm 5.196$
b. Use guess and check. 4 is too small and 5 is to big.

CC-42. Did you use $b^{1/x} = \sqrt[x]{b}$ and $b^{1} = \frac{1}{b^{1}}$?

CC-43. **a.** 4, 20, 100
b. Did you use $312{,}500 = 4(5^n)$?
c. No, the solution is not an integer.

CC-50. **a.** x: all numbers; $y > -3$
b. no
c. Did you let $x = 0$ and then let $y = 0$?
d. $y + a = 2^x$, where $a \le 0$.

CC-51. $g(x) = \frac{x^2 - 10}{5}$; be sure the domain and range are properly restricted.

CC-52. Something like $y = 2^x + 15$.

CC-53. Did you start by subtracting ay from both sides?

CC-54. Yes, they are inverses since the x and y are just interchanged. No, the second is not a function. The graphs are reflections of each other across the line $y = x$.

CC-55. **a.** The sequence is neither arithmetic nor geometric. Since the results are all squares, it is called a quadratic sequence.
b. $t(n) = n^2 - 3n + 4$

CC-56. You might want to begin by graphing $x - 1$ and $x - 2$.

CC-57. **a.** x^3 raised to what power equals x?
b. $x = 5$

CC-63. Did you set up a system of equations? A taco is $0.66.

CC-64. $2x + 5y \le 500$; $6x + 3y \le 450$; no, since $6(60)+3(70) > 450$.

CC-65. **a.** Inflation
b. How can you use $25.25 = 1.50(m^{60})$?

CC-66. Be sure the y-values for the functions are getting smaller as they approach either the x-axis or the line $y = 5$.

CC-67. **a.** Did you substitute 2 for x?
b. 12
c. Did you substitute 12 for x in $f(x)$?
d. 54
e. You don't need a calculator for this one. You could use prime factorization to rewrite 81 with a base of 3.

CC-68. $h(k(x)) = k(h(x)) = x$

CC-69. **a.** Did you use $y = k(m^t)$?
b. Use guess and check or the equation you developed in part *a*?

CC-70. $y = 2x \pm 2\sqrt{30}$

CC-71. Did you set up an equation using two ratios, such as $\frac{3}{4} = \frac{x}{8}$?

CC-72. $y = \log_3 x$

CC-78. **a.** base 12
b. Base 12 implies how many fingers?

CC-79. $x \approx 23.450$

CC-80. Did you try to solve $3x = x - 8$?

CC-81. **a.** $\frac{1}{2} < x < \infty$; $3 \le y < \infty$
b. $g(x) = \frac{(x-3)^2 + 1}{2}$
c. $3 < x < \infty; \frac{1}{2} \le y < \infty$
d. 6
e. Did you substitute 6 into $f(x)$ first?

CC-82. **a.** true
b. Did you substitute values for x? Did you graph each side of the equation?

CC-83. **a.** Did you multiply each term of the equation by $x + 4$?
b. $\frac{-1 \pm \sqrt{5}}{2}$, or 0.61803 and -1.61803

CC-84. **a.** $-0.889, -0.962, -0.988, -0.996$
c. The values get closer to -1.

CC-85. Did you raise each side of the equation to the 1/5 power?

CC-86. **a.** Did you use $A = \pi \cdot r^2$?
b. 1/3

CC-93. **a.** 25
b. 2
c. 343
d. $\sqrt{3}$
e. 3
f. 4

CC-94. **b.** $a \log k = \log (k^a)$

CC-95. Did you use logarithms this time? $x \approx 23.4498$

CC-96. **a.** Did you substitute values for x and y? What values for x and y make each equation equal 0?
b. (0,0)

CC-97. No; $\log_3 2 < 1$ and $\log_2 3 > 1$.

CC-98. **a.** $k = \frac{y}{m^x}$
b. m is the xth root of $\frac{y}{k}$.

CC-99. **a.** The second is just the first shifted up 10 units.
b. $y = k \cdot mx + b$

CC-100. Did you think of a parabola facing downward? With the y-value of the vertex between 10 and 11?

CC-101. a. $y=\sqrt{\frac{x+7}{3}}-2$, assuming you restricted the original domain to $x \geq -2$.
b. $x \geq -7$; $y \geq -2$

CC-108. Be sure you mentioned taking the eighth root.

CC-109. a. 1/8
b. $\frac{1}{a}$
c. $m = \pm 1.586\cdots$
d. Did you take the sixth root and obtain two solutions?
e. $x = b^{1/a}$

CC-110. $x = 17$

CC-111. a. $x = -3$, $y = 5$, $z = 10$
b. infinitely many solutions
c. The planes intersect in a line.

CC-112. $0 < b < 1$

CC-118. a. Have you tried taking the log of both sides of the equation?
b. 11.228

CC-119. Did you use $y = k(mx)$? 16.5 months; 99.2 months

CC-123. a. $\left(\frac{-3}{4}, \frac{5}{2}\right)$
b. Multiplying each side of the top equation by 5, and multiplying each side of the bottom equation by 3 will help.

CC-124. a. Did you begin by squaring 9?
b. $c(x) = x^2 - 5$

CC-125. a. no
b. not necessarily.

CC-126. a. Did you use $y = k(m^x)$, and let $m = 0.5$?
b. $\approx$ 6640 years
c. Never!

CC-131. a. Decreasing by 30% means multiplying by 0.7 each time. "Multiplier" implies exponential.
b. Did you use $y = k(m^x)$?
c. \$8060.50
d. $\approx$3.83 years
e. $\approx$\$61,560.64

CC-135. a. 1/2
b. any number except 0
c. 1.0×10^{23}

CC-136. a. Did you take the sixth root or raise each side of the equation to the one-sixth power?
b. 4.230
c. 0.316
d. 2.021
e. Did you divide by 4 first?
f. 3.659

CC-137. a. Did you divide by 3 first to undo this function?
b. $f(x)=\sqrt[3]{\frac{x-6}{3}}$
c. $f(x)=\frac{x+1}{x-1}$
d. In order to find the inverse did you begin by multiplying by $3 - x$?

CC-138. a. Did you use 1.025 as the multiplier?
b. $\approx$24.2 years
c. Did you use 0.95 as the multiplier?

CHAPTER 8

CF-5. a. $x = 2, 4$
b. $x = 3$
c. $x = -2, 0, 2$

CF-7. What operations are polynomials restricted to?

CF-8. What would be the highest power of x if you were to multiply the factors together?

CF-9. a. 0 or 1
b. 0, 1, or 2
c. 0, 1, 2, 3, or 4
d. 0, 1, 2, 3, or 4. (1 and 3 require the parabola to be tangent to the circle.)

CF-10. $(-2, -1)$, $(3, 4)$

CF-12. Parts c, f, and g are not polynomials

CF-13. The second graph is the same as the first graph shifted up 5 units.

CF-20. a. two
b. $\sqrt{7}$ and $-\sqrt{7}$

CF-21. $x = -1 \pm \sqrt{6}$
a. 2
b. $\left(-1+\sqrt{6}, 0\right)$ and $\left(-1-\sqrt{6}, 0\right)$
c. at $x \approx 1.45$ and $x \approx -3.45$

CF-23. at (74, 0), a double root, and at (–29, 0)

CF-24. a. 2
b. 5
c. 3
d. 6

CF-25. First find p(x) by seeing 4 as 2^2. Then substitute and solve; $x = -1$ or 5

CF-26. a. $y = 3^x - 4$
b. $y = 3^{(x-7)}$

CF-27 There are two solutions for the equation, –2 and 0. Therefore, the x-coordinates of the points of intersection are -2 and 0. To determine the y-coordinates substitute these two numbers for x. The points of intersection are $(-2, -6)$ and $(0, 0)$.

CF-28. The first statement is correct, but the second is false. Substituting numbers would be one way to demonstrate this.

CF-34. **a.** 3, 0, −3

CF-35. Solve the equation $\frac{1}{2} = \frac{16}{x^2 - 4}$. (6, 1/2), (−6, 1/2)

CF-36. **a.** $\frac{x+3}{2}$

b. $\sqrt{x-2}+3$

CF-37. **b.** $f^{-1}(x) = \left(\frac{x-3}{2}\right)^2 + 1;\ g^{-1}(x) = \sqrt{3(x+2)} - 1$

CF-38. 50π ft ≈ 157 ft

CF-39. **a.** 3

b. $x = 2$, $x = 4$, and $x \approx -0.767$

CF-40. **a.** 1/6

b. 3/16

CF-48. **a.** nowhere

b. It has no real solution; therefore, there is no number for x that gives $y = 0$, and the graph cannot cross the x-axis.

c. We get a square root of a negative number.

CF-49. **a.** 13, 17, 21

b. arithmetic

CF-51. Actually, it is not!

CF-52. **a.** See Appendix A, Fraction Busters, and multiply every term in the equation by $(x + 2)(x - 2)$. Solution: $x = -26$.

b. $x = \frac{15}{2}$. Did you get $x^2 - 7x + 6x - 30 = x^2 - 5x$ after multiplying by the common denominator?

CF-53. **a.** $4n - 23$

b. at least 2506 times.

CF-54. **b.** $x^2 - 6x + 9$

CF-55. **a.** $\frac{2}{20} = \frac{1}{10}$

b. 1/19

CF-59. **a.** $x = 3$

b. $y = 2$

CF-60. **a.** Try making a table for each company.

b. SC: $y = 20x + 180$,
AH: $y = 80(1.15^x)$. They are equal in about 11.8 years.

CF-61. Substitute some numbers for x, a, and b in order to decide. Or show how to use a common denominator to add $\frac{a}{x} + \frac{a}{b}$.

CF-62. Draw and label the triangle. AC = 10 inches

CF-63. $-1 \pm \sqrt{2}$

CF-64. 20 days, arithmetic. You will need to use a table to work this out.

CF-65. **a.** 1/2

c. Use the fact that there are 360° in a circle.

d. 1/2

e. 1/16

f. $1 - \frac{\pi}{4}$

CF-73. **a.** −18 − 5i

b. $1 \pm 2i$

CF-74. $1 \pm 2i$

CF-75. 1

CF-76. Think of 16 as 2^4, and 8 as 2^3.

CF-77. What happens when you multiply $(x - 5)(x + 5)$ and compare the result to the product of $(5 - x)(5 + x)$?

CF-78. **i.** 7 **ii.** 18.3 **iii.** d

a. −1

b. −1

CF-79. **a.** $7i$

b. $\sqrt{2}i$

c. −16

CF-80. **b.** $h(x) = (x + 2)^3 - 7$;
$h^{-1}(x) = \sqrt[3]{x+7} - 2$

CF-86. **a.** 5

b. 34

c. 17

d. 53

CF-87. **a.** $4 - i$

CF-88. **i.** 2

ii. 3

iii. 0

iv. 1

CF-90. **a.** $x = \pm$ i; therefore, it has no real roots and cannot cross the x-axis.

CF-91. **a.** three real linear factors (one repeated); therefore, 2 real (1 single, 1 double) and 0 imaginary roots

b. one linear and one quadratic factor; therefore 1 real and 2 imaginary roots

d. two linear and one quadratic factor; 2 real and 2 imaginary roots

CF-92. **e.** a) 5, b) 5, c) 4, d) 6

CF-93. Use the quadratic formula and consider possible numbers for b that would lead to real or imaginary roots. $b \geq 20$ or $b \leq -20$

CF-94. **a.** −21

b. $-10 + 7i$

CF-95. $(1 + 2i, 2 + 4i)$, $(1 - 2i, 2 - 4i)$

CF-101. **a.** about 980,000 with the cost over $19,600

b. The new tank will be $1.2m \times 5.85m \times 3.98m$. About 367,000 balls will fill it to a depth of 0.8 m at a cost of $7355

CF-102. about 0.004

CF-108. (0, 0), (3, 0), and (−0.5, 0)

CF-110. Figure out a by substituting 2 for x and 12 for y.
$y=-\frac{2}{3}x(x-3)(x+1)^2$

CF-111. **a.** real
b. imaginary
c. imaginary
d. real
e. real
f. imaginary

CF-112. **b.** Since the graphs do not intersect, the system has no real number solution. You can think of the intersection as being imaginary. The solutions are imaginary: $(1 \pm 2i, -3 \pm 4i)$

CF-113. **a.** $y = \log_b x$
b. (2, 0). The x-intercept of the asymptote which is (2, 0) could be the locator point, or the x-intercept of the graph, which is (3, 0), could be used as a locator point.
c. $y = \log_2(x-2)$ is one possibility.

CF-114. **a.** $x:\left(-\frac{5}{2},0\right),(0,0),\left(\frac{7}{2},0\right); y:(0,0)$
b. $x:(-3,0),\left(\frac{15}{2},0\right)$ (double root); y: (0, 675)

CF-116. **a.** (2, 8), (4, 4)
b. $(3 + i, 6 - 2i), (3 - i, 6 + 2i)$

CF-117. $y = 600 + 5x$; $y = 3(1.15^x)$; in 40 months

CF-118. The function could have three real roots (one x-intercept must be a double root) and two imaginary roots. Or it could have one real triple root, one real double root, and zero imaginary roots. Or it could have one quadruple real root, another single real root, and no imaginary roots.

CF-119. **a.** The roots are imaginary numbers.
b. $y = (x - 3)^2 + 4$ or $y = x^2 - 6x + 13$
c. $x = 3 \pm 2i$

CF-120. The graphs do not intersect in the real plane.

CF-121. $\frac{[2(x-5)]^2}{3}+1$

CF-122. **a.** Repeat $1, i, -1, -i, \ldots$
b. $1, i, -1, -i$

CF-123. **a.** 1
b. i
c. -1

CF-125. Remember, if you cannot figure out how to write an equation, you can always start with a guess and check table. Make a few guesses for the number of adult tickets, then use x as the next guess to get the equation, or use x and y to get two equations. Answer: 450

CF-126. **a.** $2x^2 + 12x + 18$
b. $12 + 5x - 2x^2$

CF-129. 8/12

CF-130. $y = -0.5(x - 2)(x + 4)$ or
$y = -0.5x^2 - x + 4$

CF-131. **a.** 1, 3
b. $4 \pm 3i$
d. $\pm 10i$

CF-133. $\approx$ (5.492, 50)

CF-134. **a.** You should get the result $x = 1$, but read part b.
b. This equation has *no* solutions, which means not real and not imaginary solutions. Some would say it's not even an equation. Why?

CF-135. 1, 2, or 3
a. $x = 2, \frac{1\pm i\sqrt{3}}{2}$, one real, two imaginary

APPENDIX B

B-1. **a.** $(-4, -6)$

B-2. **a.** 2
b. 4.25
e. $(x^2 + 5x + 2 + 4.25) - 4.25$
f. $(x + 2.5)^2 - 4.25$

B-3. **a.** $f(x) = (x + 3)^2 - 2$
b. $f(x) = (x - 5)^2 - 25$

B-4. $(b/2)^2$, square half of the middle term to determine what you need.

B-6. **a.** $(x + 1)^2 - 1$
b. $(x-2)^2 - 2$
c. $(x-5)^2 - 4$
d. $(x - 3.5)^2 - 10.25$

B-7. $(x + 3)^2 = 19$
$x + 3 = \pm\sqrt{19}$
$x = -3 \pm\sqrt{19}$

APPENDIX C

C-1. **a.** four points: (5, 0), (0, 5), (−5, 0), (0, −5)

C-2. **a.** $x^2 + y^2 = 9$
b. $x^2 + y^2 = 49$
c. $x^2 + y^2 = 4$
d. $x^2 + y^2 = 16$

C-5. Not—substitute some numbers. Or notice that the second equation can be written
$y = -x + 12$, which is a line, not a circle.

C-7. $(x - h)^2 + (y - k)^2 = r^2$

C-8 **a.** $(x - 9)^2 + (y + 3)^2 = 16$
b. $(x + 5)^2 + y^2 = 23$

C-10. **a.** C:(−9, 4), r: $5\sqrt{2}$
b. C:(0, −5), r: 4
c. C:(3, 7), r: 5
d. Parabola: $V(3,1)$, opening downward

C-11. **a.** $(x - 1)^2 + y^2 = 9$
b. $(x + 3)^2 + (y - 4)^2 = 4$

C-12. concentric circles with center (3, 5)

C-13. circles with radius 4, whose centers lie on a line 5 units above the x-axis

C-14. circles with radius 3

C-15. $(x + 3)^2 + (y - 5)^2 = 9$

C-16. $(x + 4)^2 + (y - 2)^2 = 16$ or $(x - 4)^2 + (y - 2)^2 = 16$

C-17. $(x + 4)^2 + (y - 7)^2 = 16$

C-18. circle of radius 5, centered at origin, and its interior

C-19.

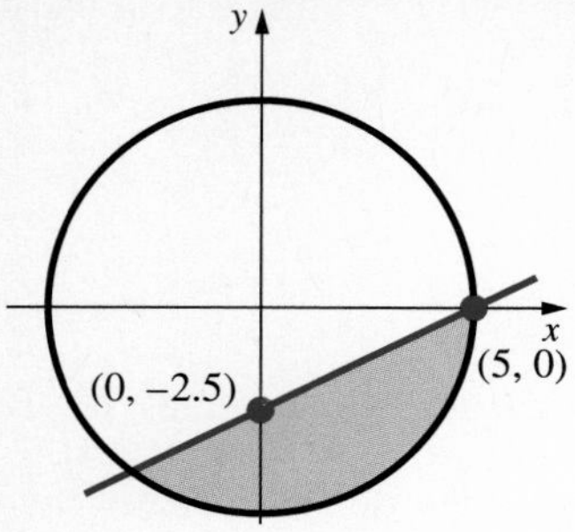

INDEX

ABOUT THE INDEX

Each entry of the index is in alphabetical order and is referenced by the problem number in which it is discussed. In some cases, a term or idea may be discussed in several of the following problems as well. Each problem is listed by the two-letter code that references the chapter followed by the problem number. Whenever possible, terms are cross-referenced to make the search process quicker and easier. For example, *common difference* can also be found under the heading *arithmetic sequence*.

TABLE OF CHAPTER LABELS

Chapter 1	PS	Problem Solving: Getting Started
Chapter 2	EF	Sharpening Pencils: Exploring Functions
Chapter 3	BB	The Bouncing Ball and Other Sequences
Chapter 4	FX	Fast Cars and Depreciation: Exponential Functions
Chapter 5	PG	The Gateway Arch: Parabolas and Other Parent Graphs
Chapter 6	LS	The Toy Factory: Linear Systems
Chapter 7	CC	The Case of the Cooling Corpse: Logarithms and Other Inverses
Chapter 8	CF	At the County Fair: Polynomials and Other Systems
Appendix A	A	Solving Equations
Appendix B	B	Completing the Square
Appendix C	C	Circles
Appendix D	D	3×3 Technology